SCIENCE AND THE QUEST FOR REALITY

MAIN TRENDS OF THE MODERN WORLD

General Editors: Robert Jackall and Arthur J. Vidich

Propaganda
Edited by Robert Jackall

Metropolis: Center and Symbol of Our Times
Edited by Philip Kasinitz

Social Movements: Critiques, Concepts, Case-Studies
Edited by Stanford M. Lyman

Science and the Quest for Reality
Edited by Alfred I. Tauber

The New Middle Classes: Life-Styles, Status Claims and Political Orientations
Edited by Arthur J. Vidich

Science and the Quest for Reality

Edited by

Alfred I. Tauber
Director of the Center for Philosophy and History of Science
Boston University

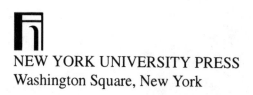

NEW YORK UNIVERSITY PRESS
Washington Square, New York

First published in the U.S.A. in 1997 by
NEW YORK UNIVERSITY PRESS
Washington Square
New York, N.Y. 10003

This book is printed on paper suitable for recycling and
made from fully managed and sustained forest sources.

Library of Congress Cataloging-in-Publication Data
Science and the quest for reality / edited by Alfred I. Tauber.
p.   cm. — (Main trends of the modern world)
Includes bibliographical references and index.
ISBN 0–8147–8218–3. — ISBN 0–8147–8220–5 (pbk.)
1. Science—Philosophy.   I. Tauber, Alfred I.   II. Series.
Q175.3.S315   1996
501—dc20   95–26709
                                                CIP

Printed in Great Britain

For Joel, Dylan, Benjamin and Hana

Contents

Series Preface

Main Trends of the Modern World is a series of books analyzing the main trends and the social psychology of our times. Each volume in the series brings together readings from social analysts who first identified a decisive institutional trend and from writers who explore its social and psychological effects in contemporary society.

The series works in the classical tradition of social theory. In this view, theory is the historically informed framing of intellectual problems about concrete social issues and the resolution of those problems through the analysis of empirical data. Theory is not, therefore, the study of the history of ideas about society, nor the abstract, ahistorical modeling of social realities, nor, as in some quarters, pure speculation often of an ideological sort unchecked by empirical reality. Theory is meaningful only when it illuminates the specific features, origins, and animating impetus of particular institutions, showing how these institutions shape experience and are linked to the social order as a whole.

Social analysts such as Karl Marx, Max Weber, Émile Durkheim, Sigmund Freud, Georg Simmel, Thorstein Veblen and George Herbert Mead, whose works we now consider classics, never consciously set out to construct paradigms, models or abstract theories of society. Instead they investigated concrete social phenomena such as the decline of feudal society and the emergence of industrial capitalism, the growth of bureaucracy, the consequences of the accelerating specialization of labor, the significance of religion in a scientific and secular age, the formation of self and the moral foundations of modern society, and the on-going rationalization of modern life. The continuing resonance of their ideas suggests the firmness of their grasp of deep-rooted structural trends in Western industrial society.

Later European and American social thinkers, deeply indebted though they were to the intellectual frameworks produced by the remarkable men who preceded them, faced a social order marked by increasing disarray, one that required fresh intellectual approaches. The social, cultural and intellectual watershed was, of course, the Great War and its aftermath. The world's first total war ravaged a whole generation of youth. In Europe it sowed the seeds of revolution, militarism, totalitarianism, fascism and state socialism; in both Europe and America it signaled the age of mass propaganda. On both continents the aftermath of the war brought economic and political turmoil, cultural frenzies, widespread disenchantment and disillusionment, and social movements of every hue and description that led eventually to the convulsions of the Second World War. These later social thinkers grappled with issues such as:

- The deepening bureaucratization of all social spheres and the ascendance of the new middle classes.
- The collapse of old religious theodicies that once gave meaning to life and the emergence of complex social psychologies of individuals and masses in a rationalized world.
- The riddles posed by modern art and culture.
- The emergence of mass communications and propaganda as well as the manufacture of cultural dreamworlds of various sorts.
- War, militarism and the advent of totalitarianism, fascism and state socialism.
- The deepening irrational consequences and moral implications of the thoroughgoing rationalization of all life spheres.

Emil Lederer, Hans Speier, Joseph Schumpeter, Kenneth Burke, Robert MacIver, Harold Lasswell, Walter Lippmann, Robert Park, W.I. Thomas, Florian Znaniecki, George Orwell, Hannah Arendt, Herbert Blumer and Hans H. Gerth are only a few of the men and women who carried forward the theoretical attitude of the great classical thinkers in the course of working on the pressing issues of their own day. In this tradition, social theory means confronting head-on the social realities of one's own times, trying to explain both the main structural drift of institutions as well as the social psychologies of individuals, groups and classes.

What then are the major structural trends and individual experiences of our own epoch? Four major trends come immediately to mind, each with profound ramifications for individuals. We pose these as groups of research problems.

BUREAUCRACY AS THE ORGANIZATIONAL FORM OF MODERNITY

- What are the social and psychological consequences of living and working in a society dominated by mass administered bureaucratic structures? How do mass bureaucratic structures affect the private lives of the men and women exposed to their influences?

- What is the structure and meaning of work in a bureaucratic society? In particular, how does bureaucracy shape moral consciousness? What are the organizational roots of the collapse of traditional notions of accountability in our society?
- What is the relationship between leaders and followers in a society dominated by a bureaucratic ethos? What are the changing roles of intellectuals, whether in the academy or public life, in defining, legitimating, challenging or serving the social order?

THE TECHNOLOGIES OF MASS COMMUNICATION AND THE MANAGEMENT OF MASS SOCIETY

- What role do public relations, advertising and bureaucratized social research play in shaping the public opinions and private attitudes of the masses?
- What is the relationship between individuals' direct life experiences (with, for example, family, friends, occupations, sex and marriage) and the definitions that the mass media suggest for these individual experiences? What illusions and myths now sustain the social order? What are the ascendant forms of this-worldly salvation in our time?
- What are the different origins, dynamics and consequences of modern political, social and cultural mass movements with their alternative visions of justice and morality?
- What social, economic and cultural trends have made many great metropolises, once the epitomes of civilization and still the centers and symbols of modern life, into new wildernesses?

THE ON-GOING SOCIAL TRANSFORMATIONS OF CAPITALISM

- What are the prospects for a transformed capitalism in a post-Marxist, post-Keynesian era?
- How has the emergence of large bureaucratic organizations in every sector of the social order transformed the middle classes? What is the social and political psychology of these new middle classes?
- What transformations of the class and status structure have been precipitated by America's changing industrial order?
- What are the social, cultural and historical roots of the pervasive criminal violence in contemporary American society? What social

factors have fostered the breakdown of traditional mechanisms of social control and the emergence of violence as a primary means for effecting individual or group goals?

THE CLASH BETWEEN WORLDVIEWS AND VALUES, OLD AND NEW

- How has science, particularly in its bureaucratized form, transformed the liberal doctrines of natural rights, individual rights and concomitant conceptions of the human person, including notions of life and death?
- How have the old middle classes come to terms with mass bureaucratic institutions and the subsequent emergence of new classes and status groups? What social forces continue to prompt the framing of complicated social issues in terms of primal antagonisms of kith, kin, blood, color, soil, gender and sexual orientation?
- What are the roots of the pervasive irrationalities evident at virtually every level of our society, despite our Enlightenment legacy of reason and rationality and our embrace of at least functional rationality in our organizational apparatus? To what extent is individual and mass irrationality generated precisely by formally rational bureaucratic structures?

In short, the modern epoch is undergoing social transformations every bit as dramatic as the transition from feudalism to industrial capitalism. The very complexity of the contemporary world impedes fixed social scientific understanding. Moreover we may lack the language and concepts necessary to provide coherent analyses of some emerging features of our social order. Nonetheless this series tries to identify and analyze the major trends of modern times. With an historical awareness of the great intellectual work of the past and with a dispassionate attitude toward contemporary social realities, the series tries to fashion grounded, specific images of our world in the hope that future thinkers will find these more useful than speculation or prophecy.

Each volume in this series addresses one major trend. The book in hand analyzes the institution of science, a dominant force in modern society since the Enlightenment, but one whose legacy is now under critical scrutiny.

ROBERT JACKALL
ARTHUR J. VIDICH

Preface

Various strategies are employed to characterize the role of science in Western civilization. These include descriptions of nature as currently understood, the developments of scientific theory in different disciplines, the growth of scientific institutions, philosophical accounts of scientific rationality, or sociological pictures of current scientific practice. We might also attempt to describe how science has provided powerful inspiration for various metaphysical pictures and even ideological agendas, such as "Nature is a garden" or "Nature is a machine," along with the attendant ramifications of such metaphors traced through their diverse cultural and intellectual effects. This complexity is integral to the subject matter, because there is no obvious or rigid demarcation separating science from general epistemology, intellectual and cultural history, sociology and psychology, art, or in fact from virtually any domain of our lives.

But beyond the particular editorial strategies that have been employed, a deeper narrative appears in these essays. The genre of anthologies offers revealing glimpses of their editors, and often succinct statements of their *Zeitgeist*. As I assembled this collection, I noted with some trepidation how peculiar some earlier enterprises of a similar nature appeared to my gaze. This historical perspective cautions, even humbles me. Perhaps my concern with present-centrism should be freely acknowledged as a guiding parameter of this work. Sensitive to the strong currents of contemporary scholarship and political reassessment that would attempt to sweep aside science's sanctimonious role in society, this project has maintained a wary eye towards what appears to be a significant revaluation of science's philosophical infrastructure, its social organization, and more far-reaching, its claims to serve as a model for intellectual discourse and objectivity. In the midst of this challenge, an attempt has been made to present the basis for defining the changing boundaries of science and the jurisdiction it offers towards defining our world view. I do so freely admitting my own bivalency: As a scientist, I cling to the objectivity which guides the laboratory scrutiny of nature; as a philosopher of science, I recognize the assumptions of those methods and the limits of its epistemology. This anthology reflects the precarious balance of these potentially conflicting views of science.

I would not have contemplated, much less embarked on this audacious project of presenting contemporary science in its cultural and philosophical matrix without the steadfast encouragement and support of Robert Jackall. I thank him for the opportunity. I also wish to acknowledge Simon Feldman for his superb assistance in preparing the manuscript. This endeavor has greatly benefited from the critical comments of Pnina Abir-Am, Burton Dreben, Alberto Cambrosio, Robert Cohen, David Kazhdan, Evelyn Fox Keller, Thomas Kuhn, Thomas Soderqvist, and especially, Eileen Crist, whose careful and astute reading is most appreciated. Although much of their respective criticisms helped mold the final presentation, they cannot be held responsible for my foibles, mistakes, and views, for I turned to each of them to test the arguments, not necessarily to verify them. I am fortunate to have such fine critics, and I appreciate their efforts to help me clarify my own perspective.

ALFRED I. TAUBER
Center for Philosophy and History of Science
Boston University

Acknowledgements

The editor and publishers acknowledge with thanks permission from the following to reproduce copyright material:

Macmillan Publishing Company, for selected excerpts from pp. 1–28 of *Science and the Modern World*, by Alfred North Whitehead in Chapter 1. © 1925, Macmillan Publishing Company, renewed by Evelyn Whitehead.

HarperCollins Publishers, Inc., for selected excerpts from pp. 115–136 of *The Question Concerning Technology and Other Essays*, by Martin Heidegger in Chapter 2. English translation by W. Lovitt © 1977, Harper & Row Publishers, Inc. Reprinted by permission. Also for selected excerpts from pp. 76–92 of *Physics and Philosophy: The Revolution in Modern Science*, by Werner Heisenberg in Chapter 4. © 1958, Werner Heisenberg. Reprinted by permission.

Cambridge University Press, for selected excerpts from pp. 1–39 of *The Ghost in the Atom, A Discussion of the Mysteries of Quantum Physics*, by P. C. W. Davies and J. R. Brown in Chapter 3. © 1974, Cambridge University Press. Reprinted by permission.

University of Notre Dame Press, Indiana, for selected excerpts from pp. 83–105 of *Science and Reality: Recent Work in the Philosophy of Science*, ed. J. T. Cushing, C. F. Delaney and G. M. Gutting in Chapter 5. © 1984, University of Notre Dame Press, Indiana. Reprinted by permission.

University of California Press, for selected excerpts from pp. 154–172 of *Scientific Realism*, ed. J. Leplin in Chapter 6. © 1984, University of California Press.

Duke University Press, for selected excerpts from pp. 313–331 of *Rethinking Objectivity*, ed. A. Megill in Chapter 7. © 1994, Duke University Press. Reprinted by permission.

University of Chicago Press, for selected excerpts from pp. 559–591 of *Darwin and the Emergence of Evolution of Theories of Mind and Behavior* in Chapter 8. © 1987, The University of Chicago Press. Reprinted by permission.

Philosophy of Science Association, for selected excerpts from pp. 3–13 of *The Road Since Structure* by Thomas Kuhn in Chapter 9. © 1991, Philosophy of Science Association. Reprinted by permission. Sage Publications, Inc., for selected excerpts from pp. 29–63 in Chapter 10 and 393–407, 424–443 in Chapter 11 of *Handbook of Science and Technology Studies*, ed. S. Jasanoff *et al.* © 1995, Sage Publications, Inc. Reprinted by permission.

D. Reidel Publishing Company, for selected excerpts from pp. 307–323 of *For Dirk Struik. Scientific, Historical and Political Essays in Honor of Dirk J. Struik*, ed. R.S. Cohen *et al.* in Chapter 13. All rights reserved. © 1974, D. Reidel Publishing Company.

Instituto de Investiaciones Filosoficas, Universidad Avtonoma de Mexico, for selected excerpts from pp. 135–141 in Chapter 14. © 1982, Universidad Nacional Avtonoma de Mexico. Reprinted by permission.

Kluwer Academic Publishers, for selected excerpts from pp. 189–200 of *The Understanding of Nature: Essays in the Philosophy of Biology*, ed. R. Cohen and M. Wartofsky, in Chapter 15. © 1974, Kluwer Academic Publishers. Reprinted by permission.

Oxford University Press, for selected excerpts from pp. 137–144; 150–156 of *From Max Weber: Essays in Sociology*, trans. and ed. G. H. Gerth and C. W. Mills in Chapter 16. © 1946, Oxford University Press, Inc., renewed 1973 by Hans G. Gerth. Reprinted by permission.

America Bar Association, for selected excerpts from pp. 13–16 and pp. 55–57 of *Natural Resources and the Environment*, 2(2) in Chapter 12. © 1986, American Bar Association. Reprinted by permission.

Notes on the Contributors

Nicholas A. Ashford is Professor of Technology and Policy at the Massachusetts Institute of Technology, where he teaches environmental occupational health law and policy. He is author of *Crisis in the Workplace*, and a co-author of *Monitoring the Worker for Exposure and Disease*, and *Technology, Law and the Working Environment*.

Julien R. Brown is affiliated with the Research Centre for English and Applied Linguistics at the University of Cambridge, England.

P.C.W. Davies is Professor of Natural Philosophy at the University of Adelaide, Australia. He is a prolific writer on astronomy, cosmology, physics, and philosophy. Among his works are *Space and Time in the Modern Universe*, *The Runaway*, *Other Worlds*, *The Edge of Infinity*, *The Cosmic Blue Print*, *The New Physics* (2 vols), *The Accidental Universe*, and *The Mind of God*.

Michel Callon is Professor of Sociology and Director of the Centre de Sociologie de l'Innovation at the Ecole Nationale Supérieure des Mines de Paris. He co-edited *Mapping the Dynamics of Science and Technology: Sociology of Science in the Real World*.

Robert S. Cohen is Professor Emeritus of the Departments of Physics and Philosophy at Boston University. He is the general editor of *Boston Studies in the Philosophy of Science* and a co-editor of *Vienna Circle Collection*.

Thomas Gieryn is Associate Professor in the Department of Sociology at Indiana University. He co-edited *Social Research and the Practicing Professionals*, and *Theories of Science in Society*. His book-in-progress, *Making Space for Science: Cultural Cartography Episodically Explored*, represents a study of pragmatic demarcations of science.

Karin A. Gregory, a lawyer and health care consultant, owns and operates Custom Solutions in Boston, Massachusetts, and serves as an Adjunct Assistant Professor at the Boston University School of Medicine.

Marjorie Grene is Professor Emeritus at the Center for the Study of Science and Society at the Virginia Polytechnic Institute and State

University. Her books include *The Anatomy of Knowledge, Approaches to Philosophical Biology, Dreadful Freedom, The Knower and the Known,* and *Knowing and Being.*

Ian Hacking is Professor of Philosophy at the University of Toronto. He is author of *Logic of Statistical Inference, Why Does Language Matter for Philosophy?, The Emergence of Probability,* and *The Taming of Chance.*

Martin Heidegger (1889–1976), a pivotal figure of twentieth century philosophy, was Professor of Philosophy at the University of Freiburg and the University of Marburg. Among his most influential philosophical texts are *Being and Time, Holzwege, What Is Philosophy?, Hegel's Phenomenology of Spirit, History of the Concept of Time, Essays in Metaphysics: Identity and Difference,* and *Kant and the Problem of Metaphysics.*

Werner Heisenberg (1901–1976) was a winner of the Nobel Prize, Professor of the University of Leipzig and director of the Max Planck Institute for Physics and Astrophysics at Munich. In 1925 he published a paper, "About the Quantum-theoretical Reinterpretation of Kinetic and Mechanical Relationships," in which he proposed a reinterpretation of the basic concepts of mechanics. Among his works are *Cosmic Radiation, Nuclear Physics, Philosophical Problems of Quantum Physics,* and *Physics and Beyond.*

Evelyn Fox Keller is Professor of History and Philosophy of Science in the Program of Science, Technology, and Sociology at the Massachusetts Institute of Technology. She is author of *A Feeling for the Organism: The Life and Work of Barbara McClintock, Reflections on Gender and Science, Secrets of Life, Secrets of Death: Essays on Language, Gender and Science,* and *Refiguring Life.*

Thomas Kuhn was Professor Emeritus of the Department of Linguistics and Philosophy at the Massachusetts Institute of Technology. Often cited as the most influential historian of science of this century, his books include *The Structure of Scientific Revolutions, The Essential Tension, The Copernican Revolution,* and *Black-Body Theory and the Quantum Discontinuity 1894–1912.*

Larry Laudan is Professor of Philosophy at the University of Hawaii. His major interests in the philosophy of science have centered around the issue of scientific rationality as illuminated by historical case studies of actual scientific practice. His books include *Progress and Its Problems, Science and Hypothesis, Science and Values,* and *Beyond Positivism and Relativism.*

Hilary Putnam is Walter Beverly Pearson Professor of Modern Mathematics and Mathematical Logic at Harvard University. He has authored *Meaning and the Moral Sciences, Philosophy of Logic, Reason, Truth and History, Philosophical Papers* (3 vols), *The Many Faces of Reality, Representation and Reality,* and *Realism with a Human Face.*

Robert J. Richards is Professor in the Departments of History, Philosophy, and Psychology at the University of Chicago. His books include *Darwin and the Emergence of Evolutionary Theories of Mind and Behavior,* and *The Meaning of Evolution.*

Alfred I. Tauber is Professor of Philosophy and Professor of Medicine at Boston University, where he directs the Center for Philosophy and History of Science. He is author of *The Immune Self, Theory or Metaphor?,* co-author of *Metchnikoff and the Origins of Immunology,* and editor of *Organism and the Origins of Selt* and *The Elusive Synthesis: Aesthetics and Science.*

Max Weber (1864–1920) was a highly influential German sociologist, historian, and philosopher, holding professorships of economics at Freiberg, Munich and at the University of Vienna. He proposed a theory of "ideal types" which found its most celebrated application in his famous *The Protestant Ethic and the Spirit of Capitalism.* Among his other books are *The Theory of Social and Economic Organization, General Economic History, On the Methodology of Social Sciences,* and *The City.*

Alfred North Whitehead (1861–1947) was a major figure of twentieth-century philosophy, holding professorships at Imperial College (London) and Harvard. Co-author with Bertrand Russell of *Principia Mathematica,* Whitehead developed wide-ranging interests in metaphysics, epistemology, and philosophies of science, religion, and education. His best known works are *An Inquiry Concerning the Principles of Natural Knowledge, Concept of Nature, Science and the Modern World,* and *Process and Reality.*

Introduction
Alfred I. Tauber

AN OVERVIEW

We live in a world dominated by scientific consciousness, not only in the practicalities of our everyday lives, but with respect to our most basic notions of reality and objectivity, not to mention how we regard ourselves as animal creatures, rational thinkers, or elements of the entire cosmos. Science has no less than created a world view, and this too has its own consequences. This anthology does not attempt to chart the course of Western science, nor even give a cursory summary of what science has wrought. Instead, these essays present broad philosophical, historical, and sociological orientations to place science in its widest intellectual and cultural context. More specifically, however, the task and focus of this volume is to come to grips with science as a cognitive activity, whose philosophical foundations are open to reconstruction and critical scrutiny, whose history suggests dissension and sometimes confusion as to the nature of scientific change, and whose sociology, quite beyond what was postulated a generation ago, argues for an integration of scientific practice in its supporting culture.

A new academic discipline has emerged called "science studies," and this anthology, as part of its larger sociological series, is heavily influenced by the particular mode of discourse its participants have adopted. This rather nondescript designation refers to the confluence of what were heretofore regarded as separate fields that examined the history, sociology, and philosophy of science, respectively. These modes of inquiry remain distinct, but a new hybrid has emerged that endeavors to understand the "network" of science as it penetrates throughout our culture, and in turn is molded by that culture. As Bruno Latour describes the phenomenon, science is "blended," by which he means to emphasize how artificial it is to attempt to dissect "science" as though it were an insulated and circumscribed social activity (Latour, 1993). Beyond the social analyses that divided complex cultural activities in order to examine them as entities, contemporary social theory avers again and again the fundamental intercontextualization of complex Western social institutions in which science partakes. Science has become the object of anthropological analysis analogous to studies of exotic cultures, where the economics, religious beliefs, kinship systems, etc. are integrated into a global view of the culture. Science studies have become another venue of such social analyses, and within this theoretical orientation certain characteristics are shared. For our purposes, perhaps the most important thesis to understand is that the study of Nature and the study of Society are inexorably linked, for not only do they reside interwoven in a trivial social sense, their very belief systems are interdependent to the point that it makes no sense to speak of Nature (as science examines her) and Culture (as historians, philosophers, or sociologists practice their studies) as independent domains. The implications of this view go well beyond characterizing science in the modern world: the project is no less than an attempt to characterize ourselves.

The interesting development in science studies is that traditional disciplinary boundaries have been broken and new challenges have been hurled at the sanctified view of science as Autonomous and Objective. What heretofore was an inquiry into the social organization and practice of science based on some normative view of scientific discovery, has evolved to encompass a sociological account of how scientific knowledge is generated, and how its validity is established. In other words, sociology has taken on epistemological concerns of *what* science describes, and *how* it does so. Current investigations go beyond *The Social Construction of Reality* written by Peter Berger and Thomas Luckman in 1966, where they attempted to develop a systematic theory for the sociology of knowledge, and thereby identify its deep penetration into all forms of sociological analyses. "Reality" and "knowledge" fall between philosophy and commonsense in the sociological context, for here the concern has shifted from questions of *what* is real or true, to a focus on what *counts* as real or true, how validation is attained, and why views of reality and knowledge differ from one culture to another or from one historical moment to the next. This problem of the "slippage" into a relativist perspective was recognized at the very birth of modern science; as Blaise Pascal noted, what is truth on one side of the Pyrenees is error on the other. The general public might well wonder what all the fuss is about. Indeed, it is often taken for granted that there is a special standing of scientific endeavors as compared to other forms of inquiry concerned with defining or comprehending the psyche, society, history, arts, or religion. These latter domains appear governed by different principles of inquiry, and criteria of truth, and they offer different *kinds* of descriptions (see for example, Barbour, 1990).

This orientation closes the door to global relativism and disallows the displacement of scientific knowledge.

Although social forces such as award systems, economic supports, and institutional pursuits may shape the progression of science, to appreciate the important role of such factors does not necessarily lead to a relativization of scientific knowledge. This is the position I hold, and this anthology is designed to present the view that science is indeed a social phenomenon, but a very special one, because of the constraints exerted by its object of study and its mode of analysis. Implicit in my own orientation is the endeavor to preserve science's special place in the pantheon of knowledge. Science has an undeniable authority when it remains within its appropriate domain of inquiry. Scientific method and verification (or falsification) are applicable within experimental systems; statements can be made and tested against nature to describe physical and biological realities. This is not to deny that scientific interpretation may on occasion be over-extended or wrong, or that nature is contrived and contorted to fit the experimental description, or that there are powerful cultural forces at play that allow prejudice to undermine objective ideals, but nevertheless, despite all these caveats, science *is* successful and we must characterize and acknowledge its accomplishments, while still realizing its limitations and false applications.

The scientific enterprise is committed to a kind of verification not found (and usually unattainable) in other domains of knowledge. Despite the obvious successes that rely on this claim, objectivity as understood in the scientific context rests on a complex philosophical foundation that remains contested. This is a highly complex issue, but in its simplest formulation, the argument concerns how criteria are chosen for successful prediction and verification. The positivist would argue for normative standards and methods; some social constructivists maintain that consensus dynamics not only are operative, in some instances they are determinative. In another venue, when referring to the "maturity" of a science, we are concerned with the differing sophistication of the various disciplines to command Prediction. Although success may depend on methodological, even technical considerations, there are fundamental concerns that certain phenomena may not be amenable to scientific analysis, and this limitation then raises the spectre of the boundaries and limitations of scientific knowledge. Finally, there is a deep philosophical issue concerning the very relationship of prediction and its encompassing theory. Here I am referring to the dissociation of prediction and theoretical validity, i.e. their logical, and in some instances, practical separation. In other words, a prediction may be correct, but its theoretical basis may be false. These are matters, and there are others, that philosophers of science wrestle with, because they seek some logical means by which theories may be judged True, some rigor by which to ascertain the validity of Scientific Method, and some overarching Rationality that explains the predictive success of science. We might easily devote our entire attention to these matters, but given the context of this series, these philosophical problems serve simply as the backdrop for the sociological debate concerning the nature of scientific practice.

These essays address this debate from two points of view. The first illustrates certain developments in twentieth century science and philosophy that together intimated that science deserved a more sophisticated appreciation than that offered by those who lauded a naive positivism as the paragon of knowledge, the standard by which science might, and should be judged. The second perspective is to describe what the contemporary sociological appreciation of science sees as the practice of science, moving beyond an idealized vision of scientists cloistered together in a community dedicated to the unprejudiced pursuit of Truth, and toward a conception of science as a thoroughly human affair, characterized by social pursuits not qualitatively different from those of other social constellations. Finally, there is an ethical, or perhaps humanist undercurrent present in the essays of this anthology. I suspect that the relativist undermining of science as absolute or final knowledge reflects a deeper moral apprehension concerning the potential tyranny of a narrowly-conceived Rationality. In general, there are those proponents who regard scientific growth as synonymous with modernity and progress, values that have not only bequeathed us the capacity increasingly to control nature and our standard of living, but also provided a powerful means for understanding our world and ourselves. But there is also a critical chorus that regards science as dominating and imperialistic. In its endeavor to seek truth and be governed by rational discourse, science from this perspective is regarded as overwhelming other modes of knowledge and perverting humanistic values and concerns to those of blind materialism. On this view, originating in the Romantic period, the relentless reductionism of the natural sciences (and those social sciences based on them) has provided narrow and distorted views of Man, Society, and Nature. I am interested then in another codifying issue, perhaps mooted by the epistemological and political concerns, namely, that science encompasses an ethical position.

I take the view that science may be regarded as a means to a grand vision of nature; not only is it enlightening in all of its full intellectual splendor, it is deeply saturated with aesthetic preoccupations (Tauber, 1993; Tauber, 1996). This is a Romantic sentiment, and may assume moral overtones, but it is also rooted in an immanent ethical stance that science assumes, one integral to its intellectual precepts: science is essentially pluralistic, accepting detracting, as well as integrating criticism as part of its very code (Popper, 1945). In this regard science is a bulwark of liberal, democratic society (Merton, 1973). In respecting that science does indeed seek truth by such principles, we must be wary of confusing its ostensible, and in my opinion largely attained, moral goals from the exceptional cases of dogmatic attitudes or fraudulent practices that threaten to subvert the ideal.

So much for the general orientation of this anthology. Note, the selections are chosen from the twentieth century, although a broader historical context is evident. I have presented the debate in its modern format, with little attention given to its historical antecedents, which date to ancient times in the challenge of the Skeptics, through its early formal expression in the sixteenth and seventeenth centuries' debates between the Rationalists and the Empiricists. Although there is a strong philosophical flavor to the anthology, the sociological approach is equally emphasized to keep this book consistent with the larger project of this series concerning the main trends of the modern world. With these general guideposts in place, we can now commence to situate these essays within their broader concerns.

SCIENCE AND ITS WORLD VIEW

Thoughtful commentators have, at least since the birth of modern science in the sixteenth century, recognized the philosophical issues that limited scientific analysis. I have chosen the classic statement (which is not to say uncontested) of Alfred North Whitehead's opening chapter of *Science and the Modern World* (1925) (Chapter 1). There, he explores the intellectual origins of modern science as simultaneously a reaction to medieval scholasticism and a progeny of its metaphysics. Obviously modern science has been an integral factor in undermining, even overthrowing religious authority. In contradistinction to the rule of an august Rationality, "irreducible and stubborn facts" would be sought in direct experience and verification. The science of the late fifteenth century was born in this attempt to know nature directly. This attempt for direct knowledge came entangled with certain accompanying assumptions. First, a critical disjuncture between the observing subject and the object of inquiry was established. This subject–object dichotomy took a deeper hold with the Cartesian mind–body duality, where a knowing mind was conceived as distinct and separate from its encompassing world. Second, an accepted universality of such descriptions rested upon an assumed inherent and autonomous order of nature and an assured utility of an analytical logic to explain and understand this order. Finally, there was the implicit assumption that there are secrets to unveil which link us as inquiring minds to Being. These cornerstone presumptions are profound metaphysical correlates to what we call science, and Whitehead demands that we examine and take seriously these foundations in order to avoid science fragmenting "into a medley of *ad hoc* hypotheses." He was concerned that the science of Galileo, Descartes, and Newton was, in the twentieth century, at a turning point. The very philosophical assumptions upon which physics was based, concerning the nature of causality, time–space, matter, and the stability and order of the organic domain now required a reorienting metaphysics. According to Whitehead, this project was crucial not only to unify and propel science, but also to delineate its broader significance as a human activity.

Whereas Whitehead was enjoining the construction of a new metaphysics based on the post-Darwinian, post-Einsteinian revolutions, at the same time, Martin Heidegger was declaring the end of metaphysics. With respect to our purposes, we need not dwell on Heidegger's general philosophical agenda beyond his assessment of science. In "The Age of the World Picture" ([1938], 1977) (Chapter 2), he argued that the essence of science was research, and in this sense the essay is an interesting complement to Whitehead's metaphysical characterization of science's birth and governing ethos. But Heidegger goes further than Whitehead as a critic of science, for he attacks the very basis of objectification and the pursuit of what he refers to as "object-ness." Beyond the common concern that an imperialistic science may unilaterally establish *the* picture of the world, he perceived that "world picture... does not mean a picture of the world but the world conceived and grasped as a picture." (Again, the isolated observing Cartesian subject is the pivotal philosophical turning point upon which the entire enterprise rests.) But in Heidegger's view, irrespective of the potential freedom the inquiring subject might now possess, science's domination was ultimately limited, because the method of science, its very essence, excludes that which cannot be incorporated into its world picture, namely, that which cannot be represented and directly related to Man in the anthropomorphic cosmic picture.

So, even beyond the overwhelming presence of science in industry, education, politics, and warfare – indeed its power to influence or shape every dominion of our lives – *"science is the theory of the real"* (Heidegger, "Science and Reflection" 1977, p. 157) and as such, according to Heidegger, science imprisons human understanding within narrow philosophical boundaries. Science cannot make a totalizing claim to knowledge, and from his perspective, we have been left unable to probe the shadows we recognize beyond the boundaries of such inquiry. Science's access is limited to only one way in which nature exhibits itself, or in other words, nature is not fully captured by objectifying the world in measurement, the Galilean criterion of validity.

Despite the divergent origins of Heidegger's and Whitehead's critiques, as well as their trajectories in later philosophy, psychology, linguistics, and sociology, each philosopher focused on the reassessment precipitated by revolutions in our concepts of the physical (i.e. quantum mechanics) and biological (i.e. post-Darwinian) worlds. As noted, their views converged on the limitations of the then current expectations of science to provide us with a world view, and a basis by which knowledge might be unified under its auspices. Beyond this skepticism, which would fairly allow different *kinds* of knowledge, an even more pressing issue has become the focus of recent debate among philosophers: On what basis can science make claim to discover Reality – that is, Nature as it Is? Profound repercussions of this discussion have been felt in the history and sociology of science, whose own concerns have seemingly converged around this issue within their own domains of inquiry. Considering the extraordinary authority, power and influence of modern science, it is perhaps ironic to those not intimately aquainted with these discussions that Arthur Fine could proclaim, "Realism is dead." (1986, p. 112) What did he mean?

THE PROBLEM OF SCIENTIFIC REALISM

To orient ourselves to the general question of 'What is science?' in the late twentieth century, we must understand the problem of scientific realism. This is a profoundly complex metaphysical and epistemological issue and underpins a seminal debate concerning the role of science in modern society. Obviously the problem of knowledge – Can we know anything for certain? – originates with the birth of philosophy.

The Greeks referred to knowledge as *epistēmē* (as opposed to *doxa*, opinion), and thus 'epistemology' concerns the theory by which we ascertain what sorts of things we can know and how we come to know them. The struggle between the earliest philosophers, who operated on the premise that we could in principle *know* the world, and the Greek Skeptics who claimed that we can only know or prove that nothing can be known or proved, has persisted in novel guises.

Before we can describe science in its cultural context, we must first attempt to understand its philosophical underpinnings in the context of a reassessment of its epistemological ideals, which have undergone a radical reassessment in this century. At least since Goethe (Tauber, 1993), we have appreciated that science is not simply governed by its own inner logic. Aesthetic judgments, deep cultural values, and social, economic, and political pressures combine to form a complex matrix in which scientific rationality and method are influenced. The view of the autonomous rational growth of scientific thinking – that is of science as logically progressing and possessing universal and unwavering objective criteria to describe Nature – is perhaps a product of conflating science's declared ideal aspirations with the more subjective and heterogeneous nature of its enterprises. The idealized description, which Philip Kitcher (1993) appositely calls The Legend, begins with the unquestioned acceptance of science as rigidly rational and objective. But twentieth century physics has bequeathed an epistemological challenge to this view, dramatically altering our very conception of nature and our relation to it. In brief, scientific certitude and objectivity have been challenged by a new understanding of probabilistic causality. The traditional Cartesian segregation of subject–object relations has been blurred and a more circumspect appreciation of potentialities and tendencies has replaced a simpler picture of a cause–effect universe. The basic reformulation of physical law offered by quantum mechanics was no less than a radical epistemological revolution. One can date the birth of this crisis precisely to 1927, when the full impact of Heisenberg's uncertainty principle was appreciated by Niels Bohr and his colleagues who embraced his Copenhagen Interpretation of physics.[1] I have chosen "The Strange World of the Quantum" by P.C.W. Davies and Julien Brown (1986) (Chapter 3) to succinctly summarize this basic statement of quantum mechanics and present the epistemological quandary we now face in attempting to correlate this description with our everyday world. Suffice it here to note simply that Bohr was referring to the inherent probabilistic nature of the measurement phenomenon. Probably the most significant debate since Newton and

Leibniz then ensued between Einstein and Bohr on the absolute versus relative conception of causality (Bohr, 1985). As Arthur Fine narrates,

> What bothered Einstein most of all was twofold. First, he could not go along with the idea that probability would play an irreducible role in fundamental physics. His famous, "God does not play with dice" is a succinct version of this idea, which he also expressed by referring to the quantum theory as a "flight into statistics." [H]is concerns... had a second focus... "Most of them [quantum theorists] simply do not see what sort of risky game they are playing with reality." (Fine, 1986, pp. 1–2)

The risk was no less than placing into jeopardy the program of physics as traditionally conceived. Because of the "measurement effect" and the consequent unpredictability of describing quantum mechanical events (referring to how physical observation precludes the simultaneous description of the position and momentum of an elementary particle), physics would abandon its earlier aspiration of modeling an observer-independent reality.

The entire basis of "factual" knowledge now required reassessment, and as a consequence the very status of the idea of "objectivity" came into question. Beyond the philosophical difficulties of the apparent incommensurability of the Newtonian versus Bohr–Heisenberg languages, the naturalistic basis for a "realistic" view of the world has been left highly problematic. The repercussions of this epistemological upheaval are still incomplete, but we already perceive that quantum reality, built on foundations that appear to defy common sense, has burst the boundaries of our everyday language. But beyond the issue of description, the basic understanding of space, time, and causality underwent a radical reorientation. I have chosen Werner Heisenberg's philosophical musings on the revolution he helped orchestrate with "The Development of Philosophical Ideas since Descartes in Comparison with the New Situation in Quantum Theory (1958) (Chapter 4)." Aside from its historical interest, the essay clearly demarcates the evolution of the Kantian *a priori* categories of time, space, and causality from their transcendental, or what Heisenberg calls "metaphysical," basis to their status as a "practical concern." Space and time are no longer given or absolute, and causality has taken on a probabilistic character far removed from a simple mechanical model. The lessons learned, with deep epistemological implications, are that our most basic concepts are not sharply defined with respect to their natural language meaning, and we cannot know the precise limits of their applicability. "Therefore, it will never be possible by pure reason to arrive at some absolute truth." This is not to reduce ourselves to complete skepticism, because Heisenberg was confident of local descriptions of systems that were definitional in a mathematical scheme, and as we will see in the respective essays of Laudan and Hacking (Chapters 5 and 6), this same line is pursued as they describe science with a view of a pragmatic sense of the real. But Heisenberg's central point has been made. Reality, with a capital R, cannot be directly observed, and as a result, the character of our basic notions of time, space, and causality have become problematic when contrasted to our "natural" view of the world.

Beyond the new language of its mathematics, and the uncertainty built into its probabilistic structure, the quantum world has revealed fundamental imbroglios. Nietzsche's aphorism, "as the circle of science grows larger it touches paradox at more places" (*The Birth of Tragedy*) certainly captures how increased understanding can make the world in fact more bewildering. The question is no less than, "What is Reality?", and if we cannot "know" what is real then how do we determine criteria for truth and objectivity? The roots of our predicament sink deeply into our post-Enlightenment history, drawing from many sources of which scientific views are only tributaries. But science does have a singular role in framing our philosophical understandings, and thus these are not merely esoteric concerns of a few philosophers and physicists, for ultimately our philosophical positions serve as an important foundation from which ethics and the derivative social standards of behaviour are built.

We will delve into the philosophical discussions with an eye towards better comprehending the meaning of the new physics revolution (namely, of quantum mechanics and relativity), and perhaps in a more general (viz. popular) vein, to appreciate its repercussive effects well beyond the debates among the immediate combatants. Ironically, as quantum physics was born, champions of "fact" and "objectivity" – the logical positivists – achieved their most dominant position. They maintained that scientific method is the only source of knowledge, and that a statement is meaningful only if it is "scientific," i.e. empirically verifiable. This strong empirical orientation was opposed not only by Heidegger and Whitehead, but also by neo-Kantians such as Ernst Cassirer who argued in *The Philosophy of Symbolic Forms* (1923, 1925, 1929) that language, myth, art and science are not ways of apprehending and recording a reality structured independently of us, but must be regarded as expressions of our own spirit. Beyond the Kantian strategy of offering us *a priori* categories by which we might know a formless and incoherent nature, "the world" is partly but inextricably the product of our self-expression (Wolterstorff, 1984). (This is what Cassirer referred to as "functional" *a priori* categories.) But perhaps an even more important source for the current reaction against metaphysical realism originates in the revolt of the later Wittgenstein against the very positivists who followed his earlier work. In *Philosophical Investigations* (1953), Wittgenstein disallowed language to serve as a "representation" of the world, discounted our ability to discern the language rules that govern even the most rigorous logical analysis (much less our daily discourse), and thus dispelled the justifications that language serves as a direct correspondence to reality. Language, and the mind, could not mirror nature. The reach of his arguments have penetrated into every current philosophical issue, and for our purposes we need only note that his orientation had a crucial influence on spawning the current criticism against naive realism.

"Scientific realism" today refers to a precise position concerning how a scientific theory is to be understood and what scientific activity really is. A simplified definition is that the picture science gives us of the world is true and faithful in all details, and that the entities postulated *really* exist. (Philosophers more broadly refer to this belief as metaphysical realism.) Thus science advances by discovery, not inventions. This position is vulnerable to the attack that science is always changing, and both facts and theories are constantly being modified or discarded. But the realist nevertheless maintains that

> science aims to give us, in its theories, a literally true story of what the world is like; and acceptance of a scientific theory involves the belief that it is true. (Van Fraassen, 1980, p. 8)

A dominant opposing position is more modest, claiming that the aims of science can be well served without encumbering it with truth criteria that cannot be met. In other words, instead of proclaiming a theory to be true, this "modest realist" would simply display it, and enumerate its virtues, which may include, for instance, empirical adequacy, simplicity, comprehensiveness, coherence, predictability, etc. (Van Fraassen, 1980) Thus the scientist is never confronted with a complete theory, but is involved in a "research program" (Lakatos, 1970), situated within the broad array of factors that determine theory construction that need not be susceptible to truth claims of the sort demanded by the scientific realist. This view, which some label "anti-realist," is what Bas van Fraassen calls "constructive empiricism" (1980), Hilary Putnam "inner realism" (1981), and Arthur Fine, "the natural ontological attitude" (1986), each of course offering his own twists to what is essentially a neo-pragmatist position. The critical stance is not that the mind manufactures the world, but rather "the mind and the world jointly make up the mind and the world." (Putnam, 1981, p. xi)[2]

Whether the realism–anti-realism debate concerns a genuine philosophical question I need not attempt to adjudicate here. Suffice it to note only that the anti-realist position is an important element in opening the philosophical space for a critique of the normative view of science. Sociologists, from this perspective, picture science immersed in, rather than riding above the needs, tribulations and power politics of its supporting culture, lurching forward by rules not rigidly formalizable through logical analysis; and because of its tight intercontextualization within a complex matrix of philosophical, historical, and cultural contingencies, they argue that we cannot expect that science should possess a singular universal and prescribed method. From this general orientation, there is a spectrum of critiques, but whether one aligns with the 'strong' program of sociology of science (Barnes, 1985; Bloor, 1991), the empirical–relativist school (Collins, 1992; Pinch, 1986), ethnomethodological studies (Lynch, 1985), actor–network theory (Latour, 1987, Callon, see Chapter 10 in this volume), feminist epistemology (Haraway, 1989a, 1989b; Harding, 1986) or symbolic interactionism (Fujimura, 1992), to name only some approaches, each would demand a circumspect assessment of how to situate science's appropriate intellectual claims. Thus we are challenged to ponder how we view the power of the scientific vision, not only of nature, but of society and ourselves. We will again encounter these themes in Part IV, concerned with sociological models of scientific knowledge and practice, but first we must consider scientific praxis in the context of this general debate.

What are the implications of these philosophical discussions for the practices of science? Larry Laudan, in "Explaining the Success of Science: Beyond Epistemic Realism and Relativism," (1984) (Chapter 5) directly attacks both the claims of a naive realism (i.e. the view that we are increasingly defining what the world really *is*), and the social relativism that regards science as nothing more than a social construct. In so doing, Laudan attempts to offer an escape from the unsatisfactory bind in which each of these two views of knowledge places us. (This essay is especially germane in linking Part II with Part IV, dealing with the boundaries of science.) The argument Laudan offers is structured on first granting that science has indeed achieved success, whether assessed by predictive or manipulative control of nature, or by the precision and parsimony of our descriptions of natural phenomena. And how is one to explain these successes? When he confronts the relativist, Laudan shows that the issue of scientific success is hardly addressed, since the sociologists of science largely remain agnostic about the respective merits of various methodological and evaluative scientific strategies. Thus the matter is not even placed on the agenda, and we are to be content with a description of the activity we call "science" as simply one among other social activities. Laudan also accuses the relativist of lacking the explanatory tools for such an analysis, for there is no provision in his epistemological armamentarium for such an account. On the other hand, Laudan also attacks realism because it disallows any coherent sense of a theory's approximate truth. For example, a theory may be approximately true, but still inaccurate in those areas where it can be tested. And conversely, over the course of history, we have witnessed theories (e.g. Newtonian optics) which have been successful, but in fact from our current vantage point we now know to be fundamentally flawed. Laudan believes that an epistemic analysis of the methods of theory testing is required (i.e. how the winnowing process operates – the evolutionary models discussed in Part III), rather than proposing accounts of theory semantics. In short, he claims that a direct examination of theory testing allows us to make comparative judgments about the reliability of various methods of inquiry and of various theories, and he thus adopts the general neo-pragmatic attitude already mentioned.

The preceding critical caveats imposed by Laudan on naive realism leave us with certain modest constraints on our expectations of science. Extending Laudan's pragmatic orientation, we next examine scientific practice at its most immediate level, namely in the laboratory of experiment. (This area has been, until recently, a relatively neglected area of scrutiny; see Gooding, Pinch and Schaffer, 1989, for comprehensive references to this literature.) Ian Hacking's "Experimentation and Scientific Realism," (1984) (Chapter 6) explores how at the level of experimental practice, scientific realism is unavoidable, for there realism is not about theories and truth, since the experimenter need only be a realist about the entities used as tools. Hacking's claim, in agreement with Max Planck, is simply that the best evidence for the reality of an inferred entity is that we can begin to measure it or predict how it behaves. This surely is the practice of the working scientist, who nonchalantly ignores epistemological theorizing and adopts pragmatic criteria as the best proof of scientific realism. But Hacking would not succumb even here to Realism in a positivist sense. As he wrote elsewhere:

To experiment is to create, produce, refine and stabilize phenomena. If phenomena were plentiful in nature:...it would be remarkable if experiments didn't work. But phenomena are hard to produce in any stable way. That is why I spoke of creating and not merely discovering phenomena...Noting and reporting readings of dials – Oxford philosophy's picture of experiment – is nothing. (Hacking, 1983, p. 230).

Given this cautionary caveat, an orientation of science as first and foremost practice, regards the core importance of science in its predictive power, an empirico – logical praxis of constructing reality, and finally its demonstrable ability to manipulate nature in its technological applications. From this same perspective, Henri Atlan draws a most paradoxical conclusion:

The need for an explanation of reality is, fundamentally, anti-scientific. The satisfactory explanation is a bonus, the aesthetic pinnacle that accompanies and sometimes completes – not necessarily, but only when possible – the result truly sought: technical performance ...The need for explanation is merely a relic of metaphysical, indeed religious wonder. (1993, p. 193)

Atlan has forcefully drawn an important distinction: Do not confuse the tentative achievements of scientific theory in its various stages of maturity with the proven power of practical prediction, namely science's 'engineering' accomplishments. These remain uncontested and must dominate any debate concerning the nature of the scientific enterprise.

Hacking's pragmatic orientation may be the common (if not minimal) denominator of scientific practice, and despite its modest assertions – given the legitimate skepticism toward both scientific realism and its claims of objectivity, and deterministic claims of social influences on scientific change (discussed shortly) – we might well have to be content with studying how conceptual boundaries are created and how the scientist *behaves*. The issue then becomes whether such analyses must remain restricted to the local setting of the laboratory or should be extended to the wider reaches of the scientist's socio–political network, and in this latter sense, what kind of construction is to be applied? This is a very important matter relative to the social studies of science discussed in Part IV, concerned with the boundaries of science. A prominent body of literature has developed since the 1970s, where the focus of inquiry has been the laboratory, both the site of experimentation proper, as well as its broader reaches, to describe the setting from which facts emerge (reviewed by Gooding, Pinch, and Schaffer, 1989; Golinski, 1990; Pickering, 1992; Jasanoff *et al.*, 1995). The classic work in this regard, *Genesis and Development of a Scientific Fact*, by Ludwig Fleck (1979 [originally published in 1935]), is acknowledged as a seminal inspiration of Thomas Kuhn's *The Structure of Scientific Revolutions* (1970) (discussed in Part III), as well as the foundational opus for current sociological descriptions concerning the manufacture of a kind of local knowledge strongly dependent upon the practices employed in its making. These local case studies should be distinguished from a second, perhaps more traditional sociological approach, namely how such knowledge is propagated (e.g. Rouse, 1987, p. 125). These two modes of sociological inquiry, one focused closely on laboratory practices and the other on the dispersion of scientific knowledge will be considered in due course, but first let us review the problem of the autonomous knower, the scientist who would be pleased to be left in the laboratory to engage nature pragmatically, or at least so he or she would aspire.

The knowing subject persists on one end of the subject–object dipole as an interpreter, and here the issue of objectivity is raised. When we attempt to assign some insularity to the practicing scientist who seeks objectivity, we find her or him working both under the auspices of pragmatic, realist demands as well as in the context of social and linguistic constraints which are difficult, when not impossible, to overcome. Current portraitures regard the scientist as a complex historical, social and philosophical entity who defies assignment to either an objectivist ideal or description as a caricature drawn by the social relativist (Pickering, 1994). I have presented Hacking's perspective as seeking an alternative ground by offering a pragmatic approach that regards knowledge as made and sustained in practical activity. Contemporary philosophical, historical, and sociological perspectives largely converge in assessing that objectivity cannot be arrived at by transcendental, timeless norms of scientific practice; yet these perspectives diverge in the degree to which social forces are allowed to determine scientific content. We would do well to delve a little deeper into the matter.

Concern with scientific objectivity focuses upon the discovery or creation of facts; the status of facts in the modern scientific context dates to Francis Bacon's endeavor, in the early seventeenth century, to replace metaphysics with the concrete, the datum of experience. As Lorraine Daston notes, the word "fact" derives from Latin *facere*, "to do," and in the sixteenth century the word still meant an action or deed (1994, p. 45). The critical Baconian distinction was that facts offered neither "consensus nor freedom from all bias, but simply freedom from theoretical bias" (1994, p. 47). Daston maintains that facts became the focus of scientific discourse because they shifted attention away from the more contentious wrangling over rival theories; thus a social norm fashioned scientific practice:

> Since the academicians believed that partiality to one's own theories and opinions was the apple of discord rolled in their midst, the kind of impartiality they sought was impartiality to theory... Therefore, the purportedly theory-free Baconian facts suited their purposes perfectly, despite their other obvious disadvantages. Thus did objectivity come to be about the impartial examination of neutral facts. (Daston, 1994, p. 57)

The recent social history of the scientific community of the seventeenth century emphasizes how the civility of English aristocracy framed the powerful legitimatization process that social authority conferred on scientific discourse, and facts were in large measure accepted by the scientific community based on the social standing of the observer (Shapin and Schaffer, 1985; Shapin, 1994). This implicit trust has in our own era been transferred to the community at large. As Steven Shapin observes,

> the very power of science to hold knowledge as collective property *and* focus doubt on bits of currently accepted knowledge is founded upon a degree and a quality of trust which are arguably unparalleled elsewhere in our culture. (1994, p.417)

Trust amongst scientists is of course based on the shared ideal of scientific objectivity. Those embracing a constructivist orientation consider that objectivity is achieved primarily as a matter of rhetorical practice (e.g. Gergen, 1994). Because the individual cannot achieve objectivity as a private mental condition, monitoring objectivity then becomes a matter of broad social policy. There is a clear historical sensibility operative here, for criteria of objectivity have varied over the past four centuries. Consensus is required among participants as to what counts as an objective account, and these accepted criteria have changed in the course of the developments of science over time; moreover, in the contemporary context, we note that different scientific disciplines employ different standards of proof. The rhetorical techniques employed to establish objectivity are wedded to the development of common and esoteric languages, as well as to the use of metaphors to structure and communicate new models and theories (a topic that particularly intrigues me; see Tauber, 1994a). Linguistic means of separating subject and object, characterizing the objective world, establishing authorial presence and absence, and "cleansing the lens of perception" are prominent devices in generating a sense of objectivity; whether these devices largely reflect a commitment to a mechanical (Cartesian) view of the knowing self that strives toward an unrealizable mirroring of nature (Gergen, 1994), or represent a dynamic and evolving science based on some universal objective ideals is grounds for rigorous debate.[3] (See Boyd *et al.*, 1991 for representative arguments.)

Another way to deal with the objectivity question is to portray the scientist as vanishing, absorbed by her machines. As a simple reporter of her instruments (already discredited by Hacking – see above), the subjective element is eliminated. This is the aspired ideal Evelyn Fox Keller perceives in "The Paradox of Scientific Subjectivity" (1994) (Chapter 7). She steps back from the persona of the scientist as a social entity (the identification that dominates most sociological analyses of scientific practice) and attempts to portray her subsumed beneath the epistemological demands of what Thomas Nagel calls the "view from nowhere" (Nagel, 1986). This refers to the ostensible goal of a completely detached observer, one independent of subjective foibles and prejudices, whose conclusions come from "somewhere else." Keller begins with noting that the invention of classical perspective in painting during the fifteenth century serves as a ready metaphor for the birth of modern science. In one sense, the idea of the detached observer was the first step in accurately assessing nature, but an even more radical rupture of subject–object was required: the task of modern science was first to standardize observation and then to eradicate the observer altogether in the quest of a complete elimination of the subjective dimension. By focusing on experimental procedures, Robert Boyle effectively propagated a shared research program, which generated a "multiplication of the witnessing experience" (Shapin and Schaffer, 1985, p. 488), with public demonstrations and the enlistment of other scientists, and a rhetoric of reports that emphasized the observed facts and described experimental procedures in great detail (virtual witnessing). The singular subjective observation was thus co-witnessed and translated into a shared public objectivity through the machine's results. The disjunction of subject from observation was hardly complete, however, and for objectivity to assume its current meaning of being "aperspectival," extensive rhetorical refinement and the development of statistical analyses in the nineteenth century were required. Standardized equipment and techniques universalized scientific practice so that the first person report could be replaced by the abstract "scientist." (Note, that the very word "scientist" is itself an abstraction and was first coined by William Whewell in 1840.)

I have selected Keller's essay not so much for its epistemological insights (Hacking's contribution is more to the philosophical point) as for its historical perspectives offered on objectivity, and the issues it introduces concerning the humanistic relationship (in some cases responsibility) of the scientist with his or her knowledge. Keller believes that the historical evolution of objectivity has left the human only as a machine among machines, a conclusion that has important moral implications. Her position should be contrasted with Heidegger's (chapter 2), whose owh moral concerns derive from the limits of Man as "measure of all things." These matters are addressed in Part V of the anthology, so we would do well to keep them well within reach. However, prior to turning to these concerns, let us review how historians regard the evolution of science and characterize its practice as a function of change.

THE NATURE OF SCIENTIFIC CHANGE

The nature of scientific change as non-incremental, but rather as proceeding in sudden leaps, or what Thomas Kuhn referred to as paradigm shifts, has led to an intense historical exploration of specific historical instances of scientific change, as well as the more general question of the nature of scientific progress. To address the question of change in science, I have chosen a comprehensive review of science historiography by Robert Richards (Appendix I of *Darwin and the Emergence of Evolutionary Theories of Mind and Behavior*) (1987) (Chapter 8) to discuss how we currently understand the evolution of science. Richards presents several historiographic models that mark the major trends in the writing of science's history. The first two represent normative views of science. (1) The Growth Model of the eighteenth century, regarded science as relatively isolated from other disciplines. This view persisted into the twentieth century in the work of George Sarton, who maintained that science advanced by rationally exact methods, reflecting the unity and continuity of knowledge. Independent of social pressures and historical contingencies, science was envisioned as progressing as it discovered the Real World. (2) The Revolutionary Model, first advocated at the end of the eighteenth century, emphasized the sudden shift of thinking inaugurated by the Scientific Revolution of Bacon, Galileo, and Copernicus. *What* that revolution was and continues to be is open to interpretation, but *that* science underwent a radical methodological and metaphysical shift in the sixteenth century is generally acknowledged (see, e.g., Koyré, 1957). Each of these views regarded science as essentially progressing within its own system of rational inquiry, fulfilling its mandate to describe Nature objectively. This normative point of view has been superseded and replaced with a variety of models emphasizing the social interdependence of scientific practice and theory, the philosophical uncertainty of scientific methods and verification, and the historical record of scientific change; this latter record was observed to exhibit no logical, progressive rule or method, but instead evolved by what appears as convulsive, sudden revolutions that can hardly be accounted for by a simple narrative that describes incremental stepwise conceptual changes.

Most commentators acknowledge Thomas Kuhn as the most notable author of this position. He emphasized that beyond the non-incremental nature of scientific change, some changes resulted in incommensurable world views, such as those exemplified by a Newtonian and an Aristotelean science. In *The Structure of Scientific Revolutions* (1962, 1970), Kuhn differentiates between "normal" science where there is progress in an accumulative fashion within a given paradigm, and those sudden conceptual changes that generate a "revolution." Richards refers to Kuhn's interpretation as a Gestalt Model, emphasizing Kuhn's analogy of scientific conceptual shifts as pictures which may appear as one image or another; as in the exemplary case of the duck–rabbit where the two images cannot be fused to appear as a conglomerate of both, so analogously, divergent world views while referring to the 'same' phenomenon, assemble those phenomena in radically divergent ways – and are thus, incommensurable. The Gestalt Model of scientific change illustrates how context, past experience, and familiar assumptions control, or at least guide, our perceptions and conceptual experiences as well. Kuhn's notion of paradigm – a term now hopelessly weakened by overuse and lost amongst its various interpretations (Masterman, 1970) – represented this perceptual – conceptual whole of fact and theory, whose gradual or incremental changes cannot account for the Gestalt shift in major scientific revolutions. From this perspective, there is no Truth, but only facts that cohere in the theoretical, technological, and methodological arrangements of the time. In other words, there are no "facts" autonomous from particular paradigms in which they are placed (see also Putnam, n. 2 below).[4]

The key concept, and the one generating the most discussion, is not Kuhn's argument concerning scientific change (i.e. a skeptical assessment of the picture of an orderly progression of scientific growth), but that such revolutions may render the vanquished paradigm incommensurable with its successor. Scientific theory thus domineers the structuring of epistemology, radically. Kuhn's argument of the incommensurability of paradigms has generated intense debate concerning the endeavour to describe nature "objectively," ranging from the technical limitations imposed by our instruments to the arbitrary conceptual boundaries science must accept in its pursuit of "truth" (see Hoyningen-Huene, 1993 for exhaustive references regarding Kuhn's work in general). As already discussed, the problem of objectivity remains a key issue unresolved in current discussions of science's philosophical foundations (e.g. Megill, 1994). Besides stimulating competing models of scientific change, *The Structure of Scientific Revolutions* spawned an entire generation of science studies inspired by its central thesis. We will again consider the fruits of those labors in Part IV, concerning boundaries of science, but first we must allow Kuhn a rebuttal to himself.

Kuhn's own views have been modified to the point of disavowing the "Kuhnians."[5] Thus I have chosen Kuhn's "The Road Since *Structure*" (1991) (Chapter 9) to present his more recent position regarding scientific knowledge, and more specifically, incommensurability. It is a most interesting endeavor to salvage *The Structure of Scientific Revolutions* from a radical relativism. Kuhn builds his case from familiar material: science firmly embedded in its culture ("The Archimedean platform outside of history, outside of time and space, is gone beyond recall," 1992, p. 14), an evolutionary epistemology ('Scientific development is like Darwinian evolution, a process driven from behind rather than pulled toward some fixed goal to which it grows ever closer,' 1992), and an anti-realist orientation for truth claims ("If the notion of truth has a role to play in scientific development, which I shall argue elsewhere that it does, then truth cannot be anything quite like correspondence to reality," 1992). From these positions, Kuhn redefines incommensurability as a conceptual disparity between specialties that have grown apart, a "sort of untranslatability, localized to one or another area in which two lexical taxonomies differ" (1991). (One might compare these views with those of Heidegger (Chapter 2). In some sense, he builds on a foundation shared with the Kuhnians, and the reader is left to weigh the reorientation he would command.

The Social–Psychological model, describes science as fueled by social interests and psychological needs to the point of driving, if not determining, theory conceptualization. Marxist or Darwinian applications are well known, but in the past twenty years, newer forays have been presented under the rubric of social constructivism (discussed below). Finally, there are a variety of Evolutionary Models, represented most prominently by Gerald Holton, David Hull, Imre Lakatos, Karl Popper, Robert Richards, and Stephen Toulmin. On the evolutionary view, theory survives not solely by appeal to evidence, but because other competitors are less fit. The intellectual environment selects ideas and restricts those that might be otherwise entertained in a complex calculus of the intellectual traditions and criteria of the discipline and the social situations of its practitioners. The epistemological basis of theory choice is historically contingent; because we see that standards of truth change, it is not surprising that those forces governing theory change reflect pragmatic choices and approximations. Science then is viewed as an ever-evolving enterprise, whose business is to attend to the picturing of the world, but as Representation it is only an approximation, and the pragmatic epistemological judgment of its practitioners is always unsteady as they probe to verify their depictions with nature as best they can.

THE BOUNDARIES OF SCIENCE

The boundaries of science fall into two large arenas. The first pertains to arguing whether social phenomena are amenable to scientific scrutiny and criteria. In other words, what are the relations between the natural sciences and the human sciences? What aspirations should the human sciences seek as science? Or, more fundamentally, are the social sciences "science" in the same sense we think of physics, chemistry, or biology? The second boundary question pertains to the social contextualization of the natural science. Before addressing this latter issue in detail, let us briefly review the problem of regarding the social sciences as we define natural ones.

The aspiration for scientific objectivity has been a central concern for the social sciences since Auguste Comte proclaimed his positivist ideals in the 1820s. Today, the positivist agenda for the social sciences is no longer generally accepted (see Proctor, 1991, pp. 163–181, for discussion), and revised notions of how the social sciences might be based on discovery of their own general laws persist as a central theoretical concern. One strategy is to view such an inquiry as falling on a contiuum of scientific explanation, where a more relaxed view of laws from that espoused in the natural sciences might suffice (e.g. Kinkaid, 1990). This line of argument often leads to an espousal for suggesting that a new *kind* of science is required, which then serves as a loose rationale for many programs that do not even attempt to cloak their efforts in the garments of scientific legitimacy. I suspect this admission serves to weaken the entire edifice, prompting some critics like Lee McIntyre (1993) to maintain a stauncher stance. He suggests that social scientists must seek new descriptive criteria that are amenable to reducing the phenomena to simplified laws. For McIntyre, the problem of the social sciences is to change what appear to be the natural presentation of social problems into a new vocabulary better susceptible to analysis: the social phenomena are perhaps not inherently resistant to scientific analysis, but just as in pre-modern inquiry, the suitable questions have not yet been posed.[6] A similar posture is assumed by Brian Fay (1983) where he assails weakening the criteria for laws of causality to accommodate the difficulty of predicting social behavior. Even in failing this agenda, the general orientation of what he calls "critical theory" fulfills the larger purpose of self-reflection and reform. In this sense, then, science again fulfills its larger ethical mandate, and it does so in the same more modest spirit of the scientist who recognizes the contingency of his own best efforts at defining the natural world. Implicit to any of these advocacy positions (see Martin and McIntyre, 1994, for an excellent presentation of various perspectives) is the firm espousal that humans and their cultures are fundamentally, and reducibly, extensions of the natural world, or as Alan Wolfe puts it, "humans have no special qualities and deserve no special place" (1993, p. 15)

We will not dwell further on the issue concerning how science, or at least scientific methodology and its critical aspirations might be applied to cultural phenomena, turning instead to another matter altogether: how do we understand science sociologically? It is well to note here the critical provisos we have already charted in regard to the professed ideals of scientific inquiry that have arisen within the philosophy and history of science, proper: The ambiguities regarding objectivity and the nature of the knowing subject relative to her object of study; the delineation of the limits of natural scientific theory; the anti-realist concern that epistemic judgment depends upon local contexts of use; the pragmatic attitude that must dominate scientific inquiry. We will complement these caveats with recent criticism arising from the social constructivist approach to science, which has emphasized the powerful forces of cultural determinism at play in scientific practice. This last issue now commands our attention, and serves as the focus of our discussion concerning the boundaries of natural science.

As already discussed, one cannot view scientific realism or the nature of scientific change as matters that simply may be maintained as internal to the scientific enterprise. In the past twenty years, a debate has emerged concerning the relative influence of external factors that may influence scientific practice. On the one hand, the "internalist" approach to science argues that science grows from its local, immanent concerns, that it is subject to and governed by rational discourse, and that the world it examines may be discerned objectively by the scientific hypothetico–deductive method (Hemple, 1966). The project to comprehend nature is thus viewed as essentially logical and self-sufficent. This is what Michel Callon (in "Four Models for the Dynamics of Science") (1994) (Chapter 10) calls the model of "Science as Rational Knowledge." This model imposes severe constraints on the social organization of scientific work, and implicitly relies on the Realist view of Nature, where science enunciates a sophisticated and formalized dialogue between man and nature. This is fundamentally a normative exercise, where social influences are minimized in the pursuit of Truth. Callon's "Competition" model attempts to regard science in a richer social context, but again normative rules govern practice, albeit in a more dynamic setting, and scientific growth continues to possess its own rationales.

On the other hand, these positions have been challenged, as already reviewed, from both historical and philosophical analyses, and beginning in the 1970s a school of science studies sociologists have presented a spectrum of alternatives to the Rational models of scientific progress. Callon refers to these studies as a model of "Science as Sociocultural Practice," to which Andrew Pickering ascribes no less than

> a new approach to thinking about science... The sociology of scientific knowledge... insisted that science was interestingly and constitutively social all the way into its technical core: scientific knowledge itself had to be understood as a social product. (1992, p. 1)

And in addition, perhaps ironically aping its subject, the approach was "determinedly empirical and naturalistic" (1992). The constructivist argument, in its broadest interpretation, is that science cannot be segregated from the complex economic and political forces that support its activities, but are also heavily indebted to them. On this view,

> scientists' decisions... can be explained by reference to various active elements of their situation. That is to say, actors' choices are con-

strained by their aims or interests and by the resources they select to advance them. (Golinski, 1990, p. 502)

This position is not identified with "externalism," for the scientist is acknowledged as operating within a methodological tradition that incorporates theoretical skills that "represent a set of vested interests *within* the scientific community" (Shapin, 1982).

Perhaps the best developed perspective of science embedded in its social matrix is offered by Callon's fourth model, "Extended Translation." In this account, science is displayed as a vastly intricate network, operating in a communication system which connects the various levels of its discourse constitutively, as well as outwards through a supporting social mileau. By and large, science is erroneously (or perhaps just superficially) regarded as having its own sophisticated domain, and ventures forth from its insularity with caution against a potentially intruding public. The lobby promoting science routinely expounds the obvious benefits of scientific progress for technology and medicine, and often treats the lay person to a dramatic presentation of Science at Work. Whether it is pictures of a comet crashing into Jupiter, or the pinpoint accuracy of smart missles, or the *in-vitro* fertilization of a sterile sixty-two year old woman, Francis Bacon's espousal of science's promise to better society is constantly reiterated. To propel that message, the boundaries of science extend well beyond the laboratory into the copy rooms of news agencies and the studios of television (e.g. Gieryn and Figert, 1990). Not solely the business of scientists, their findings and disputes become intertwined into the interstices of seemingly disinterested parties to become integrated within society as a whole.

Few would dispute the constructivist's claims regarding the intercontextualization of science into its supporting culture, but these critiques have generated heated debate when the arguments have followed a theoretical continuum that appears to conclude in radical deconstruction, whose end point leaves science reduced to politics. The ire of "scientists" (and here I am referring to those who regard science as a normative enterprise) concerns the assertion that theoretical formulations are heavily determined by ideological orientations, whether political or sexual (e.g. Haraway, 1989a, 1989b and Harding 1986), or by those that would regard science as no more than a rhetorical enterprise, where persuasion is used to overwhelm the opposition. On this view, the pursuit of knowledge seems to command interest only as a process where scientists are regarded as pitted against one another in an "agonist field," locked into a constant trial of rhetorical strength (Latour and Woolgar, 1979).[7] The sociological question: How is truth erected or arrived at? (e.g. Collins and Pinch, 1982; Collins, 1992; Latour, 1987), has on the one hand been interpreted as a circumscribed description of the pathways governing scientific discourse, but on the other hand, the interpretations of such practice have led some critics to decry what they perceive as a relativist assault on the scientific enterprise.

Just as there are diverse perspectives offered by social constructivists of different stripe, the rebuttals are similarly slanted towards different agendas, whether the argument is poised in historical and cultural (e.g. Holton, 1993), sociological (e.g. Cole, 1992), or philosophical terms (e.g. Laudan, 1990). It is too early to judge the full ramifications of this debate and assess the proper application of the various brands of social constructivism, that is, the extent of social influence determining communal scientific knowledge and the constraints of the empirical world on cognitive relativism.[8] But what remains most appealing about science studies, if a generalization is to be made, is the barbed critique of what Harry Collins has called the "ethnocentrism of the present," that is the feeling that "now" (as opposed to the past), we have finally achieved scientific "maturity" and have a method and a world view that approaches some Final Theory (e.g. Weinberg, 1992). The skepticism regarding universal and unchallenged modes of assessing scientific statements or for distinguishing between intrinsically scientific and non-scientific research programs has been forcefully reiterated by recent science studies, which *in toto* seem to share doubt that there is one singular, rational scientific methodology. This is an important lesson also learned from both philosophical and historical analyses, but hardly a novel one. Such skepticism forms the very foundation of research practice, an ethos that has ruled science since its birth. Recent sociological studies have reiterated that message, but have formulated it in a more complex and highly textured context than previously appreciated. This project is what Thomas Gieryn has summarized in "Boundaries of Science" (1994) (Chapter 11).

The "boundary problem" refers to the attempt to define the border between science and non-science, or perhaps more specifically, adjudicating claims of what comprises scientific practices and, ultimately, scientific knowledge. Gieryn poses the issue from the different perspectives of essentialism and constructivism. In a sociological argument with overtones reminiscent of the philosophical discussions already referred too, essentialists maintain the possibility and analytic advantage of identifying unique and invariant qualities that set science apart from other occupations, and the larger purpose of explaining its singular achievements; constructivists deny such demarcation principles and instead argue that science, like other intellectual disciplines, is contextually contingent and driven by pragmatic interests of its supporting political culture. Gieryn doubts that we may discern universal demarcation criteria, believing that sociology of science has settled on examining how, and to what ends, boundaries of science are drawn and defended in practice.[9] From Gieryn's perspective, boundaries result from a complex interplay of social conventions, both within the "scientific community" (however defined) and from its surrounding culture. The general critique that subsumes the particular concerns with how political ideology might color not only scientific goals, but the very research programs espoused as value neutral, derives primarily from European post-World War II social philosophy and American feminism.[10] In a section omitted here, Gieryn relates this orientation to other theoretical studies in this broader field of cultural studies, and he then concludes with a lucid review of recent constructivist studies of science, which deal with science as politics and the legitimatization of science (fraud and accreditation).[11]

The overarching issue to ask of the social constructivist is to what degree science practice, as socially determined, allows its *conceptual* work to be constructed from these sources. In other words, to what degree are scientific theories, hypotheses, even facts, mere reflections of societal mores, expectations, and underlying cultural structure, as opposed to the expressions of scientific experience in a rhetoric dependent on common language? There is an implicit vagueness in any groping towards the unknown, and the scientist must borrow from his language and everyday experience to articulate metaphors and models in the process of constructing a scientific universe based on empirical phenomena (see, e.g., Tauber, 1994a). The radical social constructivist has been read as arguing a different and damning relativist indictment: Beyond the usurpation of common language and public categories to describe nature, the scientist manufactures theories from the social residua of his hidden prejudice and preformed social consciousness so that the conceptual product is only the social world portrayed in a different, albeit "natural" guise (see e.g. Collins, 1992; Harding, 1986; Haraway, 1989a, 1989b). In other words, language is so imbued with the social construction of reality, we cannot legitimately separate scientific facts from their supporting social milieu.

Relativists have made an intriguing case. They argue for the contextuality of scientific objectivity, whose criteria are forever changing. But science *is* successful, and to conflate the contextuality of knowledge with the possibility of its objectivity is to deny the obvious accomplishments of scientific method. No doubt, scientific objectivity depends in part on its contextuality, but that is not to deny its strengths of verification, coherence, and predictability, even within the local context of its inquiry. So as we readily acknowledge that social factors do indeed play an important role, we must not lose sight of how science locks onto nature, offering us means for manipulation and powerful prediction. It is not an arbitrary description, and within its local domain, it serves as an important limit to relativist thinking. This is not to say that scientists do not choose among various research programs by what may be judged as "ideological" criteria (e.g. Tauber and Sarkar, 1992, 1993; Lewontin, 1991), and others have called "aesthetic," criteria (see Kuhn, 1970 and the respective essays by Margolis, McAllister, and Sarkar in Tauber (ed.), 1996). Because of training within particular research traditions or commitment to broader social concerns, scientists often choose a particular research program from several options. In short, there are different ways of probing the same problem, and in the process the phenomenon may appear quite differently from each perspective. For instance, whether molecular genetic research is intrinsically "better" than other kinds of biological approaches to the description of complex systems such as the immune or nervous systems is hotly disputed (Tauber, 1994a). The resultant scientific facts and theory will obviously reflect these different approaches, but this only demonstrates the obvious: Nature may be described in varying ways. The depiction then is to a certain extent locked into the particular mode of representation offered by its research program. The final choices may exhibit strong extra-curricular influences, but within the particular program a cognitive integrity tested against nature competes with other programs. In the evolution of scientific practice the fecundity of one approach may lead to its outlasting others. And this success is most directly determined by how well the science allows us to interface with nature.

At this point it might be useful to consider the heuristic model developed by Imre Lakatos to explain scientific change (1970), and adopt it to orient these epistemological and sociological issues. Let us regard the body of scientific knowledge as consisting of complex levels with a stable core, which remains unperturbed by the conceptual and social whirlwinds rushing around its more peripheral activities, the sites where new knowledge is being fashioned. Perhaps these peripheral sites are more subject to various theoretical interpretations that include the influence of factors arising from the social setting. But eventually the conceptual basis is "hardened" by further testing against nature. In short, at some point in the trials of scientific theory construction, knowledge coalesces around its pragmatic achievements and frees itself from its own theoretical vagaries and uncertainties, which may include the vicissitudes bequeathed by its political structure. No longer subject to radical development, a research program matures, and its core content is protected and codified. In other words, how scientific theories emerge may well be examined as products of a particular scientific method, the political interests of the supporting culture, the historical contingencies of the setting, etc., but the evolution of that process does result in a practical achievement of effective manipulation of nature: On the basis of Newton's laws, we have landed on the moon... and returned. Beyond some social consensus, there is an objective result related to our manipulation of nature. That "product" constitutes in some pragmatic sense an end point regardless of whether the theory that explains it is True in any final sense. In a different format, this is analogous to the epistemological differentiation between what scientific theories describe in nature and their claims as explanations. In short, there is an acknowledged hierarchy of validation applied to scientific knowledge, and the theoretical core may be regarded as quite limited when viewed critically; the rest is approximation with a different status of truth (see, e.g., Cartwright, 1983).

Scientific explanation follows a pragmatic course, and idealized truth claims of theory notwithstanding, science is driven by its practical concerns. Science asserts a powerful understanding of nature that cannot be denied if its claims are circumspect to its method of study and we are able to differentiate the validity of its range of assertions. Guided and prodded by its political environment, and implicitly structured by its underlying metaphysics, science remains a pragmatic enterprise resting on a constructive empiricism and the power of "local" descriptions (i.e. well-defined experimental systems) of reality. Nevertheless, science has enormous influence on shaping the conduct of our lives. Beyond determining the technology of our time and the pervasive formulation of our world view, there are strong ethical consequences derived from science on its culture with wide-ranging effects on how we organize our society and regard ourselves. To these influences, we now turn.

SCIENCE AND VALUES

After considering the problem of defining boundaries of science, we can now turn to the relationship between science and morality. This is a complex matter, because we must disentangle two often confused issues: First, what are science's ostensible values and how do they govern scientific practice? And secondly, how does science influence our broader morality? Each of these concerns relate to the boundary question, because the ethics of science arise within broader cultural ideals, and the culture may in turn be influenced by institutions of science, whose moral code may serve as a model of behavior extended to other domains. Despite the intimate historical links between scientific and moral discourses (Shapin, 1994), we must recall how science ostensibly was born in the effort to free itself from the conflations between fact and value, between natural and supernatural, between body and spirit. Even at the dawn of modern science, this issue of mapping the domain of knowledge proper was recognized as crucial, given the rise of political agendas concerned with how science could be used to serve particular social and economic interests. And it was in the recognition of science's power that its early institutionalization was structured to guarantee its independence from meddling politicians, who in turn extracted the promise that scientists would remain disinterested in worldly concerns (e.g. Shapin and Schaffer, 1985). After all, in search for Truth and Reality, the scientist should be inured from the messy debate of how the fruits of his labor were to be applied (whether in warfare, medicine, or technology-at-large) or the possible dire consequences of his discoveries for the environment and for the individual. From Science's lofty laboratories, only unperturbed truth seekers would explore Nature's secrets, oblivious to the political, social, and economic needs of the supporting culture. Some critics, such as Theodore Roszak, writing in *Where the Wasteland Ends* (1972), have deplored the moral consequences of this posture:

> Bacon went in search of a philosophy of alienation. They [scientists] broke faith with their environment by establishing between it and themselves the alienative dichotomy called "objectivity." By that means they sought to increase their power, with nothing – no sensitivity to others or the environment – to bar their access to 'the delicate mysteries of man and nature." The cult of objectivity has led scientists and the general public, to think of everything around us – people and biosphere – as "mere things on which we exercise power." Objectivity is in practice a cloak for callousness. (p. 169)

As patient spectators, the "common" people would reap the material harvest, although some would argue it was a Faustian pact. Objectivity is a value, and we will now look at its various manifestations.

The neutrality of science depends on regarding Nature as holding no value; value is rooted in human needs and desires, whereas nature is stripped of qualities, teleology, and meaning, leaving it devalorized, secularized, and disenchanted. The crucial philosophical distinction is between "what ought to be" and "what is." The attempt to free facts from value was to liberate science from its Medieval theological roots, and remains the linchpin for scientists pleading autonomy under the rubric of "objectivity," as well as for their critics, who decry the violation of neutrality of science, which obviously serves particular social agendas. But as Robert Proctor has cogently observed,

> [N]eutrality and objectivity are not the same thing. Neutrality refers to whether science takes a stand; objectivity, to whether science merits claims to reliability. The two need not have anything to do with each other. Certain sciences may be completely "objective" – that is, valid – and yet designed to serve certain political interests. Geologists know more about oil-bearing shales than about many other rocks, but the knowledge is thereby no less reliable. Counterinsurgency theorists know how to manipulate populations in revolt, but the fact that their knowledge is goal-directed does not mean it doesn't work.
>
> The appropriate critique of these sciences is not that they are not "objective" but that they are partial, or narrow, or directed towards ends which one opposes. In general, knowledge is no less objective (that is true, or reliable) being in the service of interests. (1991, p. 10)

Given these distinctions, science nevertheless can hardly be separated from its political support, and scrutiny of that political context has increasingly raised a critical chorus since World War II. Citizens maintaining a vigilant watch over scientific aspirations and purported successes no longer accept as gospel the claims and promises of a growing scientific lobby. Critics have successfully halted or modified multi-billion dollar Big Science projects in the 1990s, in what some regard as an anti-scientific conservatism, and others the appropriate constraint of a ravenously imperialistic Science. Controversies surrounding public policy concerning investment in major scientific projects that are touted as the penultimate, if not the ultimate, climax of scientific progress (e.g. Gilbert, 1992; Weinberg, 1992) are largely connected with three domains: The disappointment in past, similar programs, such as the failed War on Cancer and other overly optimistic projects, promising to deliver solutions that were unrealistic; the growing concern that resources should be more carefully allocated towards directed application and more modestly achievable goals; and finally the recognition that the naive positivist ideals have been abandoned. Further, despite the reiterated disavowals of a value-laden science, critics have exposed this innocent view.

This critical stance is based on the assertion that science as practiced is not a free-standing enterprise, but is socially based and subject to the needs and values of its supporting culture. This public domain of science refers not only to the renewal and support that our society gives scientific institutions, but the recognition that science serves in a political culture, supporting various economic and political interests. From a political perspective, the relevant issue, beyond defining the social origins of knowledge, is the requirement for a philosophy that focuses on the forms of power in and around the sciences:

> Why do we know what we know and why don't we know what we don't know? What *should* we know and what shouldn't we know? How might we know differently? (Proctor, 1991, p. 13)

In short, a political philosophy of science is emerging. Integral to this pursuit is the necessity of preserving the openness of scientific inquiry and its pluralistic attitude towards knowledge.

Nicholas Ashford and Karin Gregory; in "Ethical Problems in Using Science in the Regulatory Process" (1986) (Chapter 12) reflect on how investigators move from the laboratory into society-at-large by advising and directing public policy. Here we witness the makings of a political philosophy of science, for in the judicial or regulative advisory position, the scientist plays to both scientific and broader humane concerns. The expert must offer his or her best professional opinion regarding risk, but at the same time recognize that any ostensibly objective judgment may be biased. These authors' first caveat is to recognize that interpretations of scientific "facts" are not necessarily impartial, and that scientists carry complex personae. Prejudicial judgment has many sources, some intrinsic to the science *per se* (e.g. a commitment to a particular theory that may be in dispute, or confidence in a certain methodology that may be questioned), and other influences that are extrinsic to the narrow confines of the laboratory (e.g. a political affiliation or religious belief that might seek scientific support). Dispute, whether about the safety of a drug or the environmental impact of a chemical, generally reflects a continuum of scientific certainty, where facts may assume various meanings. This point reiterates Proctor's concerns, and further serves as a vivid illustration of a theme developed in this anthology, namely science's own quest for an elusive certainty. We see here that the doubts raised by philosophers, historians, and sociologists concerning the epistemological foundations of scientific inquiry have social consequences.

Ashford and Gregory are specifically concerned with offering guidelines to policy makers, whereas activists, such as Paul Feyerabend (1978), argue for more radical citizen participation in judging science, whether in regards to evaluating its technological products or its testimony that may have broad social consequences. The growth of the environmental and anti-nuclear movements reflect in large measure a broadly held sentiment that science must not be deified into an unassailable ideology of the state. Not only would such critics regard science with a skeptical eye, they would further argue that other forms of knowledge are important and offer legitimate bases to assess scientific advance.

Some regard such arguments as, at least incipiently, "anti-science," but in seeking to examine the underlying claims made by science for formulating our construction of reality, the scientific project itself must be strengthened. The issue is not scientific validity; afterall, none could reasonably deny scientific achievements, but Feyerabend and likeminded critics endeavor to place those accomplishments within their proper truth claims, and to reassert a more encompassing philosophy.[12] And when one also considers how scientific method might serve broader social values, a proper assessment of the scientist's role is demanded. Scientists are expert consultants, not infallible authorities, and their knowledge must be used critically as part of the complex fabric of social contingencies (see, e.g., Collins, 1992). We have already considered this issue within the more focused discussion of science's philosophical foundations in the section on current notions of scientific epistemology. But these matters cannot remain within a neatly defined boundary labelled "Philosophy" or even "Sociology."

The public nature of scientific practice has also raised new concerns regarding research fraud. A false sense of autonomy has increasingly been intruded upon by a public that trusts scientists less and demands more of them than a generation ago. Although there are no reliable statistics concerning scientific misconduct, there is rising public awareness of research fraud that threatens the very legitimacy of science. These new concerns for accountability have challenged the very ethos of scientific practice, and in the process we recognize the growing importance of science to modern society. The ever-increasing investment society makes in scientific pursuits has altered the relationship of esoteric knowledge and public access, with unpredictable consequences for both governing scientific practice and the direction and speed of its future growth. Thus we see a vivid example of science's blurred boundaries in recent government action imposed on research institutions to ensure trustworthy research.[13]

Reversing the ethical vector, we next consider what science may have bequeathed to moral discourse. The position of science as an idealized venture towards truth may be understood as representing a core human ideal toward which our society must aspire; this view is presented by Robert S. Cohen's "Ethics and Science" (1974) (Chapter 13). Science and ethics have often intersected in the ambiguity concerning the concepts of natural and social law. Cohen traces the origins of our moral thought from Hebrew and Greek sources, where he perceives that theories of knowledge and theories of ethics were not only analogous, but closely linked. This intimacy is evident throughout Western thought, from Bacon to Kant and beyond. Science becomes a dominating influence in its molding of epistemology, bestowing its conceptions of the boundaries between what is possible and impossible, and of course the derivative notions concerning human nature, social structure and the world-at-large. And in turn, moral thought has profoundly influenced science; as Cohen notes for instance, the idea of causality so centered in scientific thinking has an important origin in the ethical concept of retribution.

Obviously the role of science in ethics is complex. The most salutatory view represented by Cohen's essay is that scientific methods may provide a logic for moral discovery; scientific investigations offer "the data of ethics;" scientific achievements determine, in the broadest sense, the scope and limits of responsible moral choice; the ethics of scientific investigation serves as a model for societal behavior and truth-seeking; and ultimately, science poses for us the frontier of new problems and new circumstances for old problems. The analytic, descriptive, behavioral and problematic domains science poses for ethics each require their own scrutiny and integration into the broader matrix of science in culture. The deeper lesson from what Hans Reichenbach called "scientific philosophy" (*The Rise of Scientific Philosophy*, 1964) is that the struggle to achieve a scientific world view, with all of its problems concerning objectivity and realism, has provided us with rigorous criteria for the limits of knowledge, and thus with the basis of logical choices in the ethical domain. These claims for how science might contribute a valuable core of objectivity to ethics have hardly been accepted in all quarters, where the attempt to attribute a special truth ethic to science has itself been subject to reassessment. Those who see the fall of science from its domineering pedestal, where a naive positivism commands Truth and Reality, argue that science's posture relative to directing moral inquiry must be regarded more circumspectly: how can science be neutral if scientists are self-evidently social creatures with political biases?

I believe that how we situate science and ethics ultimately depends on recognizing the legitimate rationality of each. Leaving science within its own pragmatic domain, but not beyond, is a position that assumes that distinctive forms of thinking govern different dimensions of human experience, and that the legitimacy of different rationalities is not to advocate relativism, but to acknowledge that there is no single epistemology which may lay exclusive claim to all domains of experience. Each form of knowledge explores its own world. The common mistake is not to identify the limits of a particular logic, and to appreciate that there is no *relative* merit in scientific versus other Reasons, e.g. religion. Rather than seek a meta-theory to encompass each mode of knowledge, respect for intellectual and social experience of each practice must sustain the rightful claims of the ethical and the scientific.

> Ethics is not born of knowledge. It comes from somewhere else, and this elsewhere is what can be provisionally denominated... the "social imaginary." (Atlan, 1993, p. 276)[14]

This orientation is dependent on, what Hilary Putnam refers to as our ability to go "Beyond the Fact/Value Dichotomy" (1982, Chapter 14). Again, as we have noted previously, science itself is subject to assuming value judgments regarding its own practice. After all, science can hardly be said to proceed by any formal method. When theory and fact conflict, sometimes one is given up, sometimes the other, and the choice as often as not is made "aesthetically," by adopting what appears to be the simplest, the most parsimonious, or possessing the highest coherence, each of which themselves are *values*. These are what Putnam calls *action guiding* terms, the vocabulary of justification, which is of course historically conditioned and subject to the very debates concerning the conception of rationality. The attempt to restrict coherence and simplicity to predictive theories is self-refuting, for the very logic required to even argue such a case depends on intellectual interests unrelated to prediction as such. Putnam concludes that if coherence and simplicity are values, albeit the objective values governing science, then the classic argument against the objectivity of ethical values is undercut, for *all* values suffer of the same "subjective softness." The point is to dispel the intellectual hubris of the scientific attitude and allow "that all values, including the cognitive ones, derive their authority from our idea of human flourishing and our idea of reason."

On this note, the anthology closes with two comments stimulated by Heidegger's challenge regarding the power of science in defining not only our world view, but defining who we *are*. Perhaps the most profound influence of science on contemporary life – beyond the accrual of its medical and technological benefits – is its penetrating definition of Man and Woman. Much criticism has focused on how science distorts our humanism by imposing a reality principle that both drives and commits us to a mechanical view of nature and ourselves. There are many sources for this perspective, of which the continuity of human nature with the natural world is perhaps the most profound. With the general acceptance of cosmological and biological evolution, humans have been placed firmly in natural history with other primates. The repercussions of this reorientation still require responses to unresolved issues in ethics, religion, medicine, psychology, and sociology. "Darwin and Philosophy" by Marjorie Grene (1974) (Chapter 15) succinctly explains the enormous impact of evolutionary thinking on our modern consciousness. Beyond the revolution in biological thinking *per se*, Darwinism has had an enormous influence on our very metaphysics, namely, it has forced us to recognize our animal nature, and even more broadly the omnipotence of mutability and change. To accept the contingency of nature has had far-reaching philosophical consequences. Briefly put, "fallibility follows from the acceptance of metaphysical contingency" and from this a strengthened critical epistemology is derived: looking back to our epistemological discussion, by what standards do we now judge objectivity, truth statements, systematic relevance of what were heretofore simply unreflective normative judgments? The unsteadiness of our post-Darwinian world view has bequeathed a fundamental uncertainty.[15] Now we must additionally factor in the quagmire of understanding human nature along the opposite ends of culture and biology, or nature and nurture, along with the fundamental challenge that perhaps we are not to be distinguished from other animals. To pose this question then entertains two kinds of responses. In evolutionary studies, on the one hand, the ethicist may seek a biologically-based foundation from which moral behavior is understood to emerge (e.g. Richards, 1987; see Chapter 8 in this volume; Nitecki and Nitecki, 1993). On the other hand, the philosopher, theologian or jurist might well ponder whether our ethical aspirations in fact need be based in some biological theory: does morality in some fundamental sense transcend our animal character, offering humans a unique spiritual domain? We can hardly provide rules to ground the argument despite valiant efforts to produce them (e.g. Barbour, 1990), much less predict the resolution of a debate that has been waged for four centuries. I simply offer the outline for understanding the issues and their sources.

Already in the seventeenth century the effects of post-Galilean physics on theology would reorient the relationship of God and Man, redefine the structure of the natural world, and usher in a skeptical study of Man and Society that *in toto* we designate as the Enlightenment. Some would call this the "Age of Reason" and others the triumph of Reason's despotism, depending on whether the critical attitude regarded the enterprise as an improvement over the scholasticism and superstition of an earlier epoch, or the beginning of our general disenchantment and alienation of ourselves from both nature and our very psyche. This "disenchantment of the world" (a phrase coined by Schiller) has been a nagging echo of Romantic sentiment, and reflects what Whitehead correctly perceived as a deep discrepancy between what science provides in the way of certain, verifiable knowledge and what is referred to as meaningful (aesthetic, moral, metaphysical) existence.

Max Weber, in "Science as a Vocation," (1946 [originally published in 1922]) (chapter 16) structures his argument on "the disenchantment of the world" and draws the general lessons of that demarcation. Successful science depends on a single-minded devotion to its own methods and its own conclusions,[16] but for Weber, to be a specialist is not simply to be a calculator or tool in the scientific process, but a vital, creative agent. Having provided his conception of scientific practice, Weber presents a powerful problem by observing that "scientific work is chained to the course of progress" (which today might be more properly viewed as an evolutionary process), and such change is one of the prime elements of "intellectualization," and of the trend towards the "disenchantment" of the world. Disenchantment here comprises the belief that there are no mysterious forces at play, that "one can, in principle, master all things by calculation." At one level this is clearly an optimistic outlook which elevates science as the bearer of all *seemingly* esoteric knowledge. Because the process of intellectualization denies any possibility of a metaphysical/religious/"enchanted" solution to these problems, they remain unresolved. An inevitable vacuum is left where religion would once have been invoked: science cannot address, on this view, the problem of meaning. Each individual is left to seek value on his or her own.

Or speaking directly, the ultimately possible attitudes towards life are irreconcilable, and hence their struggle can never be brought to a final conclusion.

Meaning comes from outside of science and it is an act of will and of choice. Weber appreciated the opportunity for concrete intellectual achievement which has the capacity to satisfy a human enthusiasm and provide the thrill of inspiration, imagination and ideas, while at the same time rejecting the enticing, yet deceptive notion that science ought to provide meaning. To do science is, practically and pragmatically, to believe in its methods and its results.[17] Consequently, to ask science to answer moral questions is to break the rules of the game. The scientific language does not allow this sort of question to be asked *within the confines of its own grammar.* Science does not "partake of the contemplation of sages and philosophers about meaning in the universe." The extent to which science is dichotomized ogainst other human experience is a matter that I will explore again in the Epilogue.

Scientific pursuit thus has a value in and of itself as a calling, but not in answering ethical questions beyond its purview. Such a perspective has been employed as a persistant rejoinder against the hegemony of science and its unbridled growth. Consider this indictment by Edmund Mishan:

> Like some ponderous multi-purpose robot that is powered by its own insatiable curiosity, science lurks onward irresistably, its myriad feelers peeling away the flesh of nature, probing ever deeper beneath the surface of things, forcing entry into every sanctuary, moving a transmuted humanity forward to the day when every throb in the universe has been charted, every manifestation of life dissected to the nth particle, and nothing more remains to be discovered – except, perhaps, the road back. (1967, p. 144)

Although defenders of science seem to explicitly argue against a conflation of values and a misreading of science's perceived metaphysical dominance (e.g. Holton, 1993), given the recent history of discord, the neo-romantic critique must be taken as a serious challenge. Critics too often have not differentiated the purported crimes of science from the social uses of technology (Proctor, 1991). The efforts to discern the complex social factors guiding science, along with the more charged task to determine the contribution that science has made in what the neo-romantic critics argue is an assault on nature have resulted in on-going controversies which are unavoidably ideological in character.

One cannot underestimate the power and urgency of the anti-science lobby: The discord reflects an increasingly obvious discrepancy between what science provides in the way of verifiable knowledge and technological advances, and its crucial role in contributing to what is regarded as dehumanized industrial and post-industrial societies. As Leo Marx notes (1979), widespread public trust in the inherently beneficial character of science has been shaken since the end of World War II. The current anti-science chorus has many sources ranging from the fears regarding nuclear power to the potential untoward effects of genetic engineering. This discontent should be differentiated from the persistant lines of criticism plaguing modern science since its emergence in the seventeenth century. To be sure, much of present dissension is either continued theological/metaphysical opposition, often disguised in secularized cloak, but Marx appropriately focuses upon the criticism born in the Romantic era that called into question the legitimacy of science both as a mode of cognition and as a social institution. The key thread, however, the one that seems to connect each element of these critiques is Schiller's theme of romantic disenchantment.

In this anthology, the neo-romantic germ was introduced in the essays by Whitehead and Heidegger, implicitly raised by Keller, and critically examined by Weber. This is the protest against a mechanized science in favor of an organic view of nature: unified nature is torn asunder by a reductionist and radically objectified (the figurative metaphor for dehumanized) science, that cannot put the fragmented parts back into the original whole. The consequences are in this view, dire for both Man and Society, and now with a renewed ecological consciousness, Nature itself. In some sense these concerns express a profound remorse for a lost innocence. Science is indicted with the responsibility of wrenching us away from a sheltered niche, where we resided unique in Nature, a privileged creature in communication with God. Our metaphysical disjointedness is laid at the feet of an imperialistic scientific world-view, which not only defines nature and ourselves in an anti-spiritual language, it calls into question other modes of knowing the world, ourselves, and the Beyond. But we need not becomea party to some awed interpretation of either a positivistic description of scientific method, nor an uncritical view of its achievement. Recall how Whitehead admonished us to examine the metaphysical foundations of science. In critically examining the philosophical basis of scientific method and logic, we empower ourselves not only to appreciate its achievements, but its limits. To offer an honest appraisal in such terms is to buttress against attacks science might better withstand if it more clearly understood its epistemological claims, its social and historical character, and the limits of its methods. At the same time, we are better able to place the role of scientific insight within the larger pantheon of knowledge and experience, hopefully to compose a humane world view. That is our responsibility, for ultimately we must interpret science morally. This anthology, adopting a pluralistic theme, reiterating science's legitimate, albeit limited, claims to knowledge and a circumspect jurisdiction of its practice, has endeavored to offer a broad perspective from which scientist and lay person alike might ponder these profoundly perplexing matters.

NOTES

1. At a meeting in Como (September 16, 1927) Bohr enunciated for the first time the principle of complementarity:

   The very nature of the quantum theory . . . forces us to regard the space–time coordination and the claim of causality, the union of which characterizes the classical theories, as complementary but exclusive features of the description, symbolizing the idealization of observation and definition, respectively. (Quoted by Pais, 1982, p. 444)

2. I am most sympathetic to Hilary Putnam and his fellow travelers. He challenges the position of the metaphysical realist regarding truth to be radically non-epistemic. Putnam's argument is that the realist in stating the truth conditions for sentences, says nothing about *knowing* whether those conditions are satisfied, because even the best-confirmed theories may still be false. Thus he argues that Truth is "some sort of [idealized] rational acceptability" (1981, p. 49) or essentially an epistemic notion. He traces his position to Kant, of whom he writes:

   We can view him as rejecting the idea of truth as correspondence (to a mind-independent reality) and as saying that the only sort of truth we can have an idea of, or use for, is *assertibility* (by creatures with our rational natures) *under optimal conditions* (as determined by our sensible natures). Truth becomes a radically epistemic notion. (1983, p. 210)

   Putnam continues to modify his views (1978; 1981; 1983; 1990; 1994), but essentially affirms our common picture of the world and the everyday language by which we deal with it, and in so doing readily acknowledges the impossibility of attaining either direct correspondence with reality or independent facts concerning it. These are profound philosophical conundrums concerning the relation of mind with the material world, and the difficulties with a correspondence theory of truth. He discards the two traditional avenues that sought to overcome the problem. One offers the mind access to "'forms" or direct access to the things-in-themselves, thus obviat-

ing the problem of correspondence. The other strategy is to postulate a
built-in structure of the world, a set of essences, allowing a correspondence
between the signs and their objects. Putnam would offer instead

> a species of pragmatism... "internal" realism: a realism which recog-
> nizes a difference between "p" and "I think that p," between being right,
> and merely thinking one is right without locating that objectivity in
> either transcendental correspondence or mere consensus. ("Why there
> Isn't a Ready-Made World," 1983, pp. 225–226)

Or, as he wrote elsewhere:

> The time has come for a moratorium on the kind of ontological specula-
> tion that seeks to describe the Furniture of the Universe and to tell us
> what is Really There and what is Only a Human Projection, and for a
> moratorium on the kind of epistemological speculation that seeks to tell
> us the One Method by which all our beliefs can be appraised. ("Why is
> a Philosopher?" 1990, p. 118)

Critics abound (e.g. Sellars, 1963; Trigg, 1980; Boyd, 1984; Wolterstorff,
1984; Harre, 1987; Rescher, 1987) and there are efforts towards constructive
dialogue (e.g. Gutting, 1983; Laudan, 1990). See recent anthologies
concerning the "realism problem" (e.g. Cushing, Delaney and Gutting,
1984; Leplin, 1984; Moser, 1990; Boyd *et al.*, 1991, as well as Goldstein and
Goldstein, 1978) for a pragmatic introduction to science as practice.
3. Gergen's case is based on the structure of the knowing self as dualistic, where
an inner psychological state of "knowing" is contrasted to an external world
to be "known." The objective mind "sees things for what they are" and is
thus "in touch with reality," and there is a correspondence of this reality with
our language that depicts it. Gergen rejects this view, and argues that objectivity rests on an unsteady psychological foundation (1994). How to determine the accuracy of one's internal identifications receives Gergen's major
attention, for it is in the collective agreement concerning perception that
the limits of individual knowledge are subsumed to consensus.
4. Although the contingency of facts, and the problem of their theory-ladenness, were dominant concerns of Kuhn (1970) and Norwood Russell Hanson (1970), the genealogy of this problem dates most immediately to
Ludwig Fleck ([1935]1979) and more distantly to Goethe at the end of the
eighteenth century (Tauber, 1993).
5. As Kuhn plaintively complained to me in the summer of 1994, "I *do* believe
in objectivity!" and elsewhere he has written,

> I think their ["Kuhnians"] viewpoint damagingly mistaken, have been
> pained to be associated with it, and have for years attributed that asso-
> ciation to misunderstanding. (1992, p. 3)

(This retrenchment had already begun in the Appendix of the second edition of *Structures*, 1970.)
6. McIntyre argues against the notion that the subject matter of social science
is too complex for scientific inquiry, persuasively noting that complexity is
only a function of the level of examination (that is, the construction of the

subject matter under investigation may be made in a different fashion to afford universalized explanation). Assuming that such boundaries or definitions may be changed, the analysis could then become better articulated and more susceptible to scientific study. This reorientation of the subject matter, no less than the very redefinition of the problems-at-hand, would then enable the social scientist to discover universal laws. He uses as a correlate the complexity of natural phenomena and their successful reduction to descriptions that have fulfilled our criteria of natural laws.

7. Beneath the sociological dissection of the local fights for dominance in the evolution of scientific knowledge, lurks another corollary that arises from the broader sociological concern of how science fits into its larger cultural matrix. At the dawn of modern science 400 years ago, Francis Bacon astutely recognized that scientific knowledge confers social power. As purveyors of this power, scientists may be regarded as an instrument of political power, whether in monarchial, totalitarian or democratic societies. Further, as the scientific community has grown in this century, scientists and their supporting industry have increasingly been characterized as an interest group advocating their method and product for their own economic purposes. Unfortunately, there are dramatic historical examples of perverse political manipulation of science in the case of Stalinistic genetics (the Lysenko affair) and Nazi racial views, but these are regarded by radical critics as only more obvious examples of the political nature of even "normal" science. Does science, perhaps unknowingly, hide behind an objective mask to pursue unstated ideological goals? This is the underlying core issue of the radical constructivist critique, and the one that arouses the most hostility among defenders of Science, who would argue that in these perverse uses of science, we are witness to the political usurpation of what should be an autonomous, if not value-neutral endeavor. This matter is discussed again below.

8. The arguments have become quite vociferous. In reaction to the "strong" constructivists (see Ashmore, 1989, for a review), who focus their critique upon the program of denying science a unique rational activity, Stephan Cole mounts a vigorous defense from a sociological perspective. He attacks the "strong" position at a highly vulnerable site (if in fact that position is legitimately ascribed, a matter I cannot adjudicate here)

> Almost all constructivists have adopted a relativist epistemological position which emphasizes the underdetermination of solutions to scientific problems and deemphasizes or altogether denies the importance of the empirical world as a constraint on the development of scientific knowledge... [Further all constructivists argue that] the actual cognitive content of the natural sciences can only be understood as an outcome of social processes and as influenced by social variables. (Cole, 1992, p. 35)

9. In adopting a constructivist attitude, Gieryn attempts to discredit three essentialist positions: Karl Popper's criteria of falsifiability that regards demarcation between science and non-science as purely an epistemological matter; Thomas Kuhn's scientific paradigm concept which, despite its attempt to encompass the broadest outline of consensus, remains a nebulous and contested construction of scientific and social elements; and what is

increasingly regarded as the least defensible social theory of science, Robert Merton's normative structure of scientific practice, which most commentators regard as possessing only "surface rules that do not translate into behavior patterns in an immediate and direct way." This of course sounds very much like Wittgenstein's approach to language games, their definition in practice, and the difficulty, if not impossibility, of analytically deciphering the operative rules by which the discourse is governed (Wittgenstein, 1953). Beyond this particular application, the entire discipline that falls under the current rubric of sociology of science studies reveals a strong philosophical resonance with a reading of Wittgenstein's philosophy as a "social theory of-knowledge" (Bloor, 1983), which takes the central topics of epistemology as empirical problems for social science research. Considering the pivotal role of Wittgenstein in inspiring the anti-realist position already described, one must ponder the nature of the philosophical bridge that appears to link this epistemology with social constructivism. In this regard, the debate between Michael Lynch and David Bloor (in Pickering, 1992) is illuminating.

10. See for example Helen Longino's "Science and Ideology" (Chapter 9 of *Science as Social Knowledge*, 1990) for an excellent review of the relation of science, values and ideology based on the work of Habermas, Foucault, and the feminists Keller and Haraway. Ostensibly, science studies attempt to assume an idealized stance, outside of the phenomenon and peering in as an unobtrusive observer to characterize science anthropologically. Ironically, then, much of the recent work in this discipline has evoked strong debate: Some opponents argue that these meta-analyses in general cannot achieve the first objective priority and call into question whether a neutral constructivist project is even possible at all. Other criticism appears to be leveled against particular ideological agendas. Thus controversy concerning the degree to which science may be criticized from a feminist or Marxist perspective revolves around the explicit or implicit charges concerning the prejudice (i.e. male-dominated or capitalist-driven science) that might heavily determine, if not direct the *conceptualization* project of 'objective' science itself. These are dangerous waters, and I only warn the unwary that one must carefully dissect the various schools of sociological analyses to discern the often times hidden motives of the debate.

11. The literature concerning science studies is growing rapidly but is well reviewed by Pickering (1992) and Jasanoff *et al.* (1995). Perhaps the key text (appropriately summarized and reviewed by Gieryn) is Steven Shapin and Simon Schaffer's *Leviathan and the Air Pump: Hobbes, Boyle and the Experimental Life* (1985). This study (already referred to in Keller's essay, chapter 7 in this volume) is a seminal work in dispelling what the social constructivists maintain is a false division between epistemology and sociology. The authors argue that beyond showing how the social context affected the scientific debate between Boyle and Hobbes in seventeenth century England, what was really at stake was the very invention of a science, its social context, and the demarcation between the two. Bruno Latour lauds the book as "the real beginning of a comparative anthropology that takes science seriously" (1993, p. 15).

12. Feyerabend's point has been stated in many different formats. For instance, Edmund Husserl, in *The Crisis of European Science and Transcen*

dental Phenomenology (1970) (a work that deserves broader appreciation) regarded scientific rationality as usurping the wider project of philosophical Reason, assuming in its practical victories the place of a more comprehensive theoretical Reason. Herbert Marcuse offers a succinct description of Husserl's criticism (which in a different guise sounds similar to Whitehead's own concerns originating from a different philosophical orientation altogether):

> The new science does not elucidate the conditions and the limits of its evidence, validity, and method; it does not elucidate its inherent historical denominator. It remains unaware of its own foundation, and it is therefore unable to recognize its servitude...What happens in the developing relation between science and the empirical reality is the abrogation of the transcendence of Reason. Reason loses its philosophical power and its scientific right to define and project ideas and modes of Being beyond and against those established by the prevailing reality. I say "beyond" the empirical reality, not in any metaphysical but in a historical sense, namely, in the sense of projecting essentially different, historical alternatives. (Marcuse, 1985, p. 23)

We need not delve into Husserl's argument, nor his philosophical project to correct the limits science imposed on Philosophy, to recognize that serious criticism has confronted science in terms quite divorced from any technological influence exerted on the wider social domain.

13. Gieryn discusses fraud as part of the boundary problem. The literature is immense and easily accessible. The reader is referred to Chubin (1990) and Bulger, Heitman and Reiser (1993) for excellent orienting discussions.

14. Interestingly, Henri Atlan extends this same logic to sociological analyses of science:

> [A]n anthropology of knowledge remains possible; but instead of being an explanatory and unifying metatheory, it becomes the locus of dialogue between contradictory conceptual frameworks that determine different modes of defining what makes a fact a fact, different theories and different criteria of relevance. Even though criteria of truth can function in each of these frameworks, no single criterion traverses all of them. In terms of our own discussion, even though each game has its rules, there is no unique rule for playing with the games. (1993, p. 370)

15. One example of how a Darwinian orientation may direct our personal attitudes regards a most intimate practical concern, our health. I have argued that Darwinism pervades our organic self-conception from two defining perspectives: Our scientific theory of organic being and a metaphysical orientation regarding the nature of defining ourselves in time (1994a). Beyond the explicit influence on post-Darwinian biology directly, there is a Nietzschean vision of ourselves as perpetually evolving towards some (unrealizable) ideal (1994b) that has permeated throughout our culture to direct our medical expectations (1994c), not to speak of other matters concerning self-identity.

16. To situate science in terms of its humane function, rather than its epistemological standing or its technological application, Weber referred to the "inward calling for science," that is, he addressed the possible meaning of the enterprise for its practitioners. In a section omitted here, he suggested that the defined scope of scientific disciplines provides an opportunity for specialization, and that the fragmentation of domains of knowledge in modern society entails that genuine achievement is possible for the individual only within a narrow and confined domain of expertise. However, Weber rejected the notion that science "has become a problem in calculation." He was unwilling to accept that a "factory" method of cold calculation and methodical computation can alone yield scientific results, and he strongly maintained the necessity of intuition and inspiration – of "ideas" – in science (as in art).

17. Weber believed in the value neutrality of science and was a strong advocate of value-neutrality in the social sciences. See Proctor (1991, pp. 134–154) for discussion.

REFERENCES

Ashmore, M., *The Reflexive Thesis. Wrighting Sociology of Scientific Knowledge* (Chicago: The University of Chicago Press, 1989).

Atlan, H., *Enlightenment to Enlightenment. Intercritique of Science and Myth*, trans. L.J. Schramm. (Albany: State University of New York Press, 1993).

Barbour, I.G., "Ways of Relating Science and Religion," Chapter 1 of I.G. Barbour, *Religion in an Age of Science. The Gifford Lectures 1989–1991*, vol. 1 (San Francisco: Harper & Row. 1990), pp. 3–30.

Barnes, B., *About Science* (Oxford: Blackwell, 1985).

Berger, P.L. and Luckman, T., *The Social Construction of Reality. A Treatise in the Sociology of Knowledge.* (Garden City, NY: Doubleday, 1966).

Bloor, D., *Wittgenstein: A Social Theory of Knowledge* (New York: Columbia University Press, 1983).

Bloor, D., *Knowledge and Social Imagery*, 2nd edn (Chicago: University of Chicago Press, 1991).

Bohr, N., 'The Bohr–Einstein Dialogue,' in A.P. Frenchand P.J. Kennedy (eds.), *Niels Bohr. A Centenary Volume* (Cambridge, MA: Harvard University Press, 1985), pp. 121–140.

Boyd, R.N., 'The Current Status of Scientific Realism,' in J. Leplin (ed.), *Scientific Realism* (Berkeley: University of California Press, 1984), pp. 41–82.

Boyd, R.N., Gasper, P. and Trout, J.D. (eds.), *The Philosophy of Science* (Cambridge, MA: MIT Press, 1991).

Bulger, R.E., Heitman, E. and Reiser, S.J. (eds.), *The Ethical Dimensions of the Biological Sciences* (Cambridge: Cambridge University Press, 1993).

Callon, M., "Four Models for the Dynamics of Science," in S. Jasanoff, G.E. Markle, J.C. Petersen and T. Pinch (eds.), *Handbook of Science and Technology Studies* (Thousand Oaks, CA: Sage Publications, 1995, pp. 29–63 (Chapter 10 in this volume).

Cartwright, N., *How the Laws of Physics Lie* (Oxford: Oxford University Press, 1983).

Cassirer, E., *The Philosophy of Symbolic Forms. Volume 1: Language* [1923], trans. R. Manheim (New Haven: Yale University Press, 1953).

Cassirer, E. *The Philosophy of Symbolic Forms. Volume 2: Mythical Thought* [1925], trans. R. Manheim (New Haven: Yale University Press, 1955).

Cassirer, E. *The Philosophy of Symbolic Forms. Volume 3: The Phenomenology of Knowledge* [1929], trans. R. Manheim (New Haven: Yale University Press, 1957).

Chubin, D.E., "Scientific Malpractice and the Contemporary Politics of Knowledge," in S.E. Cozzens and T.F. Gieryn (eds.), *Theories of Science in Society*, (Bloomington: Indiana University Press, 1990), pp. 144–163.

Cohen, R.S., 'Ethics and Science,' in R.S. Cohen, J.J. Stachel and M.W. Wartofsky (eds.), *For Dirk Struik. Scientific, Historical and Political Essays in Honor of Dirk J. Struik* (Dordrecht: D. Reidel Publishing Co., 1974), pp. 307–323 (chapter 13 in this volume).

Cole, S., *Making Science. Between Nature and Society* (Cambridge, MA: Harvard University Press, 1992).

Collins, H.M., *Changing Order. Replication and Induction in Scientific Practice* (Chicago: The University of Chicago Press. 1992).

Collins, H.M. and Pinch, T., *Frames of Meaning. The Social Construction of Extraordinary Science* (London: Routledge & Kegan Paul, 1982).

Comte, A., 'Philosophical Considerations on the Sciences and Savants' (1825), in *The Crisis of Industrial Civilization. The Early Essays of Auguste Comte*, ed. and trans. R. Fletcher (New York: Crane, Rusak & Co., 1974), pp. 182–213.

Cushing, J.T., Delaney, C.F. and Gutting, G.M. (eds.), *Science and Reality. Recent Work in the Philosophy of Science* (Notre Dame: University of Notre Dame Press, 1984).

Davies P.C.W. and J.R. Brown, 'The Strange World of the Quantum,' in P.C.W. Davies and J.R. Brown (eds.), *The Ghost in the Atom*, (Cambridge: Cambridge University Press, 1986), pp. 1–39 (chapter 3 in this volume).

Daston, L., "Baconian Facts, Academic Civility, and the Prehistory of Objectivity," in A. Megill (ed.), in *Rethinking Objectivity* (Durham: Duke University Press, 1994), pp. 37–64.

Fay, B., "General Laws and Explaining Human Behavior," in D.R. Sabia and J. Wallulis (eds.), *Changing Social Science* (Albany: State University of New York Press, 1983), pp. 103–128.

Feyerabend, P., *Science in a Free Society* (London: NLB, 1978), pp. 73–86, 96–107.

Fine, A., "The Natural Ontological Attitude," Chapter 7 in *The Shaky Game. Einstein Realism and the Quantum Theory* (Chicago: The University of Chicago Press, 1986), pp. 112–135.

Fleck, L. *Genesis and Development of a Scientific Fact* (Chicago: The University of Chicago Press, [1935] 1979)

Fox Keller, E., "The Paradox of Scientific Subjectivity," in A. Megill (ed.), *Rethinking Objectivity* (Durham: Duke University Press, 1994), pp. 313–331 (chapter 7 in this volume).

Fujimura, J., "Crafting Science: Standardized packages, boundary objects, and 'translation'," in A. Pickering (ed.), *Science as Practice and Culture* (Chicago: University of Chicago Press, 1992), pp. 168–211.

Gergen, K.J., 'The Mechanical Self and the Rhetoric of Objectivity,' in A. Megill (ed.), *Rethinking Objectivity* (Durham: Duke University Press, 1994), pp. 265–288.

Gieryn, T.F., "Boundaries of Science," in S. Jasanoff, G.E. Markle, J.C. Petersen and T. Pinch (eds.), *Handbook of Science and Technology Studies* (Thousand Oaks, CA: Sage Publications, 1995), pp. 393–443 (chapter 11 in this volume).

Gieryn, T.F. and Figert, A.E., "Ingredients for the Theory of Science in Society: O-rings, Ice Water, C-Clamp, Richard Feynman and the Press," in Susan E. Cozzens and Thomas Grieryn (eds.), *Theories of Science in Society* (Bloomington: Indiana University Press), pp. 67–97.

Gilbert, W., "A Vision of the Grail," in D.J. Kelves and L. Hood (eds.), *The Code of Codes. Scientific and Social Issues in the Human Genome Project* (Cambridge, MA: Harvard University Press, 1992), pp. 83–97.

Goldstein, M. and Goldstein, I.F.G., *How We Know. An Exploration of the Scientific Process* (New York: Plenum Press, 1978).

Golinski, J., "The Theory of Practice and the Practice of Theory: Sociological Approaches in the History of Science," *ISIS*, 81 (1990), pp. 492–505.

Gooding, D., Pinch, T. and Schaffer, S. (eds.), *The Uses of Experiment. Studies in the Natural Sciences* (Cambridge: Cambridge University Press, 1989).

Grene, M., "Darwin and Philosophy," *Proceedings of the Colloquium "Connaissance scientifique et Philosophie,"* (Brussels, 1974), R. Cohen and M. Wartofsky (eds.), *The Understanding of Nature: Essays in the Philosophy of Biology* (Dordrecht: D. Reidel, 1974), pp. 189–200 (chapter 15 in this volume).

Gutting, G., "Scientific Realism vs. Constructive Empiricism: A Dialogue," *Monist*, 65 (1983), pp. 336–349.

Hacking, I., "Experimentation and Scientific Realism," in J. Leplin (ed.), *Scientific Realism* (Berkeley: University of California Press, 1984), pp. 154–172 (chapter 6 in this volume).

Hacking, I., *Representing and Intervening. Introductory Topics in the Philosophy of Natural Science* (Cambridge: Cambridge University Press, 1983).

Hanson, N.R., *Patterns of Discovery* (Cambridge: Cambridge University Press, 1970).

Haraway, D., *Primate Visions: Gender, Race, and Nature in the World* (London: Routledge & Kegan Paul, 1989a).

Haraway, D., "The Biopolitics of Postmodern Bodies: Determinations of Self in Immune Systems Discourse," *differences: A Journal of Feminist Cultural Studies*, 1 (1989b pp. 3–43).

Harding, S., *The Science Question in Feminism* (New York: Cornell University Press, 1986).

Harré, R., *Varieties of Realism* (Oxford: Basil Blackwell, 1987).

Heidegger, M., *The Question Concerning Technology and Other Essays*, trans W. Lovitt (New York: Harper & Row. 1977), pp. 115–136 (chapter 2 in this volume).

Heisenberg, W., "The Development of Philosophical Ideas since Descartes in Comparison with the New Situation in Quantum Theory," in W. Heisenberg, *Physics and Philosophy. The Revolution in Modern Science* (New York: Harper & Row, 1958), pp. 76–92 (chapter 4 in this volume).

Hemple, C.G., *Philosophy of Natural Science* (Englewood Cliffs, NJ: Prentice Hall, 1966).

Holton, G., *Science and Anti-science* (Cambridge, MA: Harvard University Press, 1993).

Hoyningen-Huene, P., *Reconstructing Scientific Revolutions. Thomas S. Kuhn's Philosophy of Science*, trans. A.T. Levine (Chicago: The University of Chicago Press, 1993).

Husserl, E., *The Crisis of European Science and Transcendental Phenomenology. An Introduction to Phenomenological Philosophy*, trans. D. Carr (Evanston: Northwestern University Press, 1970).

Jacob, M.C., *The Newtonians and the English Revolution 1689–1720* (New York: Gordon & Breach, 1990), (reprint of Cornell University Press edition, 1976).

Jasanoff, S., Markle, G.E., Petersen, J.C. and Pinch, T. (eds.), *Handbook of Science and Technology Studies* (Thousand Oaks, CA: Sage Publications, 1995).

Kincaid, H., "Defending Laws in Social Sciences," *Philosophy of Social Science* (1990), 20, pp. 56–83.

Kitcher, P., *The Advancement of Science. Science without Legend, Objectivity without Illusions* (New York and Oxford: Oxford University Press, 1993).

Koyré, A., *From the Closed World to the Infinite Universe* (Baltimore: The Johns Hopkins University Press, 1957).

Kuhn, T.S., *The Structure of Scientific Revolutions*, 2nd edn (Chicago: The University of Chicago Press, 1970) [1st edn 1962].

Kuhn, T.S., "The Road Since *Structure*," *PSA 1990* vol. 2 (1991), pp. 3–13 (chapter 9 in this volume).

Kuhn, T.S., "The Trouble with the Historical Philosophy of Science," Cambridge: Department of the History of Science, Harvard University (1992).

Lakatos, I., "Falsification and the Methodology of Scientific Research Programmes," in I. Lakatos and A. Musgrave (eds.), *Criticism and the Growth of Knowledge* (Cambridge: Cambridge University Press, 1970), pp. 91–196.

Latour, B., *Science in Action. How to Follow Scientists and Engineers through Society* (Cambridge, MA: Harvard University Press, 1987).

Latour, B., *We Have Never Been Modern*, trans. C. Porter (Cambridge, MA: Harvard University Press, 1993).

Latour, B. and Woolgar, S., *Laboratory Life: The Construction of Scientific Facts* (Princeton: Princeton University Press, 1979).

Laudan, L., "Explaining the Success of Science: Beyond Epistemic Realism and Relativism," in J.T. Cushing, C.F. Delaney and G.M. Gutting (eds.), *Science and Reality. Recent Work in the Philosophy of Science* (Notre Dame: University of Notre Dame Press, 1984), pp. 83–105 (chapter 5 in this volume).

Laudan, L., *Science and Relativism. Some Key Controversies in the Philosophy of Science* (Chicago: The University of Chicago Press, 1990).

Leplin, J. (ed.), *Scientific Realism* (Berkeley: University of California Press, 1984).

Lewontin, R.C., *Biology as Ideology. The Doctrine of DNA* (New York: Harper Collins, 1991).

Longino, H.E., "Science and Ideology," Chapter 9 in H.E. Longino, *Science as Social Knowledge* (Princeton: Princeton University Press, 1990), pp. 187–214.

Lynch, M., *Art and Artifact in Laboratory Science: A Study of Shop Work and Shop Talk in a Research Laboratory* (London: Routledge & Kegan Paul, 1985).

Marcuse, H., "On Science and Phenomenology," in R.S. Cohen and M.W. Wartofsky (eds.), *A Portrait of Twenty-five Years. Boston Colloquium for the Philosophy of Science 1960–1985* (Dordrecht: D. Reidel Publishing Co., 1985), pp. 19–30.

Martin, M. and McIntyre, L.C. (eds.), *Readings in the Philosophy of Social Science* (Cambridge, MA: MIT Press, 1994).

Marx, L., "Reflections on the Neo-romantic Critique of Science," in G. Holton and R.S. Morison (eds.), *Limits of Scientific Inquiry* (New York: W.W. Norton, 1979), pp. 61–74.

Masterman, M., "The Nature of a Paradigm," in I. Laxatos and A. Musgrave (eds.), *Criticism and the Growth of Knowledge* (Cambridge: Cambridge University Press, 1970), pp. 59–90.

McIntyre, L.C., "Complexity and Social Scientific Laws," *Synthèse*, 97 (1993) pp. 209–227, reprinted in M. Martin and L.C. McIntyre (eds.), *Readings in the Philosophy of Social Science* (Cambridge, MA: MIT Press, 1994), pp. 131–143.

Megill, A. (ed.), *Rethinking Objectivity* (Durham: Duke University Press, 1994).

Merton, R.K., *The Sociology of Science: Theoretical and Empirical Investigations*, ed. N.W. Storer (Chicago: The University of Chicagoo Press, 1973).

Mishan, E.J., *The Costs of Economic Growth* (London: Staples Press, 1967).

Moser, P.K. (ed.), *Reality in Focus* (Englewood Cliffs, NJ.: Prentice-Hall, 1990).

Nagel, T., *The View from Nowhere* (New York and Oxford: Oxford University Press, 1986).

Nitecki, M.H. and Nitecki, D.V. (eds.), *Evolutionary Ethics* (Albany: State University of New York Press, 1993).

Pais, A., *"Subtle is the Lord." The Science and the Life of Albert Einstein* (Oxford: Oxford University Press, 1982).

Pickering, A., 'Objectivity and the Mangle of Practice,' *Annals of Scholarship*, 8 (1991), 9 (1992); reprinted in A. Megill (ed.), *Rethinking Objectivity* (Durham: Duke University Press, 1994), pp. 109–124.

Pickering, A. (ed.), *Science as Practice and Culture* (Chicago: The University of Chicago Press, 1992).

Pinch, T., *Confronting Nature: The Sociology of Neutrino Detection* (Dordrecht: D. Reidel Publishing Co., 1986).

Popper, K.R., *The Open Society and its Enemies* (London: Routledge & Kegan, 1945).

Proctor, R.N., *Value-Free Science? Purity and Power in Modern Knowledge* (Cambridge, MA: Harvard University Press, 1991).

Putnam, H., *Meaning and the Moral Sciences* (London: Routledge & Kegan Paul, 1978).

Putnam, H., *Reason, Truth and History* (Cambridge: Cambridge University Press, 1981).

Putnam, H., "Beyond the Fact/Value Dichotomy," *Critica*, 14 (1982) pp. 3–12; reprinted in J. Conant (ed.), *Realism with a Human Face* (Cambridge, MA: Harvard University Press, 1990), pp. 135–141 (Chapter 14 in this volume).

Putnam, H., "Why there Isn't a Ready-Made World?" in *Realism and Reason, Philosophical Papers, Volume 3* (Cambridge: Cambridge University Press, 1983).

Putnam, H. "Why is a Philosopher?," in J. Conant (ed.), *Realism with a Human Face* (Cambridge, MA: Harvard University Press, 1990).

Putnam, H., *Words and Life*, J. Conant (ed.) (Cambridge, MA: Harvard University Press, 1994).

Reichenbach, H., *The Rise of Scientific Philosophy.* (Berkeley: University of California Press, 1964).

Rescher, N., *Scientific Realism* (Dordrecht: D. Reidel Publishing Co., 1987).

Richards, R.J., *Darwin and the Emergence of Evolutionary Theories of Mind and Behavior.* Appendix I (Chicago: The University of Chicago Press, 1987), pp. 559–591 (Chapter 8 in this volume).

Roszak, T., *Where the Wasteland Ends* (Garden City, NY: Doubleday, 1972).

Rouse, J., *Knowledge and Power. Toward a Political Philosophy of Science.* (Ithaca: Cornell University Press, 1987).

Sellars, W., *Science, Perception, and Reality* (New York: Humanities Press, 1963).

Shapin, S., "History of Science and Its Sociological Reconstructions," *History of Science,* 20 (1982), pp. 157–211.

Shapin, S., *A Social History of Truth. Civility and Science in Seventeenth-Century England* (Chicago: The University of Chicago Press, 1994).

Shapin, S. and Schaffer, S., *Leviathan and the Air Pump: Hobbes, Boyle and the Experimental Life* (Princeton: Princeton University Press, 1985).

Tauber, A.I., "Goethe's Philosophy of Science. Modern Resonances," *Perspectives in Biology and Medicine,* 36 (1993), pp. 244–257.

Tauber, A.I., *The Immune Self: Theory or Metaphor?* (New York and Cambridge: Cambridge University Press, 1994a).

Tauber, A.I., "A Typology of Nietzsche's Biology," *Biology and Philosophy,* 9 (1994b), pp. 24–44.

Tauber, A.I., "Darwinian Aftershocks: Repercussions in Late Twentieth Century Medicine," *Journal of the Royal Society of Medicine,* 87 (1994c), pp. 27–31.

Tauber, A.I., "From Descartes' Dream to Husserl's Nightmare," in A.I. Tauber (ed.), *The Elusive Synthesis: Aesthetics and Science* (Dordrecht: Kluwer Academic Publishers, 1996), pp. 289–312.

Tauber, A.I. (ed.), *The Elusive Synthesis: Aesthetics and Science* (Dordrecht: Kluwer Academic Publishers, 1996).

Tauber, A.I. and Sarkar, S., 'The Human Genome Project: Has Blind Reductionism Gone too Far?' *Perspectives in Biology and Medicine,* 35 (1992), pp. 220–235.

Tauber, A.I. and Sarkar, S., "The Ideological Basis of the Human Genome Project," *Journal of the Royal Society of Medicine,* 86 (1993), pp. 537–540.

Trigg, R., *Reality at Risk: A Defense of Realism in Philosophy and the Sciences* (Brighton, England: Harvester Press, 1980).

van Fraassen, B., *The Scientific Image* (Oxford: Clarendon Press, 1980).

Weber, M., "Science as a Vocation" [1922] in *From Max Weber: Essays in Sociology,* trans. and ed. H.H. Gerth and C.W. Mills (New York: Oxford University Press, 1946), pp. 137–156 (Chapter 16 in this volume).

Weinberg, S., *Dreams of a Final Theory* (New York: Pantheon, 1992).

Whitehead, A.N., "The Origins of Modern Science," Chapter 1 in A.N. Whitehead, *Science and the Modern World* (New York: Macmillan, 1925), pp. 1–28 (Chapter 1 in this volume).

Wittgenstein, L., *Philosophical Investigations* (Oxford: Blackwell, 1953).

Wolfe, A., *The Human Difference. Animals, Computers, and the Necessity of Social Science.* (Berkeley: University of California Press, 1993).

Wolterstorff, N., 'Realism vs. Anti-realism,' in D.O. Dahlstrom (ed.), *Realism Proceedings and Addresses of the American Catholic Philosophical Association,* vol. 59 (Washington, DC: The American Catholic Philosophical Association, 1984), pp. 182–205; reprinted in P.K. Moser (ed.), *Reality in Focus* (Englewood cliffs, NJ: Prentice-Hall), pp. 50–64.

Part I
Science and its World View

1 The Origins of Modern Science*

Alfred North Whitehead

THE ORIGINS OF MODERN SCIENCE

The progress of civilisation is not wholly a uniform drift towards better things. It may perhaps wear this aspect if we map it on a scale which is large enough. But such broad views obscure the details on which rests our whole understanding of the process. New epochs emerge with comparative suddenness, if we have regard to the scores of thousands of years throughout which the complete history extends. Secluded races suddenly take their places in the main stream of events: technological discoveries transform the mechanism of human life: a primitive art quickly flowers into full satisfaction of some aesthetic craving: great religions in their crusading youth spread through the nations the peace of Heaven and the sword of the Lord.

The sixteenth century of our era saw the disruption of Western Christianity and the rise of modern science. It was an age of ferment. Nothing was settled, though much was opened – new worlds and new ideas. In science, Copernicus and Vesalius may be chosen as representative figures: they typify the new cosmology and the scientific emphasis on direct observation. Giordano Bruno was the martyr; though the cause for which he suffered was not that of science, but that of free imaginative speculation. His death in the year 1600 ushered in the first century of modern science in the strict sense of the term. In his execution there was an unconscious symbolism: for the subsequent tone of scientific thought has contained distrust of his type of general speculativeness. The Reformation, for all its importance, may be considered as a domestic affair of the European races. Even the Christianity of the East viewed it with profound disengagement. Furthermore, such disruptions are no new phenomena in the history of Christianity or of other religions. When we project this great revolution upon the whole history of the Christian Church, we cannot look upon it as introducing

* Chapter 1 in *Science and the Modern World* (New York: Macmillan, 1925), pp. 1–28.

a new principle into human life. For good or for evil, it was a great transformation of religion; but it was not the coming of religion. It did not itself claim to be so. Reformers maintained that they were only restoring what had been forgotten.

It is quite otherwise with the rise of modern science. In every way it contrasts with the contemporary religious movement. The Reformation was a popular uprising, and for a century and a half drenched Europe in blood. The beginnings of the scientific movement were confined to a minority among the intellectual élite. In a generation which saw the Thirty Years' War and remembered Alva in the Netherlands, the worst that happened to men of science was that Galileo suffered an honourable detention and a mild reproof, before dying peacefully in his bed. The way in which the persecution of Galileo has been remembered is a tribute to the quiet commencement of the most intimate change in outlook which the human race had yet encountered. Since a babe was born in a manger, it may be doubted whether so great a thing has happened with so little stir.

The thesis which these lectures will illustrate is that this quiet growth of science has practically recoloured our mentality so that modes of thought which in former times were exceptional are now broadly spread through the educated world. This new colouring of ways of thought had been proceeding slowly for many ages in the European peoples. At last it issued in the rapid development of science; and has thereby strengthened itself by its most obvious application. The new mentality is more important even than the new science and the new technology. It has altered the metaphysical presuppositions and the imaginative contents of our minds; so that now the old stimuli provoke a new response. Perhaps my metaphor of a new colour is too strong. What I mean is just that slightest change of tone which yet makes all the difference. This is exactly illustrated by a sentence from a published letter of that adorable genius, William James. When he was finishing his great treatise on the *Principles of Psychology*, he wrote to his brother Henry James, "I have to forge every sentence in the teeth of irreducible and stubborn facts."

This new tinge to modern minds is a vehement and passionate interest in the relation of general principles to irreducible and stubborn facts. All the world over and at all times there have been practical men, absorbed in "irreducible and stubborn facts": all the world over and at all times there have been men of philosophic temperament who have been absorbed in the weaving of general principles. It is this union of passionate interest in the detailed facts with equal devotion to abstract generalisation which forms the novelty in our present society. Previously it had appeared sporadically and as if by chance. This bal-

ance of mind has now become part of the tradition which infects culti-
vated thought. It is the salt which keeps life sweet. The main business
of universities is to transmit this tradition as a widespread inheritance
from generation to generation.

Another contrast which singles out science from among the European movements of the sixteenth and seventeenth centuries is its universality. Modern science was born in Europe, but its home is the
whole world. In the last two centuries there has been a long and confused impact of Western modes upon the civilisation of Asia. The wise
men of the East have been puzzling, and are puzzling, as to what may
be the regulative secret of life which can be passed from West to East
without the wanton destruction of their own inheritance which they so
rightly prize. More and more it is becoming evident that what the West
can most readily give to the East is its science and its scientific outlook.
This is transferable from country to country, and from race to race,
wherever there is a rational society.

In this course of lectures I shall not discuss the details of scientific
discovery. My theme is the energising of a state of mind in the modern
world, its broad generalisations, and its impact upon other spiritual
forces. There are two ways of reading history, forwards and backwards.
In the history of thought, we require both methods. A climate of opinion – to use the happy phrase of a seventeenth century writer – requires
for its understanding the consideration of its antecedents and its issues.
Accordingly in this lecture I shall consider some of the antecedents of
our modern approach to the investigation of nature.

In the first place, there can be no living science unless there is a
widespread instinctive conviction in the existence of an *Order of Things*,
and, in particular, of an *Order of Nature*. I have used the word *instinctive*
advisedly. It does not matter what men say in words, so long as their activities are controlled by settled instincts. The words may ultimately destroy the instincts. But until this has occurred, words do not count.
This remark is important in respect to the history of scientific thought.
For we shall find that since the time of Hume, the fashionable scientific
philosophy has been such as to deny the rationality of science. This
conclusion lies upon the surface of Hume's philosophy. Take, for example, the following passage from Section IV of his *Inquiry Concerning
Human Understanding:*

In a word, then, every effect is a distinct event from its cause. It
could not, therefore, be discovered in the cause; and the first invention or conception of it, *a priori*, must be entirely arbitrary.

If the cause in itself discloses no information as to the effect, so that the first invention of it must be *entirely* arbitrary, it follows at once that science is impossible, except in the sense of establishing *entirely arbitrary* connections which are not warranted by anything intrinsic to the natures either of causes or effects. Some variant of Hume's philosophy has generally prevailed among men of science. But scientific faith has risen to the occasion, and has tacitly removed the philosophic mountain.

In view of this strange contradiction in scientific thought, it is of the first importance to consider the antecedents of a faith which is impervious to the demand for a consistent rationality. We have therefore to trace the rise of the instinctive faith that there is an Order of Nature which can be traced in every detained occurrence.

Of course we all share in this faith, and we therefore believe that the reason for the faith is our apprehension of its truth. But the formation of a general idea – such as the idea of the Order of Nature – and the grasp of its importance, and the observation of its exemplification in a variety of occasions are by no means the necessary consequences of the truth of the idea in question. Familiar things happen, and mankind does not bother about them. It requires a very unusual mind to undertake the analysis of the obvious. Accordingly I wish to consider the stages in which this analysis became explicit, and finally became unalterably impressed upon the educated minds of Western Europe.

Obviously, the main recurrences of life are too insistent to escape the notice of the least rational of humans; and even before the dawn of rationality, they have impressed themselves upon the instincts of animals. It is unnecessary to labour the point, that in broad outline certain general states of nature recur, and that our very natures have adapted themselves to such repetitions.

But there is a complementary fact which is equally true and equally obvious – nothing ever really recurs in exact detail. No two days are identical, no two winters. What has gone, has gone forever. Accordingly the practical philosophy of mankind has been to expect the broad recurrences, and to accept the details as emanating from the inscrutable womb of things beyond the ken of rationality. Men expected the sun to rise, but the wind bloweth where it listeth.

Certainly from the classical Greek civilisation onwards there have been men, and indeed groups of men, who have placed themselves beyond this acceptance of an ultimate irrationality. Such men have endeavoured to explain all phenomena as the outcome of an order of things which extends to every detail. Geniuses such as Aristotle, or Archimedes, or Roger Bacon, must have been endowed with the full scientific mentality, which instinctively holds that all things great and small·are conceivable as exemplifications of general principles which reign throughout the natural order.

But until the close of the Middle Ages the general educated public did not feel that intimate conviction, and that detailed interest, in such an idea, so as to lead to an unceasing supply of men, with ability and opportunity adequate to maintain a coördinated search for the discovery of these hypothetical principles. Either people were doubtful about the existence of such principles, or were doubtful about any success in finding them, or took no interest in thinking about them, or were oblivious to their practical importance when found. For whatever reason, search was languid, if we have regard to the opportunities of a high civilisation and the length of time concerned. Why did the pace suddenly quicken in the sixteenth and seventeenth centuries? At the close of the Middle Ages a new mentality discloses itself. Invention stimulated thought, thought quickened physical speculation, Greek manuscripts disclosed what the ancients had discovered. Finally although in the year 1500 Europe knew less than Archimedes who died in the year 212 BC, yet in the year 1700, Newton's *Principia* had been written and the world was well started on the modern epoch.

There have been great civilisations in which the peculiar balance of mind required for science has only fitfully appeared and has produced the feeblest result. For example, the more we know of Chinese art, of Chinese literature, and of the Chinese philosophy of life, the more we admire the heights to which that civilisation attained. For thousands of years, there have been in China acute and learned men patiently devoting their lives to study. Having regard to the span of time, and to the population concerned, China forms the largest volume of civilisation which the world has seen. There is no reason to doubt the intrinsic capacity of individual Chinamen for the pursuit of science. And yet Chinese science is practically negligible. There is no reason to believe that China if left to itself would have ever produced any progress in science. The same may be said of India. Furthermore, if the Persians had enslaved the Greeks, there is no definite ground for belief that science would have flourished in Europe. The Romans showed no particular originality in that line. Even as it was, the Greeks, though they founded the movement, did not sustain it with the concentrated interest which modern Europe has shown. I am not alluding to the last few generations of the European peoples on both sides of the ocean; I mean the smaller Europe of the Reformation period, distracted as it was with wars and religious disputes. Consider the world of the eastern Mediterranean, from Sicily to western Asia, during the period of about 1400 years from the death of Archimedes [in 212 BC] to the irruption of the Tartars. There were wars and revolutions and large changes of religion: but nothing much worse than the wars of the sixteenth and seventeenth centuries throughout Europe. There was a great and wealthy civilisation, Pagan, Christian, Mahometan. In that period a great deal was added to science. But on the whole the progress was slow and wavering; and, except in mathematics, the men of the Renaissance practically started from the position which Archimedes had reached. There had been some progress in medicine and some progress in astronomy. But the total advance was very little compared to the marvellous success of the seventeenth century. For example, compare the progress of scientific knowledge from the year 1560, just before the births of Galileo and of Kepler, up to the year 1700, when Newton was in the height of his fame, with the progress in the ancient period, already mentioned, exactly ten times as long.

Nevertheless, Greece was the mother of Europe; and it is to Greece that we must look in order to find the origin of our modern ideas. We all know that on the eastern shores of the Mediterranean there was a very flourishing school of Ionian philosophers, deeply interested in theories concerning nature. Their ideas have been transmitted to us, enriched by the genius of Plato and Aristotle. But, with the exception of Aristotle, and it is a large exception, this school of thought had not attained to the complete scientific mentality. In some ways, it was better. The Greek genius was philosophical, lucid and logical. The men of this group were primarily asking philosophical questions. What is the substratum of nature? Is it fire, or earth, or water, or some combination of any two, or of all three? Or is it a mere flux, not reducible to some static material? Mathematics interested them mightily. They invented its generality, analysed its premises, and made notable discoveries of theorems by a rigid adherence to deductive reasoning. Their minds were infected with an eager generality. They demanded clear, bold ideas, and strict reasoning from them. All this was excellent; it was genius; it was ideal preparatory work. But it was not science as we understand it. The patience of minute observation was not nearly so prominent. Their genius was not so apt for the state of imaginative muddled suspense which precedes successful inductive generalisation. They were lucid thinkers and bold reasoners.

Of course there were exceptions, and at the very top: for example, Aristotle and Archimedes. Also for patient observation, there were the astronomers. There was a mathematical lucidity about the stars, and a fascination about the small numerable band of runaway planets.

Every philosophy is tinged with the colouring of some secret imaginative background, which never emerges explicitly into its trains of reasoning. The Greek view of nature, at least that cosmology transmitted from them to later ages, was essentially dramatic. It is not necessarily wrong for this reason: but it was overwhelmingly dramatic. It thus conceived nature as articulated in the way of a work of dramatic art, for the exemplification of general ideas converging to an end. Nature was differentiated so as to provide its proper end for each thing. There was the centre of the universe as the end of motion for those things which are heavy, and the celestial spheres as the end of motion for those things whose natures lead them upwards. The celestial spheres were for things which are impassible and ingenerable, the lower regions for things passible and generable. Nature was a drama in which each thing played its part.

I do not say that this is a view to which Aristotle would have subscribed without severe reservations, in fact without the sort of reservations which we ourselves would make. But it was the view which subsequent Greek thought extracted from Aristotle and passed on to the Middle Ages. The effect of such an imaginative setting for nature was to damp down the historical spirit. For it was the end which seemed illuminating, so why bother about the beginning? The Reformation and the scientific movement were two aspects of the historical revolt which was the dominant intellectual movement of the later Renaissance. The appeal to the origins of Christianity, and Francis Bacon's appeal to efficient causes as against final causes, were two sides of one movement of thought. Also for this reason Galileo and his adversaries were at hopeless cross purposes, as can be seen from his *Dialogues on the Two Systems of the World*.

Galileo keeps harping on how things happen, whereas his adversaries had a complete theory as to why things happen. Unfortunately the two theories did not bring out the same results. Galileo insists upon 'irreducible and stubborn facts,' and Simplicius, his opponent, brings forward reasons, completely satisfactory, at least to himself. It is a great mistake to conceive this historical revolt as an appeal to reason. On the contrary, it was through and through an anti-intellectualist movement. It was the return to the contemplation of brute fact; and it was based on a recoil from the inflexible rationality of medieval thought. In making this statement I am merely summarising what at the time the adherents of the old régime themselves asserted. For example, in the fourth book of Father Paul Sarpi's *History of the Council of Trent*, you will find that in the year 1551 the Papal Legates who presided over the Council ordered:

> That the Divines ought to confirm their opinions with the holy Scripture, Traditions of the Apostles, sacred and approved Councils, and by the Constitutions and Authorities of the holy Fathers; that they ought to use brevity, and avoid superfluous and unprofitable questions, and perverse contentions...This order did not please the Italian Divines; who said it was a novity, and a condemning of School-Divinity, which, in all difficulties, *useth reason*, and because it was not lawful [i.e., by this decree] to treat as St Thomas [Aquinas], St Bonaventure, and other famous men did.

It is impossible not to feel sympathy with these Italian divines, maintaining the lost cause of unbridled rationalism. They were deserted on all hands. The Protestants were in full revolt against them. The Papacy failed to support them, and the Bishops of the Council could not even understand them. For a few sentences below the foregoing quotation, we read:

> Though many complained here-of [i.e., of the Decree], yet it prevailed but little, because generally the Fathers [i.e., the Bishops] desired to hear men speak with intelligible terms, not abstrusely, as in the matter of Justification, and others already handled.

Poor belated medievalists! When they used reason they were not even intelligible to the ruling powers of their epoch. It will take centuries before stubborn facts are reducible by reason, and meanwhile the pendulum swings slowly and heavily to the extreme of the historical method.

Forty-three years after the Italian divines had written this memorial, Richard Hooker in his famous *Laws of Ecclesiastical Polity* makes exactly the same complaint of his Puritan adversaries.[1] Hooker's balanced thought – from which the appellation "The Judicious Hooker" is derived – and his diffuse style, which is the vehicle of such thought, make his writings singularly unfit for the process of summarising by a short, pointed quotation. But, in the section referred to, he reproaches his opponents with *Their Disparagement of Reason*; and in support of his own position definitely refers to "The greatest amongst the school-divines" by which designation I presume that he refers to St Thomas Aquinas.

Hooker's *Ecclesiastical Polity* was published just before Sarpi's *Council of Trent*. Accordingly there was complete independence between the two works. But both the Italian divines of 1551, and Hooker at the end of that century testify to the anti-rationalist trend of thought at that epoch, and in this respect contrast their own age with the epoch of scholasticism.

This reaction was undoubtedly a very necessary corrective to the unguarded rationalism of the Middle Ages. But reactions run to extremes. Accordingly, although one outcome of this reaction was the birth of modern science, yet we must remember that science thereby inherited the bias of thought to which it owes its origin.

The effect of Greek dramatic literature was many-sided so far as concerns the various ways in which it indirectly affected medieval thought. The pilgrim fathers of the scientific imagination as it exists today are the great tragedians of ancient Athens, Aeschylus, Sophocles, Euripides. Their vision of fate, remorseless and indifferent, urging a tragic incident to its inevitable issue, is the vision possessed by science. Fate in Greek Tragedy becomes the order of nature in modern thought. The absorbing interest in the particular heroic incidents, as an example and a verification of the workings of fate, reappears in our epoch as concentration of interest on the crucial experiments. It was my good fortune to be present at the meeting of the Royal Society in London when the Astronomer Royal for England announced that the photographic plates of the famous eclipse, as measured by his colleagues in Greenwich Observatory, had verified the prediction of Einstein that rays of light are bent as they pass in the neighbourhood of the sun. The whole atmosphere of tense interest was exactly that of the Greek drama: we were the chorus commenting on the decree of destiny as disclosed in the development of a supreme incident. There was dramatic quality in the very staging – the traditional ceremonial, and in the background the picture of Newton to remind us that the greatest of scientific generalisations was now, after more than two centuries, to receive its first modification. Nor was the personal interest wanting: a great adventure in thought had at length come safe to shore.

Let me here remind you that the essence of dramatic tragedy is not unhappiness. It resides in the solemnity of the remorseless working of things. This inevitableness of destiny can only be illustrated in terms of human life by incidents which in fact involve unhappiness. For it is only by them that the futility of escape can be made evident in the drama. This remorseless inevitableness is what pervades scientific thought. The laws of physics are the decrees of fate.

The conception of the moral order in the Greek plays was certainly not a discovery of the dramatists. It must have passed into the literary tradition from the general serious opinion of the times. But in finding this magnificent expression, it thereby deepened the stream of thought from which it arose. The spectacle of a moral order was impressed upon the imagination of classical civilisation.

The time came when that great society decayed, and Europe passed into the Middle Ages. The direct influence of Greek literature vanished. But the concept of the moral order and of the order of nature had enshrined itself in the Stoic philosophy. For example, Lecky in his *History of European Morals* tells us "Seneca maintains that the Divinity has determined all things by an inexorable law of destiny, which He has decreed, but which He Himself obeys." But the most effective way in which the Stoics influenced the mentality of the Middle Ages was by the diffused sense of order which arose from Roman law. Again to quote Lecky,

> The Roman legislation was in a twofold manner the child of philosophy. It was in the first place formed upon the philosophical model, for, instead of being a mere empirical system adjusted to the existing requirements of society, it laid down abstract principles of right to which it endeavoured to conform; and, in the next place, these principles were borrowed directly from Stoicism.

In spite of the actual anarchy throughout large regions in Europe after the collapse of the Empire, the sense of legal order always haunted the racial memories of the Imperial populations. Also the Western Church was always there as a living embodiment of the traditions of Imperial rule.

It is important to notice that this legal impress upon medieval civilisation was not in the form of a few wise precepts which should permeate conduct. It was the conception of a definite articulated system which defines the legality of the detailed structure of social organism, and of the detailed way in which it should function. There was nothing vague. It was not a question of admirable maxims, but of definite procedure to put things right and to keep them there. The Middle Ages formed one long training of the intellect of Western Europe in the sense of order. There may have been some deficiency in respect to practice. But the idea never for a moment lost its grip. It was preeminently an epoch of orderly thought, rationalist through and through. The very anarchy quickened the sense for coherent system; just as the modern anarchy of Europe has stimulated the intellectual vision of a League of Nations.

But for science something more is wanted than a general sense of the order in things. It needs but a sentence to point out how the habit of definite exact thought was implanted in the European mind by the long dominance of scholastic logic and scholastic divinity. The habit remained after the philosophy had been repudiated, the priceless habit of looking for an exact point and of sticking to it when found. Galileo owes more to Aristotle than appears on the surface of his *Dialogues*: he owes to him his clear head and his analytic mind.

I do not think, however, that I have even yet brought out the greatest contribution of medievalism to the formation of the scientific movement. I mean the inexpungable belief that every detailed occurrence can be correlated with its antecedents in a perfectly definite manner, exemplifying general principles. Without this belief the incredible labours of scientists would be without hope. It is this instinctive conviction, vividly poised before the imagination, which is the motive power of research: – that there is a secret, a secret which can be unveiled. How has this conviction been so vividly implanted on the European mind?

When we compare this tone of thought in Europe with the attitude of other civilisations when left to themselves, there seems but one source for its origin. It must come from the medieval insistence on the rationality of God, conceived as with the personal energy of Jehovah and with the rationality of a Greek philosopher. Every detail was supervised and ordered: the search into nature could only result in the vindication of the faith in rationality. Remember that I am not talking of the explicit beliefs of a few individuals. What I mean is the impress on the European mind arising from the unquestioned faith of centuries. By this I mean the instinctive tone of thought and not a mere creed of words.

In Asia, the conceptions of God were of a being who was either too arbitrary or too impersonal for such ideas to have much effect on instinctive habits of mind. Any definite occurrence might be due to the fiat of an irrational despot, or might issue from some impersonal, inscrutable origin of things. There was not the same confidence as in the intelligible rationality of a personal being. I am not arguing that the European trust in the scrutability of nature was logically justified even by its own theology. My only point is to understand how it arose. My explanation is that the faith in the possibility of science, generated antecedently to the development of modern scientific theory, is an unconscious derivative from medieval theology.

But science is not merely the outcome of instinctive faith. It also requires an active interest in the simple occurrences of life for their own sake.

This qualification "for their own sake" is important. The first phase of the Middle Ages was an age of symbolism. It was an age of vast ideas, and of primitive technique. There was little to be done with nature, except to coin a hard living from it. But there were realms of thought to be explored, realms of philosophy and realms of theology. Primitive art could symbolise those ideas which filled all thoughtful minds. The first phase of medieval art has a haunting charm beyond compare: its own intrinsic quality is enhanced by the fact that its message, which stretched beyond art's own self-justification of aesthetic achievement, was the symbolism of things lying behind nature itself. In this symbolic phase, medieval art energised in nature as its medium, but pointed to another world.

In order to understand the contrast between these early Middle Ages and the atmosphere required by the scientific mentality, we should compare the sixth century in Italy with the sixteenth century. In both centuries the Italian genius was laying the foundations of a new epoch. The history of the three centuries preceding the earlier period, despite the promise for the future introduced by the rise of Christianity, is overwhelmingly infected by the sense of the decline of civilisation. In each generation something has been lost. As we read the records, we are haunted by the shadow of the coming barbarism. There are great men, with fine achievements in action or in thought. But their total effect is merely for some short time to arrest the general decline. In the sixth century we are, so far as Italy is concerned, at the lowest point of the curve. But in that century every action is laying the foundation for the tremendous rise of the new European civilisation. In the background the Byzantine Empire, under Justinian, in three ways determined the character of the early Middle Ages in Western Europe. In the first place, its armies, under Belisarius and Narses, cleared Italy from the Gothic domination. In this way, the stage was freed for the exercise of the old Italian genius for creating organisations which shall be protective of ideals of cultural activity. It is impossible not to sympathise with the Goths: yet there can be no doubt but that a thousand years of the Papacy were infinitely more valuable for Europe than any effects derivable from a well-established Gothic kingdom of Italy.

In the second place, the codification of the Roman law established the ideal of legality which dominated the sociological thought of Europe in the succeeding centuries. Law is both an engine for government, and a condition restraining government. The canon law of the Church, and the civil law of the State, owe to Justinian's lawyers their influence on the development of Europe. They established in the Western mind the ideal that an authority should be at once lawful, and law-enforcing, and should in itself exhibit a rationally adjusted system of organisation.

The sixth century in Italy gave the initial exhibition of the way in which the impress of these ideas was fostered by contact with the Byzantine Empire.

Thirdly, in the non-political spheres of art and learning Constantinople exhibited a standard of realised achievement which, partly by the impulse to direct imitation, and partly by the indirect inspiration arising from the mere knowledge that such things existed, acted as a perpetual spur to Western culture. The wisdom of the Byzantines, as it stood in the imagination of the first phase of medieval mentality, and the wisdom of the Egyptians as it stood in the imagination of the early Greeks, played analogous roles. Probably the actual knowledge of these respective wisdoms was, in either case, about as much as was good for the recipients. They knew enough to know the sort of standards which are attainable, and not enough to be fettered by static and traditional ways of thought. Accordingly, in both cases men went ahead on their own and did better. No account of the rise of the European scientific mentality can omit some notice of this influence of the Byzantine civilisation in the background. In the sixth century there is a crisis in the history of the relations between the Byzantines and the West; and this crisis is to be contrasted with the influence of Greek literature on European thought in the fifteenth and sixteenth centuries. The two outstanding men, who in the Italy of the sixth century laid the foundations of the future, were St Benedict and Gregory the Great. By reference to them, we can at once see how absolutely in ruins was the approach to the scientific mentality which had been attained by the Greeks. We are at the zero point of scientific temperature. But the lifework of Gregory and of Benedict contributed elements to the reconstruction of Europe which secured that this reconstruction, when it arrived, should include a more effective scientific mentality than that of the ancient world. The Greeks were over-theoretical. For them science was an offshoot of philosophy. Gregory and Benedict were practical men, with an eye for the importance of ordinary things; and they combined this practical temperament with their religious and cultural activities. In particular, we owe it to St Benedict that the monasteries were the homes of practical agriculturalists, as well as of saints and of artists and men of learning. The alliance of science with technology, by which learning is kept in contact with irreducible and stubborn facts, owes much to the practical bent of the early Benedictines. Modern science derives from Rome as well as from Greece, and this Roman strain explains its gain in an energy of thought kept closely in contact with the world of facts.

But the influence of this contact between the monasteries and the facts of nature showed itself first in art. The rise of Naturalism in the later Middle Ages was the entry into the European mind of the final ingredient necessary for the rise of science. It was the rise of interest in natural objects and in natural occurrences, for their own sakes. The natural foliage of a district was sculptured in out-of-the-way spots of the later buildings, merely as exhibiting delight in those familiar objects. The whole atmosphere of every art exhibited a direct joy in the apprehension of the things which lie around us. The craftsmen who executed the late medieval decorative sculpture, Giotto, Chaucer, Wordsworth, Walt Whitman, and, at the present day, the New England poet Robert Frost, are all akin to each other in this respect. The simple immediate facts are the topics of interest, and these reappear in the thought of science as the "irreducible stubborn facts."

The mind of Europe was now prepared for its new venture of thought. It is unnecessary to tell in detail the various incidents which marked the rise of science: the growth of wealth and leisure; the expansion of universities; the invention of printing; the taking of Constantinople; Copernicus; Vasco da Gama; Columbus; the telescope. The soil, the climate, the seeds, were there, and the forest grew. Science has never shaken off the impress of its origin in the historical revolt of the later Renaissance. It has remained predominantly an anti-rationalistic movement, based upon a naive faith. What reasoning it has wanted, has been borrowed from mathematics which is a surviving relic of Greek rationalism, following the deductive method. Science repudiates philosophy. In other words, it has never cared to justify its faith or to explain its meanings; and has remained blandly indifferent to its refutation by Hume.

Of course the historical revolt was fully justified. It was wanted. It was more than wanted: it was an absolute necessity for healthy progress. The world required centuries of contemplation of irreducible and stubborn facts. It is difficult for men to do more than one thing at a time, and that was the sort of thing they had to do after the rationalistic orgy of the Middle Ages. It was a very sensible reaction; but it was not a protest on behalf of reason.

There is, however, a Nemesis which waits upon those who deliberately avoid avenues of knowledge. Oliver Cromwell's cry echoes down the ages, "My brethren, by the bowels of Christ I beseech you, bethink you that you may be mistaken."

The progress of science has now reached a turning point. The stable foundations of physics have broken up: also for the first time physiology is asserting itself as an effective body of knowledge, as distinct from a scrap-heap. The old foundations of scientific thought are becoming unintelligible. Time, space, matter, material, ether, electricity, mechanism, organism, configuration, structure, pattern, function, all require reinterpretation. What is the sense of talking about a mechanical explanation when you do not know what you mean by mechanics?

The truth is that science started its modern career by taking over ideas derived from the weakest side of the philosophies of Aristotle's successors. In some respects it was a happy choice. It enabled the knowledge of the seventeenth century to be formularised so far as physics and chemistry were concerned, with a completeness which has lasted to the present time. But the progress of biology and psychology has probably been checked by the uncritical assumption of half-truths. If science is not to degenerate into a medley of *ad hoc* hypotheses, it must become philosophical and must enter upon a thorough criticism of its own foundations.

In the succeeding lectures of this course, I shall trace the successes and the failures of the particular conceptions of cosmology with which the European intellect has clothed itself in the last three centuries. General climates of opinion persist for periods of about two to three generations, that is to say, for periods of sixty to a hundred years. There are also shorter waves of thought, which play on the surface of the tidal movement. We shall find, therefore transformations in the European outlook, slowly modifying the successive centuries. There persists, however, throughout the whole period the fixed scientific cosmology which presupposes the ultimate fact of an irreducible brute matter, or material, spread throughout space in a flux of configurations. In itself such a material is senseless, valueless, purposeless. It just does what it does do, following a fixed routine imposed by external relations which do not spring from the nature of its being. It is this assumption that I call "scientific materialism." Also it is an assumption which I shall challenge as being entirely unsuited to the scientific situation at which we have now arrived. It is not wrong, if properly construed. If we confine ourselves to certain types of facts, abstracted from the complete circumstances in which they occur, the materialistic assumption expresses these facts to perfection. But when we pass beyond the abstraction, either by more subtle employment of our senses, or by the request for meanings and for coherence of thoughts, the scheme breaks down at once. The narrow efficiency of the scheme was the very cause of its supreme methodological success. For it directed attention to just those groups of facts which, in the state of knowledge then existing, required investigation.

The success of the scheme has adversely affected the various currents of European thought. The historical revolt was anti-rationalistic, because the rationalism of the scholastics required a sharp correction by contact with brute fact. But the revival of philosophy in the hands of Descartes and his successors was entirely coloured in its development by the acceptance of the scientific cosmology at its face value. The success of their ultimate ideas confirmed scientists in their refusal to modify them as the result of an enquiry into their rationality. Every philosophy was bound in some way or other to swallow them whole. Also the example of science affected other regions of thought. The historical revolt has thus been exaggerated into the exclusion of philosophy from its proper role of harmonising the various abstractions of methodological thought. Thought is abstract; and the intolerant use of abstractions is the major vice of the intellect. This vice is not wholly corrected by the recurrence to concrete experience. For after all, you need only attend to those aspects of your concrete experience which lie within some limited scheme. There are two methods for the purification of ideas. One of them is dispassionate observation by means of the bodily senses. But observation is selection. Accordingly, it is difficult to transcend a scheme of abstraction whose success is sufficiently wide. The other method is by comparing the various schemes of abstraction which are well founded in our various types of experience. This comparison takes the form of satisfying the demands of the Italian scholastic divines whom Paul Sarpi mentioned. They asked that *reason* should be used. Faith in reason is the trust that the ultimate natures of things lie together in a harmony which excludes mere arbitrariness. It is the faith that at the base of things we shall not find mere arbitrary mystery. The faith in the order of nature which has made possible the growth of science is a particular example of a deeper faith. This faith cannot be justified by any inductive generalisation. It springs from direct inspection of the nature of things as disclosed in our own immediate present experience. There is no parting from your own shadow. To experience this faith is to know that in being ourselves we are more than ourselves: to know that our experience, dim and fragmentary as it is, yet sounds the utmost depths of reality: to know that detached details merely in order to be themselves demand that they should find themselves in a system of things: to know that this system includes the harmony of logical rationality, and the harmony of aesthetic achievement: to know that, while the harmony of logic lies upon the universe as an iron necessity, the aesthetic harmony stands before it as a living ideal moulding the general flux in its broken progress towards finer, subtler issues.

NOTE

1. *Cf.* Book III, Section viii.

2 The Age of the World Picture*

Martin Heidegger

In metaphysics reflection is accomplished concerning the essence of what is and a decision takes place regarding the essence of truth.[1] Metaphysics grounds an age, in that through a specific interpretation of what is and through a specific comprehension of truth it gives to that age the basis upon which it is essentially formed.[2] This basis holds complete dominion over all the phenomena that distinguish the age. Conversely, in order that there may be an adequate reflection upon these phenomena themselves, the metaphysical basis for them must let itself be apprehended in them. Reflection is the courage to make the truth of our own presuppositions and the realm of our own goals into the things that most deserve to be called in question.[3]

One of the essential phenomena of the modern age is its science. A phenomenon of no less importance is machine technology. We must not, however, misinterpret that technology as the mere application of modern mathematical physical science to praxis. Machine technology is itself an autonomous transformation of praxis, a type of transformation wherein praxis first demands the employment of mathematical physical science. Machine technology remains up to now the most visible outgrowth of the essence of modern technology, which is identical with the essence of modern metaphysics.

A third equally essential phenomenon of the modern period lies in the event of art's moving into the purview of aesthetics. That means that the art work becomes the object of mere subjective experience, and that consequently art is considered to be an expression of human life.[4]

A fourth modern phenomenon manifests itself in the fact that human activity is conceived and consummated as culture. Thus culture is the realization of the highest values, through the nurture and cultivation of the highest goods of man. It lies in the essence of culture, as

* From *The Question Concerning Technology and Other Essays*, trans. W. Lovitt (New York: Harper & Row, 1977), pp. 115–136.

such nurturing, to nurture itself in its turn and thus to become the politics of culture.

A fifth phenomenon of the modern age is the loss of the gods.[5] This expression does not mean the mere doing away with the gods, gross atheism. The loss of the gods is a twofold process. On the one hand, the world picture is Christianized in as much as the cause of the world is posited as infinite, unconditional, absolute. On the other hand, Christendom transforms Christian doctrine into a world view (the Christian world view), and in that way makes itself modern and up to date. The loss of the gods is the situation of indecision regarding God and the gods. Christendom has the greatest share in bringing it about. But the loss of the gods is so far from excluding religiosity that rather only through that loss is the relation to the gods changed into mere "religious experience." When this occurs, then the gods have fled. The resultant void is compensated for by means of historiographical and psychological investigation of myth.

What understanding of what is, what interpretation of truth, lies at the foundation of these phenomena?

We shall limit the question to the phenomenon mentioned first, to science [*Wissenschaft*].

In what does the essence of modern science lie?

What understanding of what is and of truth provides the basis for that essence? If we succeed in reaching the metaphysical ground that provides the foundation for science as a modern phenomenon, then the entire essence of the modern age will have to let itself be apprehended from out of that ground.

When we use the word "science" today, it means something essentially different from the *doctrina* and *scientia* of the Middle Ages, and also from the Greek *epistēmē*. Greek science was never exact, precisely because, in keeping with its essence, it could not be exact and did not need to be exact. Hence it makes no sense whatever to suppose that modern science is more exact than that of antiquity. Neither can we say that the Galilean doctrine of freely falling bodies is true and that Aristotle's teaching, that light bodies strive upward, is false; for the Greek understanding of the essence of body and place and of the relation between the two rests upon a different interpretation of beings and hence conditions a correspondingly different kind of seeing and questioning of natural events. No one would presume to maintain that Shakespeare's poetry is more advanced than that of Aeschylus. It is still more impossible to say that the modern understanding of whatever is, is more correct than that of the Greeks. Therefore, if we want to grasp

the essence of modern science, we must first free ourselves from the habit of comparing the new science with the old solely in terms of degree,
from the point of view of progress.

The essence of what we today call science is research. In what does
the essence of research consist?

In the fact that knowing [*das Erkennen*] establishes itself as a procedure within some realm of what is, in nature or in history. Procedure
does not mean here merely method or methodology. For every procedure already requires an open sphere in which it moves. And it is precisely the opening up of such a sphere that is the fundamental event in
research. This is accomplished through the projection within some
realm of what is – in nature, for example – of a fixed ground plan[6] of
natural events. The projection sketches out in advance the manner in
which the knowing procedure must bind itself and adhere to the sphere
opened up. This binding adherence is the rigor of research.[7] Through
the projecting of the ground plan and the prescribing of rigor, procedure makes secure for itself its sphere of objects within the realm of
Being. A look at that earliest science, which is at the same time the normative one in the modern age, namely, mathematical physics, will
make clear what we mean. Inasmuch as modern atomic physics still remains physics, what is essential – and only the essential is aimed at
here – will hold for it also.

Modern physics is called mathematical because, in a remarkable
way, it makes use of a quite specific mathematics. But it can proceed
mathematically in this way only because, in a deeper sense, it is already
itself mathematical. *Ta mathēmata* means for the Greeks that which
man knows in advance in his observation of whatever is and in his
intercourse with things: the corporeality of bodies, the vegetable character of plants, the animality of animals, the humanness of man.
Alongside these, belonging also to that which is already-known, i.e., to
the mathematical, are numbers. If we come upon three apples on the
table, we recognize that there are three of them. But the number three,
threeness, we already know. This means that number is something
mathematical. Only because numbers represent, as it were, the most
striking of always-already-knowns, and thus offer the most familiar instance of the mathematical, is "mathematical" promptly reserved as a
name for the numerical. In no way, however, is the essence of the mathematical defined by numberness. Physics is, in general, the knowledge
of nature, and, in particular, the knowledge of material corporeality in
its motion; for that corporeality manifests itself immediately and universally in everything natural, even if in a variety of ways. If physics

takes shape explicitly, then, as something mathematical, this means that, in an especially pronounced way, through it and for it something is stipulated in advance as what is already-known. That stipulating has to do with nothing less than the plan or projection of that which must henceforth, for the knowing of nature that is sought after, *be* nature: the self-contained system of motion of units of mass related spatiotemporally. Into this ground plan of nature, as supplied in keeping with its prior stipulation, the following definitions among others have been incorporated: Motion means change of place. No motion or direction of motion is superior to any other. Every place is equal to every other. No point in time has preference over any other. Every force is defined according to – i.e., *is* only – its consequences in motion, and that means in magnitude of change of place in the unity of time. Every event must be seen so as to be fitted into this ground plan of nature. Only within the perspective of this ground plan does an event in nature become visible as such an event. This projected plan of nature finds its guarantee in the fact that physical research, in every one of its questioning steps, is bound in advance to adhere to it. This binding adherence, the rigor of research, has its own character at any given time in keeping with the projected plan. The rigor of mathematical physical science is exactitude. Here all events, if they are to enter at all into representation as events of nature, must be defined beforehand as spatiotemporal magnitudes of motion. Such defining is accomplished through measuring, with the help of number and calculation. But mathematical research into nature is not exact because it calculates with precision; rather it must calculate in this way because its adherence to its object-sphere has the character of exactitude. The humanistic sciences, in contrast, indeed all the sciences concerned with life, must necessarily be inexact just in order to remain rigorous. A living thing can indeed also be grasped as spatiotemporal magnitude of motion, but then it is no longer apprehended as living. The inexactitude of the historical humanistic sciences is not a deficiency, but is only the fulfillment of a demand essential to this type of research. It is true, also, that the projecting and securing of the object-sphere of the historical sciences is not only of another kind, but is much more difficult of execution than is the achieving of rigor in the exact sciences.

Science becomes research through the projected plan and through the securing of that plan in the rigor of procedure. Projection and rigor, however, first develop into what they are in methodology. The latter constitutes the second essential characteristic of research. If the sphere that is projected is to become objective, then it is a matter of bringing it to encounter us in the complete diversity of its levels and interweavings. Therefore procedure must be free to view the changeableness in whatever encounters it. Only within the horizon of the incessant-otherness of change does the plenitude of particularity – of facts – show itself. But the facts must become objective [*gegenständlich*]. Hence procedure must represent [*vorstellen*] the changeable in its changing,[8] must bring it to a stand and let the motion be a motion nevertheless. The fixedness of facts and the constantness of their change as such is "rule." The constancy of change in the necessity of its course is "law." It is only within the purview of rule and law that facts become clear as the facts that they are. Research into facts in the realm of nature is intrinsically the establishing and verifying of rule and law. Methodology, through which a sphere of objects comes into representation, has the character of clarifying on the basis of what is clear – of explanation. Explanation is always twofold. It accounts for an unknown by means of a known, and at the same time it verifies that known by means of that unknown. Explanation takes place in investigation. In the physical sciences investigation takes place by means of experiment, always according to the kind of field of investigation and according to the type of explanation aimed at. But physical science does not first become research through experiment; rather, on the contrary, experiment first becomes possible where and only where the knowledge of nature has been transformed into research. Only because modern physics is a physics that is essentially mathematical can it be experimental. Because neither medieval *doctrina* nor Greek *epistēmē* is science in the sense of research, for these it is never a question of experiment. To be sure, it was Aristotle who first understood what *empeiria* (*experientia*) means; the observation of things themselves, their qualities and modifications under changing conditions, and consequently the knowledge of the way in which things as a rule behave. But an observation that aims at such knowledge, the *experimentum*, remains essentially different from the observation that belongs to science as research, from the research experiment; it remains essentially different even when ancient and medieval observation also works with number and measure, and even when that observation makes use of specific apparatus and instruments. For in all this, that which is decisive about the experiment is completely missing. Experiment begins with the laying down of a law as a basis. To set up an experiment means to represent or conceive [*vorstellen*] the conditions under which a specific series of motions can be made susceptible of being followed in its necessary progression, i.e., of being controlled in advance by calculation. But the establishing of a law is accomplished with reference to the ground plan of the object-sphere. That ground plan furnishes a criterion and constrains the anticipatory representing of the conditions. Such representing in and through which the experiment begins is no random imagining. That is why Newton said, *hypothesis non fingo,* "the bases that are laid down are not arbitrarily invented." They are developed out of the ground plan of nature and are sketched into it. Experiment is that methodology which, in its planning and execution, is supported and guided on the basis of the fundamental law laid down, in order to adduce the facts that either verify and confirm the law or deny it confirmation. The more exactly the ground plan of nature is projected, the more exact becomes the possibility of experiment. Hence the much-cited medieval Schoolman Roger Bacon can never be the forerunner of the modern experimental research scientist; rather he remains merely a successor of Aristotle. For in the meantime, the real locus of truth has been transferred by Christendom to faith – to the infallibility of the written word and to the doctrine of the Church. The highest knowledge and teaching is theology as the interpretation of the divine word of revelation, which is set down in Scripture and proclaimed by the Church. Here, to know is not to search out; rather it is to understand rightly the authoritative Word and the authorities proclaiming it. Therefore, the discussion of the words and doctrinal opinions of the various authorities takes precedence in the acquiring of knowledge in the Middle Ages. The *componere scripta et sermones,* the *argumentum ex verbo,*[9] is decisive and at the same time is the reason why the accepted Platonic and Aristotelian philosophy that had been taken over had to be transformed into scholastic dialectic. If, now, Roger Bacon demands the *experimentum* – and he does demand it – he does not mean the experiment of science as research; rather he wants the *argumentum ex re* instead of the *argumentum ex verbo,* the careful observing of things themselves, i.e., Aristotelian *empeiria,* instead of the discussion of doctrines.

The modern research experiment, however, is not only an observation more precise in degree and scope, but is a methodology essentially different in kind, related to the verification of law in the framework, and at the service, of an exact plan of nature. Source criticism in the historical humanistic sciences corresponds to experiment in physical research. Here the name "source criticism" designates the whole gamut of the discovery, examination, verification, evaluation, preservation, and interpretation of sources. Historiographical explanation, which is based on source criticism, does not, it is true, trace facts back to laws and rules. But neither does it confine itself to the mere reporting of facts. In the historical sciences, just as in the natural sciences, the methodology aims at representing what is fixed and stable and at making history an object. History can become objective only when it is past. What is stable in what is past, that on the basis of which historiographical explanation reckons up the solitary and the diverse in history, is the always-has-been-once-already, the comparable. Through the constant comparing of everything with everything, what is intelligible is found by calculation and is certified and established as the ground plan of history. The sphere of historiographical research extends only so far as historiographical explanation reaches. The unique, the rare, the simple – in short, the great – in history is never self-evident and hence remains inexplicable. It is not that historical research denies what is great in history; rather it explains it as the exception. In this explaining, the great is measured against the ordinary and the average. And there is no other historiographical explanation so long as explaining means reduction to what is intelligible and so long as historiography remains research, i.e., an explaining. Because historiography as research projects and objectifies the past in the sense of an explicable and surveyable nexus of actions and consequences, it requires source criticism as its instrument of objectification. The standards of this criticism alter to the degree that historiography approaches journalism.

Every science is, as research, grounded upon the projection of a circumscribed object-sphere and is therefore necessarily a science of individualized character. Every individualized science must, moreover, in the development of its projected plan by means of its methodology, particularize itself into specific fields of investigation. This particularizing (specialization) is, however, by no means simply an irksome concomitant of the increasing unsurveyability of the results of research. It is not a necessary evil, but is rather an essential necessity of science as research. Specialization is not the consequence but the foundation of the progress of all research. Research does not, through its methodology, become dispersed into random investigations, so as to lose itself in them; for modern science is determined by a third fundamental event: ongoing activity.[10]

By this is to be understood first of all the phenomenon that a science today, whether physical or humanistic, attains to the respect due a science only when it has become capable of being institutionalized. However, research is not ongoing activity because its work is accomplished in institutions, but rather institutions are necessary because science, intrinsically as research, has the character of ongoing activity.

The methodology through which individual object-spheres are conquered does not simply amass results. Rather, with the help of its results it adapts [*richtet sich... ein*] itself for a new procedure. Within the complex of machinery that is necessary to physics in order to carry out the smashing of the atom lies hidden the whole of physics up to now. Correspondingly, in historiographical research, funds of source materials become usable for explanation only if those sources are themselves guaranteed on the basis of historiographical explanation. In the course of these processes, the methodology of the science becomes circumscribed by means of its results. More and more the methodology adapts itself to the possibilities of procedure opened up through itself. This having-to-adapt-itself to its own results as the ways and means of an advancing methodology is the essence of research's character as ongoing activity. And it is that character that is the intrinsic basis for the necessity of the institutional nature of research.

In ongoing activity the plan of an object-sphere is, for the first time, built into whatever is. All adjustments that facilitate a plannable conjoining of types of methodology, that further the reciprocal checking and communication of results, and that regulate the exchange of talents are measures that are by no means only the external consequences of the fact that research work is expanding and proliferating. Rather, research work becomes the distant sign, still far from being understood, that modern science is beginning to enter upon the decisive phase of its history. Only now is it beginning to take possession of its own complete essence.

What is taking place in this extending and consolidating of the institutional character of the sciences? Nothing less than the making secure of the precedence of methodology over whatever is (nature and history), which at any given time becomes objective in research. On the foundation of their character as ongoing activity, the sciences are creating for themselves the solidarity and unity appropriate to them. Therefore historiographical or archeological research that is carried forward in an institutionalized way is essentially closer to research in physics that is similarly organized than it is to a discipline belonging to its own faculty in the humanistic sciences that still remains mired in mere erudition. Hence the decisive development of the modern character of science as ongoing activity also forms men of a different stamp. The scholar disappears. He is succeeded by the research man who is engaged in research projects. These, rather than the cultivating of erudition, lend to his work its atmosphere of incisiveness. The research man no longer needs a library at home. Moreover, he is constantly on the move. He negotiates at meetings and collects information at congresses. He contracts for commissions with publishers. The latter now determine along with him which books must be written.

The research worker necessarily presses forward of himself into the sphere characteristic of the technologist in the essential sense. Only in this way is he capable of acting effectively, and only thus, after the manner of his age, is he real. Alongside him, the increasingly thin and empty Romanticism of scholarship and the university will still be able to persist for some time in a few places. However, the effective unity characteristic of the university, and hence the latter's reality, does not lie in some intellectual power belonging to an original unification of the sciences and emanating from the university because nourished by it and preserved in it. The university is real as an orderly establishment that, in a form still unique because it is administratively self-contained, makes possible and visible the striving apart of the sciences into the particularization and peculiar unity that belong to ongoing activity. Because the forces intrinsic to the essence of modern science come immediately and unequivocally to effective working in ongoing activity, therefore, also, it is only the spontaneous ongoing activities of research that can sketch out and establish the internal unity with other like activities that is commensurate with themselves.

The real system of science consists in a solidarity of procedure and attitude with respect to the objectification of whatever is – a solidarity that is brought about appropriately at any given time on the basis of planning. The excellence demanded of this system is not some contrived and rigid unity of the relationships among object-spheres, having to do with content, but is rather the greatest possible free, though regulated, flexibility in the shifting about and introducing of research apropos of the leading tasks at any given time. The more exclusively science individualizes itself with a view to the total carrying on and mastering of its work process, and the more realistically these ongoing activities are shifted into separate research institutes and professional schools, the more irresistibly do the sciences achieve the consummation of their modern essence. But the more unconditionally science and the man of research take seriously the modern form of their essence, the more unequivocally and the more immediately will they be able to offer themselves for the common good, and the more unreservedly too will they have to return to the public anonymity of all work useful to society.

Modern science simultaneously establishes itself and differentiates itself in its projections of specific object-spheres. These projection-plans are developed by means of a corresponding methodology, which is made secure through rigor. Methodology adapts and establishes itself at any given time in ongoing activity. Projection and rigor, methodology and ongoing activity, mutually requiring one another, constitute the essence of modern science, transform science into research.

We are reflecting on the essence of modern science in order that we may apprehend in it its metaphysical ground. What understanding of what is and what concept of truth provide the basis for the fact that science is being transformed into research?

Knowing, as research, calls whatever is to account with regard to the way in which and the extent to which it lets itself be put at the disposal of representation. Research has disposal over anything that is when it can either calculate it in its future course in advance or verify a calculation about it as past. Nature, in being calculated in advance, and history, in being historiographically verified as past, become, as it were, "set in place" [*gestellt*].[11] Nature and history become the objects of a representing that explains. Such representing counts on nature and takes account of history. Only that which becomes object in this way *is* – is considered to be in being. We first arrive at science as research when the Being of whatever is, is sought in such objectiveness.

This objectifying of whatever is, is accomplished in a setting-before, a representing, that aims at bringing each particular being before it in such a way that man who calculates can be sure, and that means be certain, of that being. We first arrive at science as research when and only when truth has been transformed into the certainty of representation. What it is to be is for the first time defined as the objectiveness of representing, and truth is first defined as the certainty of representing, in the metaphysics of Descartes. The title of Descartes's principal work reads: *Meditationes de prima philosophia* [*Meditations on First Philosophy*]. *Prōtē philosophia* is the designation coined by Aristotle for what is later called metaphysics. The whole of modern metaphysics taken together, Nietzsche included, maintains itself within the interpretation of what it is to be and of truth that was prepared by Descartes.

Now if science as research is an essential phenomenon of the modern age, it must be that that which constitutes the metaphysical ground of research determines first and long beforehand the essence of that age generally. The essence of the modern age can be seen in the fact that man frees himself from the bonds of the Middle Ages in freeing himself to himself. But this correct characterization remains, nevertheless, superficial. It leads to those errors that prevent us from comprehending the essential foundation of the modern age and, from there, judging the scope of the age's essence. Certainly the modern age has, as a consequence of the liberation of man, introduced subjectivism and individualism. But it remains just as certain that no age before this one has produced a comparable objectivism and that in no age before this has the non-individual, in the form of the collective, come to acceptance as having worth. Essential here is the necessary interplay between subjectivism and objectivism. It is precisely this reciprocal conditioning of one by the other that points back to events more profound.

What is decisive is not that man frees himself to himself from previous obligations, but that the very essence of man itself changes, in that man becomes subject. We must understand this word *subiectum*, however, as the translation of the Greek *hypokeimenon*. The word names that-which-lies-before, which, as ground, gathers everything onto itself. This metaphysical meaning of the concept of subject has first of all no special relationship to man and none at all to the I.

However, when man becomes the primary and only real *subiectum*, that means: Man becomes that being upon which all that is, is grounded as regards the manner of its Being and its truth. Man becomes the relational center of that which is as such. But this is possible only when the comprehension of what is as a whole changes. In what does this change manifest itself? What, in keeping with it, is the essence of the modern age?

When we reflect on the modern age, we are questioning concerning the modern world picture [*Weltbild*].[12] We characterize the latter by throwing it into relief over against the medieval and the ancient world pictures. But why do we ask concerning a world picture in our interpreting of a historical age? Does every period of history have its world picture, and indeed in such a way as to concern itself from time to time about that world picture? Or is this, after all, only a modern kind of representing, this asking concerning a world picture?

What is a world picture? Obviously a picture of the world. But what does "world" mean here? What does "picture" mean? "World" serves here as a name for what is, in its entirety. The name is not limited to the cosmos, to nature. History also belongs to the world. Yet even nature and history, and both interpenetrating in their underlying and transcending of one another, do not exhaust the world. In this designation the ground of the world is meant also, no matter how its relation to the world is thought.

With the word "picture" we think first of all of a copy of something. Accordingly, the world picture would be a painting, so to speak, of what is as a whole. But "world picture" means more than this. We mean by it the world itself, the world as such, what is, in its entirety, just as it is normative and binding for us. "Picture" here does not mean some imitation, but rather what sounds forth in the colloquial expression, "We get the picture" [literally, we are in the picture] concerning something. This means the matter stands before us exactly as it stands with it for us. "To get into the picture" [literally, to put oneself into the picture] with respect to something means to set whatever is, itself, in place before oneself just in the way that it stands with it, and to have it fixedly before oneself as set up in this way. But a decisive determinant in the essence of the picture is still missing. "We get the picture" concerning something does not mean only that what is, is set before us, is represented to us, in general, but that what is stands before us – in all that belongs to it and all that stands together in it – as a system. "To get the picture" throbs with being acquainted with something, with being equipped and prepared for it. Where the world becomes picture, what is, in its entirety, is juxtaposed as that for which man is prepared and which, correspondingly, he therefore intends to bring before himself and have before himself, and consequently intends in a decisive sense to set in place before himself. Hence world picture, when understood essentially, does not mean a picture of the world but the world conceived and grasped as picture. What is, in its entirety, is now taken in such a way that it first is in being and only is in being to the extent that it is set up by man, who represents and sets forth.[13] Wherever we have the world picture, an essential decision takes place regarding what is, in its entirety. The Being of whatever is, is sought and found in the representedness of the latter.

However, everywhere that whatever is, is *not* interpreted in this way, the world also cannot enter into a picture; there can be no world picture. The fact that whatever is comes into being in and through representedness transforms the age in which this occurs into a new age in contrast with the preceding one. The expressions "world picture of the modern age" and "modern world picture" both mean the same thing and both assume something that never could have been before, namely, a medieval and an ancient world picture. The world picture does not change from an earlier medieval one into a modern one, but rather the fact that the world becomes picture at all is what distinguishes the essence of the modern age [*der Neuzeit*].[14] For the Middle Ages, in contrast, that which is, is the *ens creatum*, that which is created by the personal Creator-God as the highest cause. Here, to be in being means to belong within a specific rank of the order of what has been created – a rank appointed from the beginning – and as thus caused, to correspond to the cause of creation (*analogia entis*). But never does the Being of that which is consist here in the fact that it is brought before man as the objective, in the fact that it is placed in the realm of man's knowing and of his having disposal, and that it is in being only in this way.

The modern interpretation of that which is, is even further from the interpretation characteristic of the Greeks. One of the oldest pronouncements of Greek thinking regarding the Being of that which is runs: *To gar auto noein estin te kai einai.*[15] This sentence of Parmenides means: The apprehending of whatever is belongs to Being because it is demanded and determined by Being. That which is, is that which arises and opens itself, which, as what presences, comes upon man as the one who presences, i.e., comes upon the one who himself opens himself to what presences in that he apprehends it. That which is does not come into being at all through the fact that man first looks upon it, in the sense of a representing that has the character of subjective perception. Rather, man is the one who is looked upon by that which is; he is the one who is – in company with itself – gathered toward presencing, by that which opens itself. To be beheld by what is, to be included and maintained within its openness and in that way to be borne along by it, to be driven about by its oppositions and marked by its discord – that is the essence of man in the great age of the Greeks. Therefore, in order to fulfill his essence, Greek man must gather (*legein*) and save (*sōzein*), catch up and preserve,[16] what opens itself in its openness, and he must remain exposed (*alētheuein*) to all its sundering confusions. Greek man *is* as the one who apprehends [*der Vernehmer*] that which is,[17] and this is why in the age of the Greeks the world cannot become picture. Yet, on the other hand, that the beingness of whatever is, is defined for Plato as *eidos* [aspect, view] is the presupposition, destined far in advance and long ruling indirectly in concealment, for the world's having to become picture.

In distinction from Greek apprehending, modern representing, whose meaning the word *repraesentatio* first brings to its earliest expression, intends something quite different. Here to represent [*vor-stellen*] means to bring what is present at hand [*das Vorhandene*] before oneself as something standing over against, to relate it to oneself, to the one representing it, and to force it back into this relationship to oneself as the normative realm. Wherever this happens, man "gets into the picture" in precedence over whatever is. But in that man puts himself into the picture in this way, he puts himself into the scene, i.e., into the open sphere of that which is generally and publicly represented. Therewith man sets himself up as the setting in which whatever is must henceforth set itself forth, must present itself [*sich . . . präsentieren*], i.e., be picture. Man becomes the representative [*der Repräsentant*] of that which is, in the sense of that which has the character of object.

But the newness in this event by no means consists in the fact that now the position of man in the midst of what is, is an entirely different one in contrast to that of medieval and ancient man. What is decisive is that man himself expressly takes up this position as one constituted by himself, that he intentionally maintains it as that taken up by himself, and that he makes it secure as the solid footing for a possible development of humanity. Now for the first time is there any such thing as a 'position' of man. Man makes depend upon himself the way in which he must take his stand in relation to whatever is as the objective. There begins that way of being human which mans the realm of human capability as a domain given over to measuring and executing, for the purpose of gaining mastery over that which is as a whole. The age that is determined from out of this event is, when viewed in retrospect, not only a new one in contrast with the one that is past, but it settles itself firmly in place expressly as the new. To be new is peculiar to the world that has become picture.

When, accordingly, the picture character of the world is made clear as the representedness of that which is, then in order fully to grasp the modern essence of representedness we must track out and expose the original naming power of the worn-out word and concept "to represent" [*vorstellen*]: to set out before oneself and to set forth in relation to oneself. Through this, whatever is comes to a stand as object and in that way alone receives the seal of Being. That the world becomes picture is one and the same event with the event of man's becoming *subiectum* in the midst of that which is.

Only because and insofar as man actually and essentially has become subject is it necessary for him, as a consequence, to confront the explicit question: Is it as an "I" confined to its own preferences and freed into its own arbitrary choosing or as the "we" of society; is it as an individual or as a community; is it as a personality within the community or as a mere group member in the corporate body; is it as a state and nation and as a people or as the common humanity of modern man, that man will and ought to be the subject that in his modern essence he *already is?* Only where man is essentially already subject does there exist the possibility of his slipping into the aberration of subjectivism in the sense of individualism. But also, only where man *remains* subject does the positive struggle against individualism and for the community as the sphere of those goals that govern all achievement and usefulness have any meaning.

The interweaving of these two events, which for the modern age is decisive – that the world is transformed into picture and man into *subiectum* – throws light at the same time on the grounding event of modern history, an event that at first glance seems almost absurd. Namely, the more extensively and the more effectually the world stands at man's disposal as conquered, and the more objectively the object appears, all the more subjectively, i.e., the more importunately, does the *subiectum* rise up, and all the more impetuously, too, do observation of and teaching about the world change into a doctrine of man, into anthropology. It is no wonder that humanism first arises where the world becomes picture. It would have been just as impossible for a humanism to have gained currency in the great age of the Greeks as it would have been impossible to have had anything like a world picture in that age. Humanism, therefore, in the more strict historiographical sense, is nothing but a moral–aesthetic anthropology. The name "anthropology" as used here does not mean just some investigation of man by a natural science. Nor does it mean the doctrine established within Christian theology of man created, fallen, and redeemed. It designates that philosophical interpretation of man which explains and evaluates whatever is, in its entirety, from the standpoint of man and in relation to man.

The increasingly exclusive rooting of the interpretation of the world in anthropology, which has set in since the end of the eighteenth century, finds its expression in the fact that the fundamental stance of man in relation to what is, in its entirety, is defined as a world view (*Weltanschauung*). Since that time this word has been admitted into common usage. As soon as the world becomes picture, the position of man is conceived as a world view. To be sure, the phrase "world view" is open to misunderstanding, as though it were merely a matter here of a passive contemplation of the world. For this reason, already in the nineteenth century it was emphasized with justification that 'world view' also meant and even meant primarily 'view of life.' The fact that, despite this, the phrase 'world view' asserts itself as the name for the position of man in the midst of all that is, is proof of how decisively the world became picture as soon as man brought his life as *subiectum* into precedence over other centers of relationship. This means: whatever is, is considered to be in being only to the degree and to the extent that it is taken into and referred back to this life, i.e., is lived out, and becomes life-experience. Just as unsuited to the Greek spirit as every humanism had to be, just so impossible was a medieval world view, and just as absurd is a Catholic world view. Just as necessarily and legitimately as everything must change into life-experience for modern man the more unlimitedly he takes charge of the shaping of his essence, just so certainly could the Greeks at the Olympian festivals never have had life-experiences.

The fundamental event of the modern age is the conquest of the world as picture. The word "picture" [*Bild*] now means the structured image [*Gebild*] that is the creature of man's producing which represents and sets before.[18] In such producing, man contends for the position in which he can be that particular being who gives the measure and draws up the guidelines for everything that is. Because this position secures, organizes, and articulates itself as a world view, the modern relationship to that which is, is one that becomes, in its decisive unfolding, a confrontation of world views; and indeed not of random world views, but only of those that have already taken up the fundamental position of man that is most extreme, and have done so with the utmost resoluteness. For the sake of this struggle of world views and in keeping with its meaning, man brings into play his unlimited power for the calculating, planning, and molding of all things. Science as research is an absolutely necessary form of this establishing of self in the world; it is one of the pathways upon which the modern age rages toward fulfillment of its essence, with a velocity unknown to the participants. With this struggle of world views the modern age first enters into the part of its history that is the most decisive and probably the most capable of enduring.

A sign of this event is that everywhere and in the most varied forms and disguises the gigantic is making its appearance. In so doing, it evidences itself simultaneously in the tendency toward the increasingly small. We have only to think of numbers in atomic physics. The gigantic presses forward in a form that actually seems to make it disappear – in the annihilation of great distances by the airplane, in the setting before us of foreign and remote worlds in their everydayness, which is produced at random through radio by a flick of the hand. Yet we think too superficially if we suppose that the gigantic is only the endlessly extended emptiness of the purely quantitative. We think too little if we find that the gigantic, in the form of continual not-ever-having-been-here-yet, originates only in a blind mania for exaggerating and excelling. We do not think at all if we believe we have explained this phenomenon of the gigantic with the catchword "Americanism".

The gigantic is rather that through which the quantitative becomes a special quality and thus a remarkable kind of greatness. Each historical age is not only great in a distinctive way in contrast to others; it also has, in each instance, its own concept of greatness. But as soon as the gigantic in planning and calculating and adjusting and making secure shifts over out of the quantitative and becomes a special quality, then what is gigantic, and what can seemingly always be calculated completely, becomes, precisely through this, incalculable. This becoming incalculable remains the invisible shadow that is cast around all things everywhere when man has been transformed into *subiectum* and the world into picture.

By means of this shadow the modern world extends itself out into a space withdrawn from representation, and so lends to the incalculable the determinateness peculiar to it, as well as a historical uniqueness. This shadow, however, points to something else, which it is denied to us of today to know. But man will never be able to experience and ponder this that is denied so long as he dawdles about in the mere negating of the age. The flight into tradition, out of a combination of humility and presumption, can bring about nothing in itself other than self-deception and blindness in relation to the historical moment.

Man will know, i.e., carefully safeguard into its truth,[19] that which is incalculable, only in creative questioning and shaping out of the power of genuine reflection. Reflection transports the man of the future into that "between" in which he belongs to Being and yet remains a stranger amid that which is. Hölderlin knew of this. His poem, which bears the superscription "To the Germans," closes:

> How narrowly bounded is our lifetime,
> We see and count the number of our years.
> But have the years of nations
> Been seen by mortal eye?
>
> If your soul throbs in longing
> Over its own time, mourning, then
> You linger on the cold shore
> Among your own and never know them.[20]

NOTES BY WILLIAM LOVITT

1. "Reflection" translates *Besinnung*. "Essence" will be the translation of the noun *Wesen* in most instances of its occurrence in this essay. Occasionally the translation "coming to presence" will be used. *Wesen* must always be understood to allude, for Heidegger, not to any mere "whatness," but to the manner in which anything, *as* what it is, takes its course and "holds sway" in its ongoing presence, i.e., the manner in which it endures in its presenc-

ing. 'What is' renders the present participle *seiend* used as a noun, *das Seiende.*

2. *der Grund seines Wesensgestalt.* Heidegger exemplifies the statement that he makes here in his discussion of the metaphysics of Descartes as providing the necessary interpretive ground for the manner in which, in the subjectness of man as self-conscious subject, being and all that is and man – in their immediate and indissoluble relation – come to presence in the modern age.

3. Heidegger's explanatory appendixes have not been included.

4. *Erlebnis*, translated here as "subjective experience"and later as "life-experience," is a term much used by life philosophers such as Dilthey and generally connotes adventure and event. It is employed somewhat pejoratively here. The term *Erfahrung*, which is regularly translated as "experience," connotes discovery and learning, and also suffering and undergoing. Here and subsequently (i.e., "mere religious experience"), "mere" is inserted to maintain the distinction between *Erlebnis* and *Erfahrung*.

5. *Entgötterung*, here inadequately rendered as "loss of the gods," actually means something more like "degodization."

6. *Grundriss.* The verb *reissen* means to tear, to rend, to sketch, to design, and the noun *Riss* means tear, gap, outline. Hence the noun *Grundriss*, first sketch, ground plan, design, connotes a fundamental sketching out that is an opening up as well.

7. "Binding adherence" here translates the noun *Bindung.* The noun could also be rendered "obligation." It could thus be said that rigor is the obligation to remain within the realm opened up.

8. Throughout this essay the literal meaning of *vorstellen*, which is usually translated with "to represent," is constantly in the foreground, so that the verb suggests specifically a setting-in-place-before that is an objectifying, i.e., a bringing to a stand as object. Heidegger frequently hyphenates *vorstellen* in this essay and its appendixes so as to stress the meaning that he intends.

9. "The comparing of the writings with the sayings, the argument from the word." *Argumentum ex re*, which follows shortly, means "argument from the thing."

10. "Ongoing activity" is the rendering of *Betrieb*, which is difficult to translate adequately. It means the act of driving on, or industry, activity, as well as undertaking, pursuit, business. It can also mean management, or workshop or factory.

11. The verb *stellen*, with the meanings to set in place, to set upon (i.e., to challenge forth), and to supply, is invariably fundamental in Heidegger's understanding of the modern age.

12. The conventional translation of *Weltbild* would be "conception of the world" or "philosophy of life." The more literal translation, "world picture," is needed for the following of Heidegger's discussion; but it is worth noting that "conception of the world" bears a close relation to Heidegger's theme of man's representing of the world as picture.

13. *durch den vorstellenden-herstellenden Menschen gestellt ist.*

14. *Die Neuzeit* is more literally "the new age." Having repeatedly used this word in this discussion, Heidegger will soon elucidate the meaning of the "newness"of which it speaks.

15. The accepted English translation of this fragment is, "For thought and being are the same thing."

16. "Preserve" translates *bewahren*. The verb speaks of a preserving that as such frees and allows to be manifest.

17. The noun *Vernehmer* is related to the verb *vernehmen* (to hear, to perceive, to understand). *Vernehmen* speaks of an immediate receiving, in contrast to the setting-before (*vor-stellen*) that arrests and objectifies.

18. *Gebild* is Heidegger's own word. The noun *Gebilde* means thing formed, creation, structure, image. *Gebild* is here taken to be close to it in meaning, and it is assumed – with the use of "structured" – that Heidegger intends the force of the prefix *ge-*, which connotes a gathering, to be found in the word. "Man's producing which represents and sets before" translates *des vorstellenden Herstellens*.

19. *Wissen, d.h., in seine Wahrheit verwahren, wird der Mensch . . .* Here the verb *wissen* (to know), strongly emphasized by its placement in the sentence, is surely intended to remind of science (*Wissenschaft*) with whose characterization this essay began. On such knowing – an attentive beholding that watches over and makes manifest – as essential to the characterizing of science as such.

20. Wohl ist enge begrenzt unsere Lebenzeit,
    Unserer Jahre Zahl sehen und zählen wir,
    Doch die Jahre der Völker,
    Sah ein sterbliches Auge sie?

    Wenn die Seele dir auch über die eigene Zeit
    Sich die sehnende schwingt, trauernd verweilest du
    Dann am kalten Gestade
    Bei den Deinen und kennst sie nie.

Part II
The Problem of Scientific Realism

3 The Strange World of the Quantum*

P. C.W. Davies and J. R. Brown

WHAT IS QUANTUM THEORY?

The word 'quantum' means 'a quantity' or 'a discrete amount'. On an everyday scale we are accustomed to the idea that the properties of an object such as its size, weight, colour, temperature, surface area, and motion are all qualities which can vary from one object to another in a smooth and continuous way. Apples, for example, come in all manner of shapes, sizes and colours without any noticeable gradations in between.

On the atomic scale, however, things are very different. The properties of atomic particles such as their motion, energy and spin do not always exhibit similar smooth variations, but may instead differ in discrete amounts. One of the assumptions of classical Newtonian mechanics was that the properties of matter are continuously variable. When physicists discovered that this notion breaks down on the atomic scale they had to devise an entirely new system of mechanics – quantum mechanics – to take account of the lumpiness which characterizes the atomic behaviour of matter. Quantum theory, then, is the underlying theory from which quantum mechanics is derived.

Considering the success of classical mechanics in describing the dynamics of everything from billiard balls to stars and planets, it is not surprising that its replacement by a new system of mechanics on the atomic scale was considered to be a revolutionary departure. Nevertheless, physicists rapidly proved the value of the theory by explaining a wide range of otherwise incomprehensible phenomena, so much so that today quantum theory is often cited as the most successful scientific theory ever produced.

* Chapter 1 in P. C.W. Davies and J. R. Brown (eds), *The Ghost in the Atom. A Discussion of the Mysteries of Quantum Physics* (Cambridge: Cambridge University Press, 1986), pp. 1–39.

ORIGINS

Quantum theory had its first faltering beginnings in the year 1900, with the publication of a paper by the German physicist Max Planck. Planck addressed himself to what was still an unsolved problem of nineteenth-century physics, concerning the distribution of radiant heat energy from a hot body among various wavelengths. Under certain ideal conditions the energy is distributed in a characteristic way, which Planck showed could only be explained by supposing that the electromagnetic radiation was emitted from the body in discrete packets or bundles, which he called quanta. The reason for this jerky behaviour was unknown, and simply had to be accepted *ad hoc*.

In 1905 the quantum hypothesis was bolstered by Einstein, who successfully explained the so-called photoelectric effect in which light energy is observed to displace electrons from the surfaces of metals. To account for the particular way this happens, Einstein was compelled to regard the beam of light as a hail of discrete particles later called photons. This description of light seemed utterly at odds with the traditional view, in which light (in common with all electromagnetic radiation) consists of continuous waves which propagate in accordance with Maxwell's celebrated electromagnetic theory, firmly established half a century before. Indeed, the wave nature of light had been demonstrated experimentally as long ago as 1801 by Thomas Young using his famous 'two-slit' apparatus.

The wave–particle dichotomy, however, was not restricted to light. Physicists were at that time also concerned about the structure of atoms. In particular, they were puzzled by how electrons could go round and round a nucleus without emitting radiation, since it was known from Maxwell's electromagnetic theory that when charged particles move along curved paths they radiate electromagnetic energy. If this were to occur continuously, the orbiting atomic electrons would rapidly lose energy and spiral into the nucleus (see Fig. 3.1).

In 1913 Niels Bohr proposed that atomic electrons are also 'quantized', in that they can reside without loss of energy in certain fixed energy levels. When electrons jump between the levels, electromagnetic energy is released or absorbed in discrete quantities. These packets of energy are, in fact, photons.

The reason why the atomic electrons should behave in this discontinuous fashion was not revealed, however, until somewhat later, when the wave nature of matter was discovered. The experimental work of Clinton Davisson and others and the theoretical work of Louis de Bro-

Fig. 3.1. Collapse of the classical atom. (a) The theories of Newton and Maxwell predict that an orbiting atomic electron will steadily radiate electromagnetic waves, thereby losing energy and spiralling into the nucleus. (b) The quantum theory predicts the existence of discrete non-radiating energy levels in which the wave associated with the electron just 'fits' around the nucleus, forming standing wave patterns reminiscent of the notes on a musical instrument. (The wave must 'fit' in the radial direction too.)

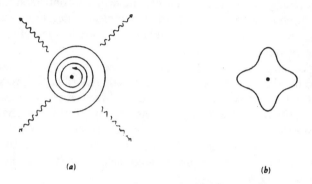

(a)   (b)

glie led to the idea that electrons as well as photons can behave both as waves and as particles, depending on the particular circumstances. According to the wave picture, the atomic energy levels Bohr proposed correspond to stationary or standing wave patterns around the nucleus. Much as a cavity can be made to resonate at different discrete musical notes, so the electron waves vibrate with certain well-defined energy patterns. Only when the patterns shift, corresponding to a transition from one energy level to another, does an electromagnetic disturbance ensue, with radiation being emitted or absorbed.

It soon became apparent that not only electrons but all subatomic particles are subject to similar wavelike behaviour. Evidently the traditional laws of mechanics as formulated by Newton, as well as Maxwell's laws of electromagnetism, fail completely in the microworld of atoms and subatomic particles. By the mid-1920s, a new system of mechanics – quantum mechanics – had been developed independently by Erwin Schrödinger and Werner Heisenberg to take account of this wave–particle duality.

The new theory was spectacularly successful. It rapidly helped scientists to explain the structure of atoms, radioactivity, chemical bonding and the details of atomic spectra (including the effects of electric and magnetic fields). Further elaborations of the theory by Paul Dirac, Enrico Fermi, Max Born and others eventually led to satisfactory explanations of nuclear structure and reactions, the electrical and thermal properties of solids, super-conductivity, the creation and annihilation of elementary particles of matter, the prediction of the existence of antimatter, the stability of certain collapsed stars and much else. Quantum mechanics also made possible major developments in practical hardware, including the electron microscope, the laser and the transistor. Exceedingly delicate atomic experiments have confirmed the existence of subtle quantum effects to an astonishing degree of accuracy. No known experiment has contradicted the predictions of quantum mechanics in the last 50 years.

This catalogue of triumphs singles out quantum mechanics as a truly remarkable theory – a theory that correctly describes the world to a level of precision and detail unprecedented in science. Nowadays, the vast majority of professional physicists employ quantum mechanics, if not almost unthinkingly, then with complete confidence. Yet this magnificent theoretical edifice is founded on a profound and disturbing paradox that has led some physicists to declare that the theory is ultimately meaningless.

The problem, which was already readily apparent in the late 1920s and early 1930s, concerns not the technical aspects of the theory but its interpretation.

WAVES OR PARTICLES?

The peculiarity of the quantum is readily apparent from the way that an object such as a photon can display both wave-like and particle-like properties. Photons can be made to produce diffraction and interference patterns, a sure test of their wave-like nature. On the other hand, in the photoelectric effect, photons knock electrons out of metals after the fashion of a coconut-shy. Here, the particle model of light seems to be more appropriate.

The co-existence of wave and particle properties leads quickly to some surprising conclusions about nature. Let us take a familiar example. Suppose that a beam of polarized light encounters a piece of polarizing material (see Fig. 3.2). Standard electromagnetic theory predicts that if the plane of polarization of the light is parallel to that of the material, all the light is transmitted. On the other hand, if the angles are perpendicular, no light is transmitted. At intermediate angles some light is transmitted; for example, at 45° the transmitted light has precisely half the intensity of the original beam. Experiment confirms this.

Fig. 3.2. Breakdown of predictability. (a) Classically, the polarized light wave will pass through the polarizer with a reduced intensity $\cos^2\theta$, emerging polarized in the 'vertical' direction. Viewed as a flux of identical photons, this phenomenon can be explained only by supposing that some photons are passed and others blocked, unpredictably, with probabilities $\cos^2\theta$ and $\sin^2\theta$, respectively. (b) Note that the incident wave could be regarded as a superposition of 'vertically' and 'horizontally' polarized waves.

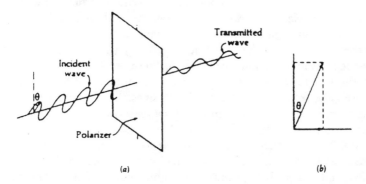

Now, if the intensity of the incoming beam is reduced so that only one photon at a time passes through the polarizer, we are faced with a puzzle. Because a photon cannot be divided up, any given photon must be either passed or blocked. With the angle set at 45°, on average half the photons must get through, while the other half are blocked. But which photons get through and which do not? As all photons of the same energy are supposed to be identical and hence indistinguishable, we are forced to conclude that the passage of photons is a purely random process. Although any given photon has a 50–50 chance (a probability of $^1/_2$) of getting through, it is impossible to predict in advance which particular ones will do so. Only the betting odds can be given. As the angle is varied so the probability can range from zero to one.

The conclusion is intriguing and yet disconcerting. Before the discovery of quantum physics the world was thought to be completely predictable, at least in principle. In particular, if identical experiments were performed, identical results were expected. But, in the case of the photons and the polarizer, one might very well find that two identical experiments produced different results, as one photon passed through the polarizer while another identical photon was blocked. Evidently the world is not wholly predictable after all. Generally we cannot know until after an observation has been made what the fate of a given photon will be.

These ideas imply that there is an element of uncertainty in the microworld of photons, electrons, atoms, and other particles. In 1927 Heisenberg quantified this uncertainty in his famous uncertainty principle. One expression of the principle concerns attempts to measure the position and motion of a quantum object simultaneously. Specifically, if we try to locate an electron, say, very precisely, we are forced to forgo information about its momentum. Conversely, we can measure the electron's momentum accurately, but then its position becomes indeterminate. The very act of trying to pin down an electron to a specific place introduces an uncontrollable and indeterminate disturbance to its motion, and vice versa. Furthermore, this inescapable constraint on our knowledge of the electron's motion and location is not merely the result of experimental clumsiness; it is inherent in nature. Apparently the electron simply *does not possess* both a position and a momentum simultaneously.

It follows that there is an intrinsic fuzziness in the microworld that is manifested whenever we attempt to measure two incompatible observable quantities, such as position and momentum. Among other things, this fuzziness demolishes the intuitive idea of an electron (or photon, or whatever) moving along a distinct path or trajectory in space. For a particle to follow a well-defined path, at each instant it must possess a location (a point on the path) and a motion (tangent vector to the path). But a quantum particle cannot have both at once.

In daily life we take it for granted that strict laws of cause and effect direct the bullet to its target or the planet in its orbit along a precisely defined geometrical path in space. We would not doubt that when the bullet arrives at the target its point of arrival represents the end-point of a continuous curve which started at the barrel of the gun. Not so for electrons. We can discern a point of departure and a point of arrival, but we cannot always infer that there was a definite route connecting them.

Seldom is this fuzziness more apparent than in the famous two-slit experiment of Thomas Young (see Fig. 3.3). Here a beam of photons (or electrons) from a small source travels towards a screen punctured by two narrow apertures. The beam creates an image of the holes on a second screen. The image consists of a distinct pattern of bright and dark 'interference fringes', as waves passing through one hole encounter those from the other hole. Where the waves arrive in step, reinforcement occurs; where they are out of step, cancellation occurs. Thus is the wave-like nature of photons or electrons clearly demonstrated.

But the beam can instead be considered as consisting of particles. Suppose the intensity is again reduced so much that only one photon

Fig. 3.3. Waves or particles? In this two-slit experiment electrons or photons from the source pass through two nearby apertures in screen *A* and travel on to strike screen *B*, where their rate of arrival is monitored. The observed pattern of varying intensity indicates a wave interference phenomenon.

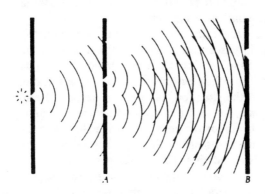

or electron traverses the apparatus at a time. Naturally each arrives at a definite point on the image screen. It can be recorded as a little speck. Other particles arrive elsewhere leaving their own specks. The effect at first seems random. But a pattern begins to build up in a speckled kind of way. Each particle is directed not by an imperative to a particular place on the image screen but by the 'law of averages'. When a large number of particles has traversed the system, an organized pattern is created. This is the interference pattern. Thus, any given photon or electron does not make a pattern; it makes only a single spot. Yet each electron or photon, while apparently free to go anywhere, cooperates in such a way as to build up the pattern in a probabilistic fashion.

Now, if one of the two apertures is blocked, the average behaviour of the electrons or photons changes dramatically; indeed, the interference pattern disappears. Nor can it be reconstructed by superimposing the two patterns obtained by recording the images from each individual slit acting alone. Interference only presents itself when both apertures are open simultaneously. Hence, each photon or electron must somehow *individually* take account of whether both or only one hole is open. But how can they do this if they are indivisible particles? On the face of it, each particle can only go through one slit. Yet somehow the particle 'knows' about the other slit. How?

One way of answering this question is to recall that quantum particles do not have well-defined paths in space. It is sometimes convenient to think of each particle as somehow possessing an infinity of different paths, each of which contributes to its behaviour. These paths, or routes, thread through both holes in the screen, and encode information about each. This is how the particle can keep track of what is happening throughout an extended region of space. The fuzziness in its activity enables it to 'feel out' many different routes.

Suppose a disbelieving physicist were to station detectors in front of the two holes to ascertain in advance towards which hole a particular electron was heading. Could not the physicist then suddenly block the other hole without the electron 'knowing', leaving its motion unaltered? If we analyse the situation, taking into account Heisenberg's uncertainty principle, then we can see that nature outmanoeuvres the wily physicist. In order for the position of each electron to be measured accurately enough to discern the hole it is approaching, the electron's motion is so disturbed that the interference pattern defiantly vanishes! The very act of investigating where the electron is going ensures that the two-hole cooperation fails. Only if we decide not to trace the electron's route will its 'knowledge' of both routes be displayed.

A further intriguing consequence of the above dichotomy has been pointed out by John Wheeler. The decision either to perform the experiment to determine the electron's route, or to relinquish this knowledge and experiment instead with an interference pattern, can be left until *after* any given electron has already traversed the apparatus! In this so-called 'delayed-choice' experiment, it appears that what the experimenter decides now can in some sense influence how quantum particles shall have behaved in the past, though it must be emphasized that the inherent unpredictability of all quantum processes forbids this arrangement from being used to send signals backwards in time or to in any way 'alter' the past.

An idealized arrangement designed to carry out a related delayed-choice experiment (with photons rather than electrons) is shown in Fig. 3.4, and forms the basis of a practical experiment performed ... by Caroll Alley and his colleagues at the University of Maryland. Laser light incident on a half-silvered mirror A divides into two beams analogous to the two paths through the slits in Young's experiment. Further reflections at mirrors M redirect the beams so that they cross and enter photon detectors 1 and 2, respectively. In this arrangement a detection of a given photon by either 1 or 2 suffices to determine which of the two alternative routes the photon will have travelled.

If, now, a second half-silvered mirror B is inserted at the crossing point the two beams are recombined, part along the route into 1 and

Fig. 3.4. Schematic diagram showing the layout of a practical version of Wheeler's delayed-choice experiment.

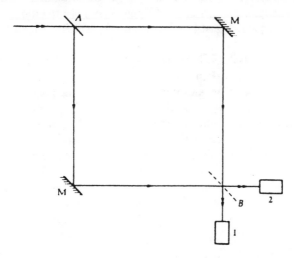

part along the route into 2. This will cause wave interference effects, and the strengths of the beams going into 1 and 2 respectively will then depend on the relative phases of the two beams at the point of recombination. These phases can be altered by adjusting the path lengths, thereby essentially scanning the interference pattern. In particular it is possible to arrange the phases so that destructive interference leads to zero beam strength going into 1, with 100% of the light going into 2. With this arrangement the system is analogous to the original Young experiment, for which it is not possible to specify which of the two routes has been taken by any given photon. (Loosely speaking, each photon takes both routes.)

Now the crucial point is that the decision of whether or not to insert the second half-silvered mirror *B* can be left until a given photon has almost arrived at the cross-over point. In other words, whether the photon *shall* have traversed the system either by one route or 'both routes' is determined only *after* the traverse has taken place.

WHAT DOES IT ALL MEAN?

The fact that electrons, photons and other quantum objects behave sometimes like waves and sometimes like particles often prompts the question of what they 'really' are. The conventional position regarding questions of this sort draws upon the later work of Bohr, who believed he had discovered a consistent interpretation of quantum mechanics. This is usually referred to as the Copenhagen interpretation, so named after Bohr's physics institute in Denmark, which he founded in the 1920s.

According to Bohr, it is meaningless to ask what an electron 'really' is. Or at least, if you ask the question, physics cannot supply the answer. Physics, he declared, tells us not about what *is*, but what we can *say* to each other concerning the world. Specifically, if a physicist carries out an experiment on a quantum system, provided a full specification of the experimental set-up is given, physics can then make a meaningful prediction about what he may observe, and thence communicate to his fellows in a well-understood language.

In Young's experiment, for example, we have a clear choice. Either we can leave the electrons or photons alone, and observe an interference pattern. Or we can take a peek at the particles' trajectories and wash out the pattern. The two situations are not contradictory, but complementary.

Similarly there is a position–momentum complementarity. We can choose to measure the position of a particle in which case its momentum is uncertain, or we can measure the momentum and trade-off knowledge of its position. Each quality – position, momentum – constitutes a complementary aspect of the quantum object.

Bohr elevated these ideas to a principle of *complementarity*. In wave–particle duality, for example, the wave and particle properties of a quantum object constitute complementary aspects of its behaviour. He argued that we should never encounter any experiments in which these two distinct behaviours conflict with each other.

A profound consequence of Bohr's ideas is that the traditional Western concept of the relationship between macro and micro, the whole and its parts, is radically altered. Bohr claimed that before you can make sense of what an electron is doing you have to specify the total experimental context; say what you are going to measure, how your apparatus is organized, and so on. So the quantum reality of the microworld is inextricably entangled with the organization of the macroworld. In other words, the part has no meaning except in relation to the whole. This holistic character of quantum physics has found considerable favour among followers of Eastern mysticism, the philosophy embodied in such oriental religions as Hinduism, Buddhism and Taoism. Indeed, in the early days of quantum theory many physicists, including Schrödinger, were quick to draw parallels between the quantum concept of part and whole, and the traditional oriental concept of the unity and harmony of nature.

Central to Bohr's philosophy is the assumption that uncertainty and fuzziness are intrinsic to the quantum world and not merely the result of our incomplete perception of it. This is quite a subtle matter. We know of many systems which are unpredictable: the vicissitudes of the weather, stockmarkets and roulette wheels, for example, are familiar enough. Yet these do not force us to make a radical reappraisal of the laws of physics. The reason is that the unpredictability of most things in everyday life can be traced to the fact that we do not have enough information to compute their behaviour at the level of detail necessary for an accurate prediction. In the case of roulette, say, we resort to a statistical description. Likewise in classical thermodynamics, the collective behaviour of myriads of molecules can successfully be described in an average way using statistical mechanics. However, the fluctuations about computed mean values are not intrinsically indeterminate in that case because, in principle, the complete mechanical description of every participating molecule could be given (ignoring for this example quantum effects!).

When the information concerning some dynamical variables is discarded, an element of vagueness and uncertainty is introduced into our description of the system. However, we know that this fuzziness is really the result of the activity of all those variables we have chosen to ignore. We might call them 'hidden variables'. They are always there, but our observations may be too crude to reveal them. Thus the measurement of gas pressure is too coarse to reveal individual molecular motions.

Why can we not attribute quantum fuzziness to a deeper level of hidden variables? Such a theory would enable us to picture the chaotic, apparently indeterminate cavorting of quantum particles as driven by a substratum of completely deterministic forces. The fact that we seem to be unable to determine both the position and momentum of an electron simultaneously might then be attributed to the crude nature of our apparatus which is as yet unable to probe the finer level of this substratum.

Einstein was convinced that something like this must be the case; that ultimately a classical world of familiar cause and effect underlies the madhouse of the quantum. He endeavoured to construct thought experiments to test the idea. The most refined of these he presented in a now famous paper written in 1935 with Boris Podolsky and Nathan Rosen.

THE EINSTEIN–PODOLSKY–ROSEN (EPR) EXPERIMENT

The purpose of this thought experiment was to expose the profound pecularities of the quantum description of a physical system extended over a large region of space. The experiment invites us to consider cheating the Heisenberg uncertainty principle by sneaking a look at both the position and momentum of a particle simultaneously. The strategy employed is to use an accomplice particle to perform a measurement by proxy on the particle of interest.

Suppose a single stationary particle explodes into two equal fragments, A and B (see Fig. 3.5). Heisenberg's uncertainty principle apparently forbids us from knowing the position and momentum of either A or B simultaneously. However, because of the law of action and reaction (conservation of momentum), a measurement of B's momentum can be used to deduce A's momentum. Similarly, by symmetry, A will have moved a distance equal to that of B from the point of explosion, so a measurement of B's position reveals A's position.

An observer at B is free, at his whim, to observe either the momentum or the position of B. As a result he will know either the momentum or the position of A, according to his choice. Thus, a subsequent observation of either As momentum or As position will now have a predictable result.

Einstein argued

> If, without in any way disturbing a system, we can predict with certainty... the value of a physical quantity, then there exists an element of physical reality corresponding to this physical quantity.

He therefore concluded that, in the situation described, the particle A must possess a real momentum or a real position, according to the choice of the observer at B.

Now the crucial point is this. If A and B have flown a very long way apart then one would be reluctant to suppose that a measurement carried out on B can affect A. At the very least, A cannot be directly affected instantaneously, because according to the special theory of

Fig. 3.5. Two equal mass fragments flying apart from a common centre (assumed at rest) have equal and opposite momenta and are always equidistant from the centre. Hence a measurement of either the momentum or position of A reveals the momentum or position of B.

relativity no physical signal or influence can travel faster than light; so *A* cannot 'know' that a measurement has been performed on *B* until at least the light travel time between them. In principle this could be billions of years!

Bohr rejected Einstein's reasoning by reiterating his Copenhagen philosophy, that the microscopic properties of a quantum particle must be viewed against the total macroscopic context. In this case a distant but correlated accomplice article, subjected to measurements, forms an inseparable part of the quantum system. Although no direct signal or influence can travel between *A* and *B*, that does not mean, according to Bohr, that you can ignore measurements carried out on *B* when discussing the circumstances of *A*. So, although no actual physical force is transmitted between *A* and *B*, they seem to *cooperate* in their behaviour in a sort of conspiracy.

Einstein found this idea of two widely separated particles conspiring to give coordinated results of apparently independent measurements performed on each too much to swallow, deriding it as 'ghostly action at a distance'. He wanted his objective reality to be localized on each particle, and it was this locality that was eventually to bring his ideas into conflict with quantum mechanics. What was needed was a practical experimental test that could discriminate between Bohr's and Einstein's views by revealing the cooperation, or ghostly action at a distance, in action. But such a development had to wait half a century.

BELL'S THEOREM

In 1965 John Bell studied the problem of two-particle quantum systems and was able to prove a powerful mathematical theorem which turned out to be of crucial importance in setting up a practical experimental test. The theory is essentially independent of the nature of the particles or the details of the forces that act on them, and focusses instead on the rules of logic that govern all measurement processes. To give a simple example of the latter, a census of the population of Britain cannot possibly find that the number of black people is greater than the number of black men plus the number of women of all races.

Bell investigated the correlations that could exist between the results of measurements carried out simultaneously on two separated particles. These measurements might be on particle positions, momenta, spin, polarization, or other dynamical properties. Many researchers have adopted polarization as a convenient means of studying EPR correlations. Suppose a parent particle with zero angular momentum decays into two photons *A* and *B*. By the laws of conservation, one photon must have the same polarization as the other. This can be confirmed by stationing measuring devices perpendicular to the paths of the particles and measuring the polarization in a certain common direction, say 'up'. It is indeed found that, when particle *A* is passed by its polarizer, *B* is always passed too. A 100% *correlation* is found. Conversely, if the polarizers are arranged perpendicular to each other, every time *A* is passed *B* is blocked. This time there is 100% *anti-correlation*. There is nothing mysterious about this; it would also be true in ordinary classical mechanics.

The crucial test comes when the polarization measuring devices are oriented obliquely to each other (see Fig. 3.6). We would now expect some result intermediate between complete correlation and complete anti-correlation, depending on the angles chosen. These may be varied both parallel and perpendicular to the line of flight of the particles, and they could be varied at random from one measurement to the next.

Bell set out to discover the theoretical limits on the extent to which the results of such measurements can be correlated. Suppose, for example, that Einstein had been basically correct, and that quantum behaviour is really the product of a substratum of chaotic classical forces. Suppose also that faster-than-light signalling is forbidden in accordance with the rules of relativity theory. Properly formulated, the first assumption is usually what is meant by 'reality', because it affirms that quantum objects *really do* possess *all* dynamical attributes in a well-defined sense at all times. The second assumption is termed 'locality' or sometimes 'separability' because it forbids objects from instantaneously exerting physical influences on each other when they are spatially separated, i.e. not at the same location.

Subject to the double assumption of 'local reality', and further assuming that the conventional rules of logical reasoning do not founder on the rocks of quantum uncertainty, Bell was able to establish a strict limit on the possible level of correlation for simultaneous two-particle mea-

Fig. 3.6. Bell's theorem applied to two oppositely directed photons from a common source predicts a limit to the degree of correlation permitted in the results of polarization measurements performed separately on each.

surement results. The whole point of the exercise then is this. Quantum mechanics *à la* Bohr predicts that, under some circumstances the degree of cooperation should *exceed* Bell's limit. That is, the conventional view of quantum mechanics requires a degree of cooperation (or conspiracy) between separated systems in excess of that logically permitted in any 'locally real' theory. Bell's theorem thus opens the way for a direct test of the foundations of quantum mechanics, and the decisive discrimination between Einstein's idea of a locally real world, and Bohr's conception of a somewhat ghostly world full of subatomic conspiracy.

ASPECT'S EXPERIMENT

A number of experiments have been conducted in an attempt to test Bell's inequality. The most successful of these was reported by A. Aspect, J. Dalibard and G. Roger in *Physical Review Letters* (vol. 39, p. 1804) in December 1982.

The experiment consisted of polarization measurements made on pairs of oppositely moving photons emitted simultaneously in single transitions by calcium atoms. The experimental arrangement is shown in Fig. 3.7.

Fig. 3.7. Aspect's experimental arrangement. Paris of photons from the source *S* travel several metres to the acousto-optical switches. The route of a photon beyond the switch determines which of the differently oriented polarizers it will encounter. The photons are detected using photo-multipliers (PM) and coincidences between the various channels are monitored electronically.

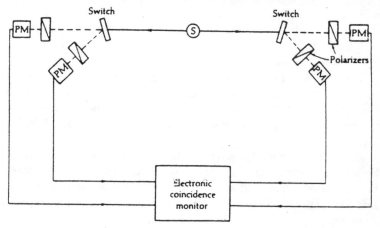

In the diagram, the source S used a beam of calcium atoms excited by a pair of lasers (i.e. two-photon excitation) to a state (S state) that could only decay again by a two-photon 'cascade'. About 6 m on either side of the source there was located an acousto-optical switching device. The principle employed was to exploit the fact that the refractive index of water will vary slightly with compression.

In the switch, an ultrasonic standing wave at about 25 MHz was established using oppositely directed transducers. By arranging for the photons to encounter the switch at near the critical angle for total internal reflection, it was possible to switch from transmission to reflection conditions at each half-cycle of the sound wave, i.e. at 50 MHz.

The photons, emerging either along the line of the incident path (after transmission) or deflected (by reflection) then encountered polarizers, which would either transmit or block them with certain definite probabilities. These polarizers were oriented at different angles relative to the polarization of the photons. The photons' fate was then monitored by stationing photomultiplier detectors beyond these polarizers. The set-up was identical on both sides of the source.

The experiment was performed by monitoring electronically the fate of each pair of photons and assessing the level of correlation. The unique and essential feature of this experiment is the ability to change at random, while the photons are in mid-flight, the subsequent path of the photons, i.e. to which polarizer they shall be directed. This is equivalent to re-orienting the polarizers on each side of the source so rapidly that no signal could have time enough to pass from one to the other, even at the speed of light.

Each switching event took about 10 ns, which should be compared with the lifetime of the photons' emission (5 ns) and the travel time of the photons (40 ns).

In practice, the switching was not strictly random. The standing waves were generated independently at different frequencies. The difference between this and truly random switching is irrelevant except in the case of the most contrived 'conspiracy' theories of hidden variables.

The authors report that a typical run lasted 12 000s, divided equally between the arrangement as described above, another in which all the polarizers were removed, and a third in which one polarizer on each side was removed. This enabled the results to be corrected for systematic errors.

THE NATURE OF REALITY

At issue in the above-mentioned test is much more than just the clarification of a technical matter between contending theories of the microworld. The debate concerns our conception of the universe and of the nature of reality.

Before the days of quantum mechanics, most Western scientists assumed that the world about us enjoys an independent existence. That is, it consists of objects such as tables, chairs, stars, atoms that are simply 'out there' whether we observe them or not. According to this philosophy, the universe is a collection of such independently existing objects that together make up the totality of things. Naturally it has to be conceded that any observation that we make of an object involves some sort of interaction with it, which implies that it will inevitably suffer a disturbance. However, the disturbance is regarded merely as an incidental perturbation on something which already possesses a concrete and well-defined existence. Indeed, in principle, the disturbance needed to measure something could be made arbitrarily small, and in any case could be computed in complete detail so that after the measurement we could deduce exactly what had happened to the observed object. If this were the true state of affairs we should not hesitate in saying that the object *really had* a complete set of dynamical attributes such as position, momentum, spin and energy both prior to and subsequent to our observation of it. Atoms and electrons would then simply be 'little things', differing only from 'big things' such as billiard balls, in the matter of *scale*. Otherwise, there is no qualitative difference in their status in reality.

This picture of the world is compelling because it is the one which most readily squares with our common-sense understanding of nature. Einstein called it 'objective reality' because the status in reality of external objects does not depend upon a conscious individual's observations. (Contrast this with the objects of our dreams, which are part of a subjective reality.) But it is precisely this common-sense view of reality that Bohr challenged with the philosophy underscoring the Copenhagen interpretation.

Bohr's position, as stated earlier, is that it is meaningless to ascribe a complete set of attributes to some quantum object prior to an act of measurement being performed on it. Thus, for example, in a photon polarization experiment we simply cannot say what polarization a photon has before we conduct a measurement. But after the measurement we may indeed be able to attribute a definite polarization state to the photon. Similarly, if we are faced with the choice of position or momentum measurements on a particle, we cannot say that the particle possesses specific values for these quantities prior to the measurement. If we decide to measure the position, we end up with a particle-at-a-place. If we choose instead to measure the momentum, we get a particle-with-a-motion. In the former case, after the measurement is complete, the particle simply does not have a momentum, in the latter case it does not have a location.

These ideas can best be illustrated with the help of a simple example (see Fig. 3.8). Consider a box into which a single electron is inserted. In the absence of an observation, the electron is equally likely to be anywhere in the box. The quantum mechanical wave corresponding to the electron is therefore spread uniformly throughout the box. Suppose now that an impenetrable screen is inserted down the middle of the box, dividing it into two chambers. Obviously the electron can only be in *either* one chamber *or* the other. However, unless we look and see which, the *wave* will still be in both chambers. On observation the electron will be revealed to be in one particular chamber. At that very instant (according to the rules of quantum mechanics) the wave abruptly disappears from the empty chamber, even if that chamber has remained isolated throughout! It is as though, prior to the observation, there are two nebulous electron 'ghosts' each inhabiting one chamber waiting for an observation to turn one of them into a 'real' electron, and simultaneously to cause the other to vanish completely.

Fig. 3.8 Collapse of a quantum wave. (a) When a single quantum particle is confined to a box its associated wave is spread uniformly throughout the interior. (b) A screen is inserted, dividing the box into two isolated chambers. (c) An observation reveals the particle to be in the right-hand chamber. Abruptly the wave in the other chamber, which represents the probability of finding the particle there, vanishes.

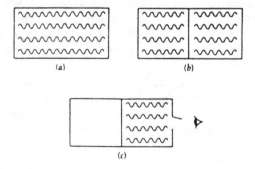

This example also nicely illustrates the non-locality of quantum mechanics. Suppose the two chambers, *A* and *B*, are disconnected and moved a long way apart (say one light year), then *A* is inspected by an observer and found to contain the particle. Instantaneously the quantum wave in *B* vanishes, even though it is a light year away. (It must be repeated, however, that this arrangement cannot be used to signal faster than light, on account of the unpredictable nature of each observation.)

In general, a quantum system will be in a state consisting of a collection (perhaps infinite in number) of quantum states superimposed. A simple example of such a superposition was given above, involving two disconnected wave patterns, one in each chamber. A more typical example is that in Young's two-slit experiment, where waves from both slits actually overlap and interfere.

We encountered this sort of superposition earlier, in the discussion of polarized light passing through an obliquely oriented polarizer. If the incoming light wave is at 45° to the polarizer, we may consider it to consist of two equal-strength waves combined coherently with polarizations at right angles to each other, as shown in Fig. 3.2. The wave parallel to the polarizer will be transmitted, the other will be blocked. We could regard a quantum state containing one photon polarized at 45° to the polarizer as a superposition of two 'ghosts' or 'potential' photons, one with parallel polarization enabling it to get through the polarizer, the other with perpendicular polarization, preventing it from getting through. When the measurement is finally made, one of these two 'ghosts' gets promoted to a 'real' photon and the other vanishes. Suppose the measurement shows that the photon passes through the polarizer. The *ghost* photon that is parallel to the polarizer prior to the measurement thus becomes the *real* photon. But we cannot say that this photon 'really existed' *prior* to the measurement. All that can be said is that the system was in a superposition of two quantum states, neither of which possessed privileged status.

The physicist John Wheeler likes to tell a delightful parable which nicely illustrates the peculiar status of a quantum particle prior to measurement. The story concerns a version of the game of 20 questions:

Then my turn came, fourth to be sent from the room so that Lothar Nordheim's other fifteen after-dinner guests could consult in secret and agree on a difficult word. I was locked out unbelievably long. On finally being readmitted, I found a smile on everyone's face, sign of a joke or a plot. I nevertheless started my attempt to find the word. 'Is it animal?' 'No.' 'Is it mineral?' 'Yes.' 'Is it green?' 'No.' 'Is it white?' 'Yes.' These answers came quickly. Then the questions began to take longer in the answering. It was strange. All I wanted from my friends was a simple yes or no. Yet the one queried would think and think, yes or no, no or yes, before responding. Finally I felt I was getting hot on the trail, that the word might be 'cloud'. I knew I was allowed only one chance at the final word. I ventured it: 'Is it cloud?' 'Yes,' came the reply, and everyone burst out laughing. They explained to me there had been no word in the room. They had agreed not to agree on a word. Each one questioned could answer as he pleased – with the one requirement that he should have a word in mind compatible with his own response and all that had gone before. Otherwise, if I challenged, he lost. The surprise version of the game of twenty questions was therefore as difficult for my colleagues as it was for me.

What is the symbolism of the story? The world, we once believed, exists 'out there' independent of any act of observation. The electron in the atom we once considered to have at each moment a definite position and a definite momentum. I, entering, thought the room contained a definite word. In actuality the word was developed step by step through the questions I raised, as the information about the electron is brought into being by the experiment that the observer chooses to make; that is, by the kind of registering equipment that he puts into place. Had I asked different questions or the same questions in a different order I would have ended up with a different word as the experimenter would have ended up with a different story for the doings of the electron. However, the power I had in bringing the particular word 'cloud' into being was partial only. A major part of the selection lay in the 'yes' and 'no' replies of the colleagues around the room. Similarly the experimenter has some substantial influence on what will happen to the electron by the choice of experiments he will do on it, 'questions he will put to nature'; but he knows there is a certain unpredictability about what any given one of his measurements will disclose, about what 'answers nature will give', about what will happen when 'God plays dice'. This comparison between the world of quantum observations and the surprise version of the game of twenty questions misses much but it makes the central point. In the game, no word is a word until that word is promoted to reality by the choice of questions asked and answers given. In the real world of quantum physics, *no elementary phenomenon is a phenomenon until it is a recorded phenomenon.*

The Copenhagen view of reality is therefore decidedly odd. It means that, *on its own* an atom or electron or whatever cannot be said to 'exist' in the full, common-sense, notion of the word.

This naturally prompts the question: 'What is an electron?' If it is not a *thing* 'out there', existing in its own right, why can we talk so confidently about electrons?

Bohr's philosophy seems to demote electrons and other quantum entities to a rather abstract status. On the other hand if we simply go ahead and apply the rules of quantum mechanics *as if* the electron were real then we still seem to get the right results; we can compute answers to all well-posed physical questions, such as how much energy does an atomic electron have, and obtain agreement with experiments.

A typical quantum calculation involving electrons consists of computing the lifetime of the excited state of an atom. If we know that the atom is excited at time t_1, then quantum mechanics enables us to compute the probability that it will no longer be excited at some later time t_2. Quantum mechanics thus provides us with an *algorithm* for relating two observations, one at t_1 and the other at t_2. The so-called 'atom' enters here as a model which enables the algorithm to predict a specific result. We never actually observe the atom directly during the decay process. All we know about it is contained in the observations of its energy at t_1 and t_2. Clearly, we do not *need* to assume anything more about the atom other than what is necessary for us to obtain satisfactory results for our predictions of actual observations. As the concept of 'atom' is only ever encountered in practice when we conduct observations on it, it could be argued that all the physicist needs to be concerned with is consistently relating the results of observations. It is unnecessary for the atom 'to exist really' as an independent thing for this consistency to be achieved. In other words, 'atom' is simply a convenient way of talking about what is nothing but a set of mathematical relations connecting different observations.

The philosophy that the reality of the world is rooted in observations is akin to what is known as logical positivism. It seems, perhaps, alien to us because, in most cases, the world still behaves *as if* it had an independent existence. It is actually only when we witness quantum phenomena that this impression looks untenable. Even then, in their practical work, many physicists continue to think of the microworld in the common-sense way.

The reason for this is that many of the purely abstract, mathematical concepts employed become so familiar that they assume a spurious air of reality in their own right. This is also true of classical physics. Take the concept of energy for example. Energy is a purely abstract quantity, introduced into physics as a useful model with which we can short-cut complex calculations. You cannot see or touch energy, yet the word is now so much part of daily conversation that people think of energy as a tangible entity with an existence of its own. In reality, energy is merely part of a set of mathematical relationships that connect together observations of mechanical processes in a simple way. What Bohr's philosophy suggests is that words like electron, photon or atom should be regarded in the same way – as useful models that consolidate in our imagination what is actually only a set of mathematical relations connecting observations.

THE PARADOX OF MEASUREMENT

Bohr's so-called Copenhagen interpretation of quantum mechanics, in spite of its strange overtones, is actually the 'official' view among professional physicists. In the practical application of quantum mechanics the physicist rarely needs to confront any epistemological problems. So long as the quantum rules are applied systematically, the theory does all that can be expected of it; that is, it correctly predicts the results of actual measurements – which is, after all, the business of physicists. Nevertheless some physicists have not been content to leave it at that, because at the heart of the Copenhagen interpretation there seems to be a devastating paradox.

Central to Bohr's view is that we can generally only speak meaningfully about the physical attributes of a quantum system after a specific measurement (or observation) has been made. Clearly this gives a crucial and special physical status to the act of measurement. As we have seen, specifying the measurement's context requires particular statements about the type and location of apparatus. Implicit in this is that we can all agree on the meaning to be attached to phrases such as 'a geiger counter placed 2 m from the source'. The trouble arises when we ask where the dividing line comes between a quantum system and a piece of macroscopic apparatus. Geiger counters are, after all, themselves made of atoms, and subject to quantum behaviour.

According to the rules of quantum mechanics, a quantum system can evolve in time in two quite distinct ways. So long as the system can be considered isolated, its temporal development is described by what mathematicians call a unitary operation. In more physical terms unitary development corresponds to something like this. Suppose the state of the system consists of several different wave patterns superimposed (see p. 97). The different component waves will continually interfere with each other and produce a complex, changing pattern, analogous to the ripples on the surface of a pond. In fact, the description of this quantum evolution is very much like that of any other wave-like system.

In contrast, suppose now that a certain sort of measurement is made. The effect is dramatic. Suddenly all but one of the contributing waves disappear, leaving only a single wave pattern corresponding to 'the answer'. Interference effects cease and the subsequent wave pattern is totally transformed. (An example of this was given on p. 108). Such a measurement-like evolution in the wave is irreversible. We cannot undo it and restore the original complex wave pattern. Mathematically, this transition is 'non-unitary'.

How can we understand these two completely different modes of behaviour in a quantum system? Obviously, the abrupt change which occurs when a measurement takes place has something to do with the fact that the quantum system is coupled to a piece of measuring apparatus with which it interacts. It is no longer isolated. The mathematician J. von Neumann was able to prove for a model system that such a coupling will indeed have the aforementioned effects. However, we here encounter once again the fundamental paradox of measurement. The measuring apparatus is itself made of atoms and so subject to the rules of quantum behaviour. In practice we do not notice any quantum effects in macroscopic devices because such effects are so small. Nevertheless, if quantum mechanics is a consistent theory, the quantum effects must be present, however large the apparatus may be. We could then choose to regard the coupled arrangement of measured object plus measuring apparatus as a single large quantum system. But, assuming the combined system can be considered isolated from yet further systems, the same rules of quantum mechanics now apply to the larger systems, including the rule of unitary development.

Why is this a problem? Suppose that the original quantum system was in a superposition of two states. Recall, for example, the case of the polarized light at 45° to the polarizer, in which the incoming state is a superposition of two possible photon states, one parallel and the other perpendicular to the polarizer. The purpose of the measurement is to see whether the photon is passed or blocked by the polarizer. The measuring apparatus will have two macroscopic states, each correlated with the two polarization states of the photon. The trouble is that, according to the laws of quantum mechanics applied to the combined system, the apparatus now passes into a superposition of states! True, if the device is properly designed, any interference effects caused by the overlap (interference) of these two states will be miniscule. But, in principle, the effects are there, and we are forced to conclude that the apparatus is now itself in the sort of indeterminate limbo state that we have come to accept for electrons, photons, etc.

Von Neumann concluded that the measuring apparatus can only be deemed to have actually accomplished an irreversible act of measurement when it too is subjected to a measurement, and thus prompted into 'making up its mind' (technically called the collapse of the wave function onto a particular eigenstate). But now we fall into an infinite regress, for this second measuring device itself requires another device to 'collapse' it into a state of concrete reality, and so on. It is as though the coupling of the apparatus to a quantum system enables the ghost-like superposition of quantum states to invade the laboratory! This ability for us to put macroscopic objects into quantum superposition dramatically demonstrates the peculiarity of quantum theory.

SCHRÖDINGER'S CAT PARADOX, AND WORSE

In 1935 Erwin Schrödinger, one of the founders of quantum mechanics, had already perceived how the philosophical problems of a quantum superposition could appear at the macroscopic level. He illustrated the issue with a touch of showmanship in a now famous thought experiment involving a cat (Fig. 3.9):

> A cat is penned up in a steel chamber, along with the following diabolical device (which must be secured against direct interference by the cat): in a Geiger counter there is a tiny bit of radioactive substance, so small, that perhaps in the course of one hour one of the atoms decays, but also with equal probability, perhaps none; if it happens, the counter tube discharges and through a relay releases a hammer which shatters a small flask of hydrocyanic acid. If one has left this entire system to itself for an hour, one would say that the cat lives if meanwhile no atom has decayed. The first atomic decay would have poisoned it.

In our own minds we are quite clear that the cat must be *either* dead *or* alive. On the other hand, according to the rules of quantum mechanics the total system within the box is in a *superposition* of two states, one with a live cat, the other a dead cat. But what sense can we make of a live–dead cat? Presumably the cat itself knows whether it is alive

Fig. 3.9. The paradox of Schrödinger's cat. The poison device is a means of amplifying a quantum superposition of states to a macroscopic scale, where a paradoxical coexistence of live and dead cats seems to be implied. (From S.B. DeWitt, 'Quantum mechanics and reality', *Physics Today,* 23, 9, 1970).

or dead; yet, accepting the reasoning of von Neumann's regress, we are obliged to conclude that the hapless creature remains in a state of suspended animation until someone peeps into the box to check on it, at which point it is either projected into full vitality or else instantly dispatched!

The paradox becomes more acute if the cat is replaced by a person, for then the friend who has been incarcerated in the box will be aware throughout as to his health or otherwise. If the experimenter opens the box to discover that the subject is still alive, he can then ask his friend how he felt prior to this, apparently crucial, observation. Obviously the friend will reply that he remained 100% alive at all times. Yet this flies in the face of quantum mechanics, which insists that the friend is in a state of live–dead superposition before the contents of the box were inspected.

The paradox of the cat demolishes any hope we may have had that the ghostliness of the quantum is somehow confined to the shadowy microworld of the atom, and that the paradoxical nature of reality in the atomic realm is irrelevant to daily life and experience. If quantum mechanics is accepted as a correct description of all matter, this hope is clearly misplaced. Following the logic of quantum theory to its ultimate conclusion, most of the physical universe seems to dissolve away into a shadowy fantasy.

Among others, Einstein could never accept this logical extreme. Surely, he once asked, the moon exists whether or not somebody is looking at it? The idea of making the observer the pivotal element in physical reality seems contrary to the whole spirit of science as an impersonal, objective enterprise. Unless there is a concrete world 'out there' for us to experiment on and conjecture about, does not all science degenerate into a game of chasing mere images?

So what is the solution to the paradox of measurement? That is really where our contributors come in because, as we shall see, they have very different views. Let us first examine some general positions.

The Pragmatic View

Most physicists do not pursue the logic of the quantum theory to the ultimate extreme. They tacitly assume that somewhere, at some level between atoms and Geiger counters, quantum physics somehow 'turns into' classical physics, in which the independent reality of tables, chairs and moons is never doubted. Bohr said that this metamorphosis required 'an irreversible act of amplification' of the quantum disturbance, leading to a macroscopically detectable result. But he left it vague as to exactly what this act entails.

Mind Over Matter

The key role that observations play in quantum physics inevitably leads to questions about the nature of mind and consciousness, and their relationship with matter. The fact that, once an observation has been made on a quantum system, its state (wave function) will generally change abruptly sounds akin to the idea of 'mind over matter'. It is as though the altered mental state of the experimenter when first aware of the result of the measurement somehow feeds back into the laboratory apparatus, and thence into the quantum system, to alter *its* state too. In short, the physical state acts to alter the mental state, and the mental state reacts back on the physical state.

In an earlier section it was mentioned how von Neumann envisaged an apparently unending chain of measuring devices, each 'observing' the preceding member of the series, but none ever bringing about the 'collapse' of the wave function. The chain can then only come to an end when a conscious individual is involved. Only with the entry of the result of measurement into somebody's consciousness will the entire pyramid of quantum 'limbo' states collapse into concrete reality.

Eugene Wigner is one physicist who has strongly advocated this version of events. According to Wigner, mind plays the fundamental part in bringing about the abrupt irreversible change in quantum state that characterizes a measurement. It is not enough to equip the laboratory with complicated automatic recording devices, video cameras and the like. Unless somebody actually looks to see where the pointer is on the meter (or actually watches the video record), the quantum state will remain in limbo.

In the last section we saw how Schrödinger employed a cat in his thought experiment. A cat is a macroscopic system that is sufficiently complex for two alternative states (live and dead) to be dramatically distinct. Yet is a cat complex enough to count as an observer, and irreversibly alter the quantum state (i.e. 'collapse the wave function')? And if a cat will do, how about a mouse? Or a cockroach? An amoeba? Where does consciousness first enter in the elaborate hierarchy of terrestrial life?

The foregoing considerations are closely associated with the vexed issue of the mind–body problem in philosophy. At one time many people adhered to what the philosopher Gilbert Ryle called the 'official view' of the relationship between mind and body (or brain), which can be traced back at least to Descartes. According to this view, mind (or soul) is a type of substance, a special type of ephemeral, intangible substance, different from, but coupled to, the very tangible sort of stuff of which our bodies are made. Mind, then, is a *thing* which can have states – mental states – that can be altered (by receiving sense data) as a result of its coupling to the brain. But this is not all. The link which couples brain and mind works both ways, enabling us to impress our will upon our brains, hence bodies.

Today, however, these dualistic ideas have fallen out of favour with many scientists, who prefer to regard the brain as a highly complex, but otherwise unmysterious electrochemical machine, subject to the laws of physics in the same way as any other machine. The internal states of the brain ought therefore to be entirely determined by its past states plus the effects of any incoming sense data. Similarly, the output signals from the brain, which control what we like to call 'behaviour', are equally fully determined by the internal state of the brain at the time.

The difficulty with this materialistic description of the brain is that it seems to reduce people to mere automata, allowing no room for an independent mind, or free will. If every nerve impulse is legislated by the laws of physics, how can mind intrude into its operation? But if mind does not intrude, how is it that we are apparently able to *control* our bodies according to personal volition?

With the discovery of quantum mechanics, a number of people, most notably Arthur Eddington, believed they had overcome this impasse.

Because quantum systems are inherently indeterministic, the mechanistic picture of all physical systems, including the brain, is known to be false. Heisenberg's uncertainty principle usually permits a range of possible outcomes for any given physical state, and it is easy to conjecture that consciousness, or mind, could have a vote in deciding which of the available alternatives is actually realized.

Picture, then, an electron in some brain cell that is critically tuned to fire. Quantum mechanics allows the electron to roam over a range of trajectories. Perhaps it only needs the mind to load the quantum dice a little, thus prodding the electron more favourably in a certain direction, for the brain cell to fire, initiating a whole cascade of dependent electrical activity, culminating in, say, the raising of an arm.

Whatever its appeal, the idea that mind finds its expression in the world by courtesy of the quantum uncertainty principle is not really taken very seriously, not least because the electrical activity of the brain seems to be more robust than that. After all, if brain cells operate at the quantum level, the entire network is vulnerable to random maverick quantum fluctuations by any one of myriads of electrons.

The whole concept of mind being an entity capable of interaction with matter has been severely criticized as a category mistake by Ryle, who derides the 'official view' of the mind as 'the ghost in the machine'. Ryle makes the point that when we talk about the brain we employ concepts appropriate to a certain level of description. On the other hand, discussion of the mind refers to an altogether different, more abstract, level of description. It is rather like the difference between the Government and the British Constitution, the former being a concrete collection of individuals, the latter an abstract set of ideas. Ryle argues that it is as meaningless to talk about communication between the Government and the Constitution as it is to talk about the mind communicating with the brain.

A better analogy, perhaps more suited to the modern era, can be found in the concepts of hardware and software in computing. Computer hardware plays the role of the brain, while the software is analogous to the mind. We can happily accept that the output of a computer is rigidly determined entirely by the laws of electric circuitry, plus whatever input is used. We seldom ask: 'How does the program manage to make all those little circuits fire in the right sequence?' Yet we are still happy to give an equivalent description in software language, using concepts like input, output, calculation, data, answer, etc.

The twin descriptions of hardware and software applied to the operation of computers are mutually complementary, not contradictory.

The situation thus closely parallels quantum mechanics, with Bohr's principle of complementarity. Indeed, the analogy is very close indeed when we consider the question of wave–particle duality. As we have seen, a quantum wave is really a description of our *knowledge* of the system (i.e. a software concept), whereas a particle is a piece of hardware. The paradox of quantum mechanics is that somehow the hardware and software levels of description have become inextricably entangled. It seems we shall not understand the ghost in the atom until we understand the ghost in the machine.

The Many-universes Interpretation

So long as one is dealing with a finite system it is possible to ignore the conceptual problems associated with the quantum measurement process. One can always rely on the interaction with the wider environment to collapse the wave function. This line of reasoning fails completely, however, when we consider the subject of quantum cosmology. If we apply quantum mechanics to the whole universe, the notion of an external measuring apparatus is meaningless. Unless mind is to be involved somehow, the physicist who wishes to make sense of quantum cosmology seems to be forced to find a meaning to the act of measurement from the quantum states themselves, as it is no longer possible for an irreversible collapse of the wave function to be brought about by an external measuring device.

Interest in quantum cosmology grew in the 1960s, with the discovery of a number of theorems concerning spacetime singularities. These are rather like boundaries or edges of spacetime at which all known physics is extinguished. Singularities are formed from intense gravitational fields, and are expected to exist inside black holes. It is also believed that the universe began with a singularity. Because singularities represent the complete breakdown of physics they are considered distasteful pathologies by some physicists. It is suspected that singularities may be an artefact of our incomplete state of knowledge about gravity, which currently fails to incorporate quantum effects satisfactorily. If quantum effects could be included, it has been argued, then the singularities might go away. To abolish the big bang singularity we have to make sense of quantum cosmology.

In 1957 Hugh Everett proposed a radical alternative interpretation of quantum mechanics that removes the conceptual obstacles to quantum cosmology. Recall that the essence of the measurement problem is to understand how a quantum system which is in a superposition of, say two or more states, jumps abruptly to one particular state with a well-defined observable, as a result of a measurement (Fig. 3.10). A good example is the Schrödinger cat experiment discussed on p. 114. There the quantum system can evolve into two very different states: live cat and dead cat. Consequently quantum mechanical ideas fail to explain how the superposition of live–dead cat switches to either the live or dead alternative.

According to Everett the transition occurs because the universe splits into two copies, one containing a live cat and the other a dead cat. Both universes contain one copy of the experimenter too, each of whom thinks he is unique. In general, if a quantum system is in a superposition of, say, n quantum states, then, on measurement, the universe will split into n copies. In most cases, n is infinite. Hence we must accept that there are actually an infinity of 'parallel worlds' co-existing alongside the one we see at any instant. Morever, there are an infinity of individuals, more or less identical with each of us, inhabiting these worlds. It is a bizarre thought.

In the original version of the theory it was supposed that every time a measurement takes place the universe branches again, though it was always vague as to exactly what constitutes a measurement. Sometimes the words 'measurement-like interaction' were used, and it seems as though splitting can occur even from the ordinary cavorting of unob-

Fig. 3.10. The branching universe. According to Everett, when a quantum system is presented with a choice of outcomes, the universe splits so that all possible choices are realized. This implies that any given universe is continually branching into a stupendous number of near-copies.

served atoms. A proponent of the many-universes interpretation, Bryce DeWitt, expresses it as follows:

> Every quantum transition taking place in every star, in every galaxy, in every remote corner of the universe is splitting our local world on Earth into myriads of copies of itself... Here is schizophrenia with a vengeance.

More recently David Deutsch has modified the theory slightly, so that the number of universes remains fixed; there is no branching. Instead, most of the universes start out completely identical. When a measurement takes place, differentiation occurs. Thus, in the Schrödinger cat experiment, two previously identical universes differentiate so that in one the cat remains alive while in the other it dies. One advantage of this new picture is that it avoids the misleading impression that something mechanical is going on as would seem to be the case if the universe actually splits.

Two major criticisms have been levelled against the many-universes theory. The first is that it introduces a preposterous amount of 'excess metaphysical baggage' into our description of the physical world. We only ever experience one universe, so to introduce an infinity of others merely to explain a subtle technical feature (the collapse of the wave function) in our own seems to be the antithesis of Occam's Razor.

In its defence, proponents of the theory argue that the theoretical 'hardware' is less relevant in a theory than the number of fundamental assumptions you have to make in formulating the theory. The other interpretations of quantum mechanics all introduce some sort of epistemological hypothesis in order to make sense out of what is at first sight a senseless theory. The many-universes theory, however, has no need of this. The interpretation, it is claimed, emerges automatically from the formal rules of quantum mechanics, without the need for any assumptions about what the theory means. There is no need to introduce a separate postulate that, on measurement, the wave function collapses. By definition, every alternative universe contains one each of the possible collapsed wave functions.

The second objection to the theory is that it is said to be untestable. If our consciousness is confined to one universe at a time, how could we ever confirm or refute the existence of all the others? Remarkably, as we shall see, it may actually be possible to test the theory after all, if one is prepared to accept the possibility of intelligent computers.

A final argument in favour of the existence of a whole ensemble of universes is that it would provide an easy explanation for the formidable range of mysterious 'coincidences' and 'accidents of nature' found in physics, biology and cosmology. For example, it turns out that, on a large scale, the universe is ordered in a remarkable way, with matter and energy distributed in a highly improbable fashion. It is hard to explain how such a fortuitous arrangement just happened to emerge from the random chaos of the big bang. If the many-universes theory were correct, however, the seemingly contrived organization of the cosmos would be no mystery. We could safely assume that *all* possible arrangements of matter and energy are represented somewhere among the infinite ensemble of universes. Only in a minute proportion of the total would things be arranged so precisely that living organisms, hence observers, arise. Consequently, it is only that very atypical fraction that ever get observed. In short, our universe is remarkable because we have selected it by our own existence!

The Statistical Interpretation

In this way of looking at things, the physicist abandons all attempts to find out what actually goes on in an individual quantum measurement event and falls back instead on statements about whole collections of measurements. Quantum mechanics correctly predicts the probabilities of the various outcomes of measurement, and as long as attention is restricted to the overall statistics, there is no case to answer as regards the measurement problems.

It may be objected that the statistical (or ensemble) interpretation does not solve the problem of measurement, it simply sidesteps it. The price paid is that there is no longer any hope of discussing what actually happens when a particular measurement takes place.

The Quantum Potential

Another, entirely different approach has evolved out of the attempts to construct a hidden variables theory of quantum mechanics. As discussed, quantum mechanics predicts that Bell's inequality is violated. If this is correct then it is necessary to relinquish one of the two physical assumptions that went into proving it. One of these is 'reality'. As we have seen, Bohr's Copenhagen interpretation adopts this stance. The other assumption is that of 'locality': roughly speaking, there should be no propagation of physical effects faster than light.

If locality is abandoned, it is possible to re-create a description of the microworld closely similar to that of the everyday world, with objects having a concrete independent existence in well-defined states and possessing complete sets of physical attributes. No need for fuzziness now.

The trade-off is, of course, that non-local effects bring their own crop of difficulties; specifically, the ability for signals to travel backwards into the past. This would open the way to all sorts of causal paradoxes.

In spite of these difficulties, some researchers, most notably David Bohm and Basil Hiley have pursued the idea of a non-local hidden variables theory, inventing something they call the 'quantum potential'. This is similar to the more familiar potentials associated with force fields such as gravity or electromagnetism, but differs in that the activity of the quantum potential depends on the holistic structure of the system. That is, it encodes information about the measuring apparatus, distant observers, and so on. Thus, the entire physical situation over a wide region of space (in principle the whole universe) is embodied in this potential.

In spite of all the strenuous efforts to make sense of quantum physics, there is still no unanimous agreement among physicists on the approach to adopt. Indeed, the brief survey given above by no means exhausts the full range of different interpretations that have been discussed in recent years. It is certainly remarkable that a theory which was otherwise more or less complete in its essential details half a century ago, and which has proved itself spectacularly successful in practical applications, nevertheless remains unfinished. This state of affairs is largely due to the fact that discussions of the foundations of quantum theory are largely theoretical. At best they tend to involve 'thought experiments'. The region of interest is so hard to probe that it is very rare that practical experiments can be performed to test the foundations of the theory. For this reason, Aspect's experimental test of Bell's inequality was received with enormous scientific interest.

EDITOR'S NOTE

This short book continues with interviews with key quantum physicists, including Alain Aspect and John Bell, whose work is described in this selection. Aspect's experiment, widely hailed as one of the most decisive tests of the foundations of quantum mechanics, concludes that the results violate Bell's inequalities, which means that a simple picture of the world, retaining Einstein's idea of separability, is not sustained.

4 The Development of Philosophical Ideas since Descartes in Comparison with the New Situation in Quantum Theory*

Werner Heisenberg

In the two thousand years that followed the culmination of Greek science and culture in the fifth and fourth centuries BC the human mind was to a large extent occupied with problems of a different kind from those of the early period. In the first centuries of Greek culture the strongest impulse had come from the immediate reality of the world in which we live and which we perceive by our senses. This reality was full of life and there was no good reason to stress the distinction between matter and mind or between body and soul. But in the philosophy of Plato one already sees that another reality begins to become stronger. In the famous simile of the cave Plato compares men to prisoners in a cave who are bound and can look in only one direction. They have a fire behind them and see on a wall the shadows of themselves and of objects behind them. Since they see nothing but the shadows, they regard those shadows as real and are not aware of the objects. Finally one of the prisoners escapes and comes from the cave into the light of the sun. For the first time he sees real things and realizes that he had been deceived hitherto by the shadows. For the first time he knows the truth and thinks only with sorrow of his long life in the darkness. The real philosopher is the prisoner who has escaped from the cave into the light of truth, he is the one who possesses real knowledge. This immediate connection with truth or, we may in the Christian sense say, with God is the new reality that has begun to

* Chapter V in W. Heisenberg, *Physics and Philosophy. The Revolution in Modern Science* (New York: Harper & Row, 1958), pp. 76–92.

become stronger than the reality of the world as perceived by our senses. The immediate connection with God happens within the human soul, not in the world, and this was the problem that occupied human thought more than anything else in the two thousand years following Plato. In this period the eyes of the philosophers were directed toward the human soul and its relation to God, to the problems of ethics, and to the interpretation of the revelation but not to the outer world. It was only in the time of the Italian Renaissance that again a gradual change of the human mind could be seen, which resulted finally in a revival of the interest in nature.

The great development of natural science since the sixteenth and seventeenth centuries was preceded and accompanied by a development of philosophical ideas which were closely connected with the fundamental concepts of science. It may therefore be instructive to comment on these ideas from the position that has finally been reached by modern science in our time.

The first great philosopher of this new period of science was René Descartes who lived in the first half of the seventeenth century. Those of his ideas that are most important for the development of scientific thinking are contained in his *Discourse on Method*. On the basis of doubt and logical reasoning he tries to find a completely new and as he thinks solid ground for a philosophical system. He does not accept revelation as such a basis nor does he want to accept uncritically what is perceived by the senses. So he starts with his method of doubt. He casts his doubt upon that which our senses tell us about the results of our reasoning and finally he arrives at his famous sentence: "cogito ergo sum." I cannot doubt my existence since it follows from the fact that I am thinking. After establishing the existence of the I in this way he proceeds to prove the existence of God essentially on the lines of scholastic philosophy. Finally the existence of the world follows from the fact that God had given me a strong inclination to believe in the existence of the world, and it is simply impossible that God should have deceived me.

This basis of the philosophy of Descartes is radically different from that of the ancient Greek philosophers. Here the starting point is not a fundamental principle or substance, but the attempt of a fundamental knowledge. And Descartes realizes that what we know about our mind is more certain than what we know about the outer world. But already his starting point with the "triangle" God-World-I simplifies in a dangerous way the basis for further reasoning. The division between matter and mind or between soul and body, which had started in Plato's philosophy, is now complete. God is separated both from the I and from the world. God in fact is raised so high above the world and men that He finally appears in the philosophy of Descartes only as a common point of reference that establishes the relation between the I and the world.

While ancient Greek philosophy had tried to find order in the infinite variety of things and events by looking for some fundamental unifying principle, Descartes tries to establish the order through some fundamental division. But the three parts which result from the division lose some of their essence when any one part is considered as separated from the other two parts. If one uses the fundamental concepts of Descartes at all, it is essential that God is in the world and in the I and it is also essential that the I cannot be really separated from the world. Of course Descartes knew the undisputable necessity of the connection, but philosophy and natural science in the following period developed on the basis of the polarity between the "res cogitans" and the "res extensa," and natural science concentrated its interest on the "res extensa." The influence of the Cartesian division on human thought in the following centuries can hardly be overestimated, but it is just this division which we have to criticize later from the development of physics in our time.

Of course it would be wrong to say that Descartes, through his new method in philosophy, has given a new direction to human thought. What he actually did was to formulate for the first time a trend in human thinking that could already be seen during the Renaissance in Italy and in the Reformation. There was the revival of interest in mathematics which expressed an increasing influence of Platonic elements in philosophy, and the insistence on personal religion. The growing interest in mathematics favored a philosophical system that started from logical reasoning and tried by this method to arrive at some truth that was as certain as a mathematical conclusion. The insistence on personal religion separated the I and its relation to God from the world. The interest in the combination of empirical knowledge with mathematics as seen in the work of Galileo was perhaps partly due to the possibility of arriving in this way at some knowledge that could be kept apart completely from the theological disputes raised by the Reformation. This empirical knowledge could be formulated without speaking about God or about ourselves and favored the separation of the three fundamental concepts God-World-I or the separation between "res cogitans" and "res extensa." In this period there was in some cases an explicit agreement among the pioneers of empirical science that in their discussions the name of God or a fundamental cause should not be mentioned.

On the other hand, the difficulties of the separation could be clearly seen from the beginning. In the distinction, for instance, between the "res cogitans" and the "res extensa" Descartes was forced to put the animals entirely on the side of the "res extensa." Therefore, the animals and the plants were not essentially different from machines, their behavior was completely determined by material causes. But it has always seemed difficult to deny completely the existence of some kind of soul in the animals, and it seems to us that the older concept of soul for instance in the philosophy of Thomas Aquinas was more natural and less forced than the Cartesian concept of the "res cogitans," even if we are convinced that the laws of physics and chemistry are strictly valid in living organisms. One of the later consequences of this view of Descartes was that, if animals were simply considered as machines, it was difficult not to think the same about men. Since, on the other hand, the "res cogitans" and the "res extensa" were taken as completely different in their essence, it did not seem possible that they could act upon each other. Therefore, in order to preserve complete parallelism between the experiences of the mind and of the body, the mind also was in its activities completely determined by laws which corresponded to the laws of physics and chemistry. Here the question of the possibility of "free will" arose. Obviously this whole description is somewhat artificial and shows the grave defects of the Cartesian partition.

On the other hand in natural science the partition was for several centuries extremely successful. The mechanics of Newton and all the other parts of classical physics constructed after its model started from the assumption that one can describe the world without speaking about God or ourselves. This possibility soon seemed almost a necessary condition for natural science in general.

But at this point the situation changed to some extent through quantum theory and therefore we may now come to a comparison of Descartes's philosophical system with our present situation in modern physics. It has been pointed out before that in the Copenhagen interpretation of quantum theory we can indeed proceed without mentioning ourselves as individuals, but we cannot disregard the fact that natural science is formed by men. Natural science does not simply describe and explain nature; it is a part of the interplay between nature and ourselves; it describes nature as exposed to our method of questioning. This was a possibility of which Descartes could not have thought, but it makes the sharp separation between the world and the I impossible.

If one follows the great difficulty which even eminent scientists like Einstein had in understanding and accepting the Copenhagen interpretation of quantum theory, one can trace the roots of this difficulty to the Cartesian partition. This partition has penetrated deeply into the human mind during the three centuries following Descartes and it will take a long time for it to be replaced by a really different attitude toward the problem of reality.

The position to which the Cartesian partition has led with respect to the "res extensa" was what one may call metaphysical realism. The world, i.e., the extended things, "exist." This is to be distinguished from practical realism, and the different forms of realism may be described as follows: We "objectivate" a statement if we claim that its content does not depend on the conditions under which it can be verified. Practical realism assumes that there are statements that can be objectivated and that in fact the largest part of our experience in daily life consists of such statements. Dogmatic realism claims that there are no statements concerning the material world that cannot be objectivated. Practical realism has always been and will always be an essential part of natural science. Dogmatic realism, however, is, as we see it now, not a necessary condition for natural science. But it has in the past played a very important role in the development of science; actually the position of classical physics is that of dogmatic realism. It is only through quantum theory that we have learned that exact science is possible without the basis of dogmatic realism. When Einstein has criticized quantum theory he has done so from the basis of dogmatic realism. This is a very natural attitude. Every scientist who does research work feels that he is looking for something that is objectively true. His statements are not meant to depend upon the conditions under which they can be verified. Especially in physics the fact that we can explain nature by simple mathematical laws tells us that here we have met some genuine feature of reality, not something that we have – in any meaning of the word – invented ourselves. This is the situation which Einstein had in mind when he took dogmatic realism as the basis for natural science. But quantum theory is in itself an example for the possibility of explaining nature by means of simple mathematical laws without this basis. These laws may perhaps not seem quite simple when one compares them with Newtonian mechanics. But, judging from the enormous complexity of the phenomena which are to be explained (for instance, the line spectra of complicated atoms), the mathematical scheme of quantum theory is comparatively simple. Natural science is actually possible without the basis of dogmatic realism.

Metaphysical realism goes one step further than dogmatic realism by saying that 'the things really exist.' This is in fact what Descartes tried to prove by the argument that "God cannot have deceived us." The statement that the things really exist is different from the statement of dogmatic realism in so far as here the word "exists" occurs, which is also meant in the other statement "cogito ergo sum" ... "I think, therefore I am." But it is difficult to see what is meant at this point that is not yet contained in the thesis of dogmatic realism; and this leads us to a general criticism of the statement "cogito ergo sum," which Descartes considered as the solid ground on which he could build his system. It is in fact true that this statement has the certainty of a mathematical conclusion, if the words "cogito" and "sum" are defined in the usual way or, to put it more cautiously and at the same time more critically, if the words are so defined that the statement follows. But this does not tell us anything about how far we can use the concepts of "thinking" and "being" in finding our way. It is finally in a very general sense always an empirical question how far our concepts can be applied.

The difficulty of metaphysical realism was felt soon after Descartes and became the starting point for the empiricist philosophy, for sensualism and positivism.

The three philosophers who can be taken as representatives for early empiricist philosophy are Locke, Berkeley and Hume. Locke holds, contrary to Descartes, that all knowledge is ultimately founded in experience. This experience may be sensation or perception of the operation of our own mind. Knowledge, so Locke states, is the perception of the agreement or disagreement of two ideas. The next step was taken by Berkeley. If actually all our knowledge is derived from perception, there is no meaning in the statement that the things really exist; because if the perception is given it cannot possibly make any difference whether the things exist or do not exist. Therefore, to be perceived is identical with existence. This line of argument then was extended to an extreme skepticism by Hume, who denied induction and causation and thereby arrived at a conclusion which if taken seriously would destroy the basis of all empirical science.

The criticism of metaphysical realism which has been expressed in empiristic philosophy is certainly justified in so far as it is a warning against the naive use of the term "existence." The positive statements of this philosophy can be criticized on similar lines. Our perceptions are not primarily bundles of colors or sounds; what we perceive is already perceived as something, the accent here being on the word "thing," and therefore it is doubtful whether we gain anything by taking the perceptions instead of the things as the ultimate elements of reality.

The underlying difficulty has been clearly recognized by modern positivism. This line of thought expresses criticism against the naive use of certain terms like "thing," "perception," "existence" by the general postulate that the question whether a given sentence has any meaning at all should always be thoroughly and critically examined. This postulate and its underlying attitude are derived from mathematical logic. The procedure of natural science is pictured as an attachment of symbols to the phenomena. The symbols can, as in mathematics, be combined according to certain rules, and in this way statements about the phenomena can be represented by combinations of symbols. However, a combination of symbols that does not comply with the rules is not wrong but conveys no meaning.

The obvious difficulty in this argument is the lack of any general criterion as to when a sentence should be considered as meaningless. A definite decision is possible only when the sentence belongs to a closed system of concepts and axioms, which in the development of natural science will be rather the exception than the rule. In some cases the conjecture that a certain sentence is meaningless has historically led to important progress, for it opened the way to the establishment of new connections which would have been impossible if the sentence had a meaning. An example in quantum theory that has already been discussed is the sentence: "In which orbit does the electron move around the nucleus?" But generally the positivistic scheme taken from mathematical logic is too narrow in a description of nature which necessarily uses words and concepts that are only vaguely defined.

The philosophic thesis that all knowledge is ultimately founded in experience has in the end led to a postulate concerning the logical clarification of any statement about nature. Such a postulate may have seemed justified in the period of classical physics, but since quantum theory we have learned that it cannot be fulfilled. The words "position" and "velocity" of an electron, for instance, seemed perfectly well defined as to both their meaning and their possible connections, and in fact they were clearly defined concepts within the mathematical framework of Newtonian mechanics. But actually they were not well defined, as is seen from the relations of uncertainty. One may say that regarding their position in Newtonian mechanics they were well defined, but in their relation to nature they were not. This shows that we can never know beforehand which limitations will be put on the applicability of certain concepts by the extension of our knowledge into the remote parts of nature, into which we can only penetrate with the most elaborate tools. Therefore, in the process of penetration we are bound sometimes to use our concepts in a way which is not justified and which carries no meaning. Insistence on the postulate of complete logical clarification would make science impossible. We are reminded here by modern physics of the old wisdom that the one who insists on never uttering an error must remain silent.

A combination of those two lines of thought that started from Descartes, on the one side, and from Locke and Berkeley, on the other, was attempted in the philosophy of Kant, who was the founder of German idealism. That part of his work which is important in comparison with the results of modern physics is contained in *The Critique of Pure Reason*. He takes up the question whether knowledge is only founded in experience or can come from other sources, and he arrives at the conclusion that our knowledge is in part "a priori" and not inferred inductively from experience. Therefore, he distinguishes between "empirical" knowledge and knowledge that is "a priori." At the same time he distinguishes between "analytic" and "synthetic" propositions. Analytic propositions follow simply from logic, and their denial would lead to self-contradiction. Propositions that are not "analytic" are called "synthetic."

What is, according to Kant, the criterion for knowledge being "a priori"? Kant agrees that all knowledge starts with experience but he adds that it is not always derived from experience. It is true that experience teaches us that a certain thing has such or such properties, but it does not teach us that it could not be different. Therefore, if a proposition is thought together with its necessity it must be "a priori." Experience never gives to its judgments complete generality. For instance, the sentence "The sun rises every morning" means that we know no exception to this rule in the past and that we expect it to hold in future. But we can imagine exceptions to the rule. If a judgment is stated with complete generality, therefore, if it is impossible to imagine any exception, it must be "a priori." An analytic judgment is always "a priori"; even if a child learns arithmetic from playing with marbles, he need not later go back to experience to know that "two and two are four." Empirical knowledge, on the other hand, is synthetic.

But are synthetic judgments a priori possible? Kant tries to prove this by giving examples in which the above criteria seem to be fulfilled. Space and time are, he says, a priori forms of pure intuition. In the case of space he gives the following metaphysical arguments:

1. Space is not an empirical concept, abstracted from other experiences, for space is presupposed in referring sensations to something

external, and external experience is only possible through the presentation of space.

2. Space is a necessary presentation a priori, which underlies all external perceptions; for we cannot imagine that there should be no space, although we can imagine that there should be nothing in space.

3. Space is not a discursive or general concept of the relations of things in general, for there is only one space, of which what we call "spaces" are parts, not instances.

4. Space is presented as an infinite given magnitude, which holds within itself all the parts of space; this relation is different from that of a concept to its instances, and therefore space is not a concept but a form of intuition.

These arguments shall not be discussed here. They are mentioned merely as examples for the general type of proof that Kant has in mind for the synthetic judgments a priori.

With regard to physics Kant took as a priori, besides space and time, the law of causality and the concept of substance. In a later stage of his work he tried to include the law of conservation of matter, the equality of "actio and reactio" and even the law of gravitation. No physicist would be willing to follow Kant here, if the term "a priori" is used in the absolute sense that was given to it by Kant. In mathematics Kant took Euclidean geometry as "a priori."

Before we compare these doctrines of Kant with the results of modern physics we must mention another part of his work, to which we will have to refer later. The disagreeable question whether "the things really exist," which had given rise to empiricist philosophy, occurred also in Kant's system. But Kant has not followed the line of Berkeley and Hume, though that would have been logically consistent. He kept the notion of the "thing-in-itself" as different from the percept, and in this way kept some connection with realism.

Coming now to the comparison of Kant's doctrines with modern physics, it looks in the first moment as though his central concept of the "synthetic judgments a priori" had been completely annihilated by the discoveries of our century. The theory of relativity has changed our views on space and time, it has in fact revealed entirely new features of space and time, of which nothing is seen in Kant's a priori forms of pure intuition. The law of causality is no longer applied in quantum theory and the law of conservation of matter is no longer true for the elementary particles. Obviously Kant could not have foreseen the new discoveries, but since he was convinced that his concepts would be "the basis of any future metaphysics that can be called science" it is interesting to see where his arguments have been wrong.

As example we take the law of causality. Kant says that whenever we observe an event we assume that there is a foregoing event from which the other event must follow according to some rule. This is, as Kant states, the basis of all scientific work. In this discussion it is not important whether or not we can always find the foregoing event from which the other one followed. Actually we can find it in many cases. But even if we cannot, nothing can prevent us from asking what this foregoing event might have been and to look for it. Therefore, the law of causality is reduced to the method of scientific research; it is the condition which makes science possible. Since we actually apply this method, the law of causality is "a priori" and is not derived from experience.

Is this true in atomic physics? Let us consider a radium atom, which can emit an α-particle. The time for the emission of the α-particle cannot be predicted. We can only say that in the average the emission will take place in about two thousand years. Therefore, when we observe the emission we do not actually look for a foregoing event from which the emission must according to a rule follow. Logically it would be quite possible to look for such a foregoing event, and we need not be discouraged by the fact that hitherto none has been found. But why has the scientific method actually changed in this very fundamental question since Kant?

Two possible answers can be given to that question. The one is: We have been convinced by experience that the laws of quantum theory are correct and, if they are, we know that a foregoing event as cause for the emission at a given time cannot be found. The other answer is: We know the foregoing event, but not quite accurately. We know the forces in the atomic nucleus that are responsible for the emission of the α-particle. But this knowledge contains the uncertainty which is brought about by the interaction between the nucleus and the rest of the world. If we wanted to know why the α-particle was emitted at that particular time we would have to know the microscopic structure of the whole world including ourselves, and that is impossible. Therefore, Kant's arguments for the a priori character of the law of causality no longer apply.

A similar discussion could be given on the a priori character of space and time as forms of intuition. The result would be the same. The a priori concepts which Kant considered an undisputable truth are no longer contained in the scientific system of modern physics.

Still they form an essential part of this system in a somewhat different sense. In the discussion of the Copenhagen interpretation of quantum theory it has been emphasized that we use the classical concepts in describing our experimental equipment and more generally in describing that part of the world which does not belong to the object of the experiment. The use of these concepts, including space, time and causality, is in fact the condition for observing atomic events and is, in this sense of the word, "a priori." What Kant had not foreseen was that these a priori concepts can be the conditions for science and at the same time can have only a limited range of applicability. When we make an experiment we have to assume a causal chain of events that leads from the atomic event through the apparatus finally to the eye of the observer; if this causal chain was not assumed, nothing could be known about the atomic event. Still we must keep in mind that classical physics and causality have only a limited range of applicability. It was the fundamental paradox of quantum theory that could not be foreseen by Kant. Modern physics has changed Kant's statement about the possibility of synthetic judgments a priori from a metaphysical one into a practical one. The synthetic judgments a priori thereby have the character of a relative truth.

If one reinterprets the Kantian "a priori" in this way, there is no reason to consider the perceptions rather than the things as given. Just as in classical physics, we can speak about those events that are not observed in the same manner as about those that are observed. Therefore, practical realism is a natural part of the reinterpretation. Considering the Kantian "thing-in-itself" Kant had pointed out that we cannot conclude anything from the perception about the "thing-in-itself." This statement has, as Weizsäcker has noticed, its formal analogy in the fact that in spite of the use of the classical concepts in all the experiments a nonclassical behavior of the atomic objects is possible. The "thing-in-itself" is for the atomic physicist, if he uses this concept at all, finally a mathematical structure; but this structure is – contrary to Kant – indirectly deduced from experience.

In this reinterpretation the Kantian "a priori" is indirectly connected with experience in so far as it has been formed through the development of the human mind in a very distant past. Following this argument the biologist Lorentz has once compared the "a priori" concepts with forms of behavior that in animals are called "inherited or innate schemes." It is in fact quite plausible that for certain primitive animals space and time are different from what Kant calls our "pure intuition" of space and time. The latter may belong to the species "man," but not to the world as independent of men. But we are perhaps entering into too hypothetical discussions by following this biological comment on the "a priori." It was mentioned here merely as an example of how the term "relative truth" in connection with the Kantian "a priori" can possibly be interpreted.

Modern physics has been used here as an example or, we may say, as a model to check the results of some important philosophic systems of the past, which of course were meant to hold in a much wider field. What we have learned especially from the discussion of the philosophies of Descartes and Kant may perhaps be stated in the following way:

Any concepts or words which have been formed in the past through the interplay between the world and ourselves are not really sharply defined with respect to their meaning; that is to say, we do not know exactly how far they will help us in finding our way in the world. We often know that they can be applied to a wide range of inner or outer experience, but we practically never know precisely the limits of their applicability. This is true even of the simplest and most general concepts like "existence" and "space and time." Therefore, it will never be possible by pure reason to arrive at some absolute truth.

The concepts may, however, be sharply defined with regard to their connections. This is actually the fact when the concepts become a part of a system of axioms and definitions which can be expressed consistently by a mathematical scheme. Such a group of connected concepts may be applicable to a wide field of experience and will help us to find our way in this field. But the limits of the applicability will in general not be known, at least not completely.

Even if we realize that the meaning of a concept is never defined with absolute precision, some concepts form an integral part of scientific methods, since they represent for the time being the final result of the development of human thought in the past, even in a very remote past; they may even be inherited and are in any case the indispensable tools for doing scientific work in our time. In this sense they can be practically a priori. But further limitations of their applicability may be found in the future.

5 Explaining the Success of Science: Beyond Epistemic Realism and Relativism[1]*

Larry Laudan

INTRODUCTION

Throughout the first half of the eighteenth century, scientific opinion concerning the structure of the cosmos was deeply polarized; numerous "systems of the world" found their advocates among prominent natural philosophers, but the two leading rival systems were those of Descartes and of Newton. Cartesian physics held sway in France and on much of the rest of the continent; Newton's reigned in England. The young Voltaire journeyed from Paris to London in the spring of 1727. He was confused by the contrasting world-views he found. With an acute case of culture shock, he wrote to a friend back home:

> A Frenchman who arrives in London finds a great shift in scientific opinion that makes the mind weary. He left the world full; he finds it empty. At Paris you see the universe composed of tiny vortices of subtle matter; in London we see nothing of the kind...With the Cartesians, all change is explained by collisions between bodies, which we don't understand very well; with the Newtonians it is done by an attraction which is even more obscure. In Paris you fancy the earth's shape like a round melon; at London it is flattened on the two sides.[2]

One gets the same dizzying and disorienting feeling in our time if one moves between circles of philosophers and sociologists of science. Many, perhaps most, philosophers in the analytic tradition (and especially philosophers of science), take it for granted that science is, at least in its essentials, largely true and substantially correct. These philosophers argue that, especially in the "mature" and well-developed

---

* Essay in J.T. Cushing, C.F. Delaney and G.M. Gutting (eds), *Science and Reality. Recent Work in the Philosophy of Science* (Notre Dame: University of Notre Dame Press, 1984), pp. 83–105.

parts of the physical sciences, scientists have come very close to discerning the way the world *really* is. Our theories about such matters are, they say, highly verisimilar. Even where science turns out not to be strictly true, most philosophers (present writer included) are still apt to consider science as our best exemplification of rationality and cognitive progress – our best guess as to how things stand.

Sociologists, by contrast, especially sociologists of knowledge, tend to see science differently. Many of them regard scientific theory, like science itself, simply as a social construct, a set of conventions which Western culture since 1700 has used for conceptualizing experience, but which has no particular purchase on reality. Every culture, they point out, has its myths and its sacred beliefs; we happen to call ours by the name 'science'; but those beliefs are no better, no more secure, objective, or rationally grounded than the guiding ideologies of other cultures. These two points of view are known as 'realism' and 'relativism' respectively. The pair of them and the injustices that each does to an understanding of science will form the foci of this paper.[3]

But before I turn to that task, one crucial qualification is in order concerning the compass of this essay. Both realism and relativism have received numerous (and often conflicting) formulations by a wide variety of writers. While there may be fewer realisms and relativisms than there are realists and relativists respectively, it is a close call. There are many species of both these groups which I shall not be discussing in this essay. Perhaps the best way of locating my concerns is to say that I will be grappling with the *specifically epistemic* formulations of realism and relativism. Equally familiar to most readers will be various *ontological* versions of both realism and relativism. I must emphasize at the outset that the latter, metaphysical theories are *not* the targets of my criticism. My preoccupation in this essay with epistemic and methodological matters should explain why I shall have little to say about many varieties of relativism (e.g., Quine's ontological relativity) which might otherwise be expected to occupy center stage in a critique of relativism.

Epistemic realism or, to be more precise, 'scientific realism', has been the reigning orthodoxy among philosophers of science for almost a generation. Philosophers as diverse in orientation as Popper, Grünbaum, McMullin, Sellars, Reichenbach, and Putnam have espoused one or other version of it. Its rival, epistemic or cognitive relativism, has found occasional philosophical advocates (e.g., Feyerabend, the later Wittgenstein, Hesse and Rorty) but cognitive relativism – at least in that variant of it which I shall treat here – is associated primarily with work in the sociology of knowledge. Mannheim, Durkheim and Kuhn have developed what are probably the three most familiar versions of this species of relativism.[4]

Both realism and relativism are theories of knowledge in the broadest sense, and both have complex ramifications for our understanding of science. To put it briefly, the realist insists that science, in the course of its development through time, provides us with an ever more accurate, an ever more nearly true, representation of the natural order. Scientific theories, if not strictly true, are nearly so; and later scientific theories are closer to the truth than earlier ones. More than that, the realist typically asserts that science or the scientific method represents the *only* (or, more weakly, the *most*) effective instrument for discovering truths about the world. The relativist, by contrast, characteristically eschews notions of truth and falsity, focussing rather on the specific and local features which shape (and, in his view, inevitably distort) the scientific image of the world. The relativist would have us believe that science is but one among indefinitely many ways in which man might represent the world; in his view, it has no special claim to validity or veracity. If we lived in a different place and time, says the relativist, we would have a fundamentally different vision of the natural order. Still worse, there is – the relativist maintains – no neutral point on which we can stand to adjudicate impartially the rival claims of these contrasting images of the world, the scientific and the nonscientific. Because we ourselves are products of a scientific culture, we cannot step outside the presuppositions of that culture to compare the legitimacy of its claims with those of nonscientific cultures. Where the realist sees the history of science as a triumphal march ever closer to the truth, the "cutting edge of objectivity" (in Gillispie's apt if notorious phrase), the epistemic relativist sees nothing more than a succession of rival and mutually incompatible representations, each reflecting various subjective and transitory interests. Where the realist sees progress in the history of science, the relativist sees only change. The realist believes that science comes as close to truth and objectivity as is humanly possible; the relativist fears that he is probably right! But they draw very different conclusions from this one point of consensus.

There is nothing especially new about this polarity. Struggles between realist and relativist perspectives span the entire history of epistemology. Precisely because of their age-old opposition, there is a tendency to see these doctrines as mutually exhaustive rivals. Any weakness in relativism (e.g., its allegedly self-indicting character) is translated into an argument for realism; while any flaw in realism (e.g., the unsatisfactory status of realist semantics) comes to be widely regarded as evidence for relativism.

A number of considerations move me to take strong exception to the view that these two doctrines more or less exhaust the range of alternatives open to us. Both seem to me to be fundamentally flawed and open to anomalies which are beyond their resources to grapple with. But more important, each fails to resolve one of the most central conceptual questions about science. In a nutshell, that question is simply: why does science work so well? In what follows, I shall seek to show:

1. That the realist recognizes the importance of this question but fails to answer it;
2. That the relativist is scarcely prepared to grant the legitimacy of the question, let alone to answer it;
3. That the question can be interestingly answered, provided that we are prepared to lay aside some of the core assumptions associated with both realism and relativism.

ESTABLISHING THE PHENOMENON: THE SUCCESS OF SCIENCE

As every student of scientific controversy understands, one man's fact is another man's fiction. (Recall, if you have any doubts, the difference between London and Paris in 1727). Nowhere is this difference in 'perception' more marked than with respect to the question of the success of science. Many of us incline to the view that nothing could be more obvious than the fact that science is a successful and effective knowledge-gathering enterprise. We may be unsure how to account for that success and our efforts to *characterize* it precisely have not been very illuminating (witness the failure of the many inductive logics and theories of confirmation); but our intuition remains unshaken that science does what it does very well indeed. It is quite another matter, however, where sceptics and relativists are concerned. It is not clear whether they positively deny that science is successful; in general, they simply do not reckon the achievements of science to be something which they are called upon to explain. Insofar as they deal with the phenomenon at all, it is to point out that the "success" attributed to science is of an ambiguous and amorphous sort. "Successful according to whom? and for what purposes?" they ask. "Successful compared to what?" "Successful by which standards?" To put the most sympathetic gloss I can on the relativists' failure to grapple with the problem of the success of science, I would say that relativists are inclined to withhold judgment on the claim that science is successful for these reasons (among others): (1) a belief that, until the notion of success is spelled out with some care, the concept is too unclear to be worthy of systematic analysis, let alone explanation; and (2) a lingering suspicion that 'success' is an evaluative rather than a descriptive term which should play no role in an empirical and naturalistically-based sociology of knowledge.[5]

What I want to do in this section is to meet that challenge by describing some notion(s) of success which should allow both relativists and realists to grant that science is, more than occasionally, successful. We can then proceed to explore the resources of realism and relativism respectively for accounting for that success.

We must begin by freeing 'success' of some of its more normative and judgmental overtones. As I shall be using the term, judgments of success in an activity imply no endorsement of that activity. There can be successful bank robbers, rapists, military campaigns, or scientific theories. One may or may not regard science and technology as forces for good; but no such evaluation is presupposed or implied by my claim that science is a successful activity. In the most general sense of the term, success in any activity always has to do with relations between ends and means and, more specifically, between aims and actions. To say that an activity is successful is simply to say that it promotes the ends of (at least some of) those engaged in it (or, and this is an important codicil, of those judging it to be successful). Just as we say before the fact that an action is rational if the actor has good reason to believe that it will achieve his goals, so do we say *post hoc* that an action is successful just insofar as it actually furthers some agent's goals.[6] Putting it this way makes it clear that success is a *relational* concept. Because agents' goals can differ, one and the same action may be unsuccessful or successful, depending upon the goals in question.

Of even greater importance is the fact that success, so conceived, is not a valuational or a normative concept. To claim that a certain action was successful is to make a contingent, empirical claim about the relation of that action and its outcomes to certain goal states. Claims about success are thus (at least in principle) as factual and as testable as any other sort of empirical claim about the world. (Although one must concede the fairness of the relativist's charge that, in practice, 'success' is too often treated as a primitive term.)[7]

Accordingly, the thesis that science is successful (or unsuccessful) amounts to the empirical assertion that the actions of scientists have in fact brought about or otherwise promoted (or failed to promote) certain goals or aims. But with respect specifically to which goals is science to be judged successful or unsuccessful? This question is both simpler and more complex than it might first appear. It is simpler because, unlike judgments about the rationality of an action, judgments of success do not require a scrutinizing of an agent's aims or motives. We can ask whether an agent's actions in fact brought about certain outcomes, quite independently of whether those outcomes were the ones which the agent intended to achieve. Just so long as we make it clear what outcomes we regard as constituting 'success', we can happily make determinations of success or failure without skating on the comparatively thin ice of attributions of intentionality to agents. Of course, we will often be interested to know whether an action is successful specifically with respect to (what we take to be) the goals of the agent who embarked on the action. But making determinations of 'success' parasitic on the agent's goals is not necessary, and in this particular case is almost certainly not desirable.[8] Through time, scientists have had a highly heterogeneous set of cognitive aims or goals.[9] There almost certainly is no such thing as *the* aims of the scientific community, any more than any other large and diverse group has universally shared and univocal aims. When we say that science has been successful (at least those of us who are prepared to venture such a conjecture), we do not usually even bother to engage in a detailed analysis of the goals or aims of all or most of the actors who have constituted the scientific community. Rather, we typically *impute* certain goals to a highly idealized caricature of the scientist (or, even more abstractly, to science as an institution) and then ascertain whether science has achieved those goals. It seems to me that it would be a more forthright and intellectually honest approach to admit that, at least for these purposes, we are not trying to ascertain whether science has managed to achieve what, as a matter of fact, working scientists have always or invariably been trying to achieve. That is certainly an interesting question, but it is not the most important question for epistemic or methodological purposes. Rather, we should say straight out that, when we are judging whether science is or has been successful, we are going to be making determinations of success with respect to the ability of science to achieve certain cognitive attributes which *we* find especially interesting. As long as we acknowledge what we are doing, we can avoid vexed questions about intentionality, incompatible goals, shifting explanatory ideals, and the host of other difficulties which confound efforts to ascertain what the aims of science have actually been through history. By taking this route, we can make our task a good deal easier than it otherwise would be.

What makes the task rather more complicated than one might expect is the necessity of spelling out clearly and precisely exactly what criterion of success we are utilizing. Since virtually any action will have some consequences or other, some outcome or other, it is always possible after the fact to find some set of descriptions under which any particular action can be made to appear to be successful. Such an approach would obviously trivialize the undertaking. Accordingly, we need to find an interesting, unusual, and demanding set of outcomes, with respect to which we will proceed to make judgments of success or failure. (There is obviously no unique set of that sort.) But there is one set of cognitive outcomes which has interested epistemologists and philosophers of science for a long time. These goal states concern themselves with certain interesting epistemic and pragmatic attributes. Consider a typical list of some of those aims:

(a) to acquire *predictive control* over those parts of one's experience of the world which seem especially chaotic and disordered;

(b) to acquire *manipulative control* over portions of one's experience so as to be able to intervene in the usual order of events so as to modify that order in particular respects;

(c) to increase the *precision* of the parameters which feature as initial and boundary conditions in our explanations of natural phenomena;

(d) to integrate and *simplify* the various components of our picture of the world, reducing them where possible to a common set of explanatory principles.

If we define cognitive 'success' along these lines, then it seems uncontroversial to say that portions of the history of science in the last 300 years have been a striking success story. For instance, we are now in a position to predict a much broader range of phenomena than we were in 1700. We can intervene in the natural order (e.g., with respect to the course of many diseases) so as to make things go more to our liking far more effectively than we could formerly. Our instruments for measuring various variables and constants are incomparably more precise than they once were. (Consider, for instance, the refinements in the last two hundred years of determinations of the velocity of light.) Finally, even if the ultimate unification of science still eludes us, it is quite clear that we can now explain a more diverse set of phenomena in terms of a smaller number of general principles than our forebears could.

In saying that science has been successful in this cognitive sense, I am certainly not claiming that science has managed to achieve all the goals of all of its practitioners. Nor am I making any judgment about whether, all things considered, science is worthy or admirable. At least for purposes of this analysis, we need make no judgment about the moral or social value of the sorts of outcomes which science has achieved. I am here simply noting certain facts, and very striking facts they are, about the diachronic development of science. Because these facts are so striking, because there was no reason *a priori* to expect man to be able to achieve such cognitive feats, because no undertaking can guarantee success *of this particular sort*, we are confronted with a genuine problem: why is science so successful? What is it about the manner in which scientists formulate and test their theories which makes this sort of success possible?[10]

Any account of science which fails to answer, or even to address, such questions is (to put the criticism in its mildest form) fundamentally *incomplete*. Whatever one's disciplinary or philosophical orientation, one cannot pretend to be accounting for, or explaining, science in a comprehensive manner unless one has an answer to such questions as these. As we shall see, such is the sorry state of both realism and relativism.

REALISM AND SUCCESS

By and large, scientific realists have recognized the pivotal importance of the problem of explaining the cognitive success of science. Indeed, especially in the last few years, numerous realists have claimed that one of the chief arguments in favor of realism is precisely that it, allegedly unique among rival epistemologies, can explain why science is successful. Hilary Putnam, for instance, asserts that "the positive argument for realism is that it is the only philosophy that doesn't make the success of science a miracle".[11] McMullin, Newton-Smith, Boyd, and Niiniluoto have made similar claims on behalf of realist epistemology.[12] In this section, I want to examine briefly the claims of contemporary realism to be able to explain the success of science.

Realists argue that scientific theories, at least in such "mature sciences" as physics, are approximately true and that the central terms in such theories genuinely refer to objects in the physical world. They go on to insist that the approximate truthlikeness of our theories (and the related authenticity of reference exhibited by their central concepts) explains why science works as well as it does. Our theories are success-ful, the realist maintains, precisely because they come close to repre-senting things as they really are.

What the realist is trading on here is the perfectly sound intuition that *if* our theories were true unqualifiedly, then all their consequences would likewise be true; and if all those consequences were true, we would indeed expect theories to exhibit just that sort of predictive accu-racy and reliability which (on the account mentioned earlier) consti-tutes "the cognitive success of science".[13] Sadly, the realist is enough of a "realist" (in the hard-headed, ordinary language sense of that term), to recognize that the relevant antecedent conditions in this intuition are unsatisfied. We have overwhelmingly good reasons to suspect that our theories about the world, even our best-tested ones, are not true *simpliciter*. Yet the realist still wants to cash in on the hunch that the 'truthlikeness' of our theories is responsible for their success. Accord-ingly, the realist maintains that, even if our best theories are only ap-proximately true, or nearly true, then he is in a position to explain the success of science. The core idea here is that an approximately true the-ory will have consequences *most* of which are true, or at least which are close to the truth.

As I have shown elsewhere in detail,[14] this argument is fundamen-tally flawed. There is as yet no coherent sense of 'approximate truth' which entails that an approximately true theory will be uniformly suc-cessful in *any* of the senses sketched above. We can put the point more strongly: there is as yet no semantic account of truthlikeness which en-tails that a theory, all of whose central explanatory claims are approxi-mately true, will be any more successful than a theory whose central explanatory claims are wildly inaccurate. It is entirely conceivable, for instance, that a theory might be approximately true, in any explored sense of the term, and still be massively inaccurate in those domains where it can be tested. Indeed, on the best-articulated sense of truth-likeness (namely Popper's theory of verisimilitude), it can be shown that a theory may have a high "truth content" yet have all its observable consequences false. To make a long story short, it can be shown that – because there is no reason to believe that approximately true theories need be (or are even likely to be) empirically successful – the near-truthlike status of theories – even assuming we had an epistemology which would warrant such attributions of truthlikeness – cannot be in-voked to explain their pragmatic success.

But the situation is even gloomier than this for the realist. As even a brief glance at the history of science will show, there are many theories which have been highly successful for long periods of time (e.g., theories postulating spontaneous generation or the aether) which clearly have not been approximately true in terms of the deep-structure claims they have made about the world. Thus, Newtonian optics – which predicted a wide variety of phenomena, and which inspired the construction of a plethora of successful instruments and measuring devices – was committed to a basic ontology of light which (so we now believe) is desperately wide of the mark. Because there seems to be nothing in the world which even approximately corresponds to Newtonian light corpuscles, it is clear that Newton's theory was not, indeed could not have been, approximately true. (Assuming that, as I have argued elsewhere,[15] the scientific realist's notion of approximate truthlikeness presupposes genuineness of reference.) How then is the realist to account for the success of Newtonian optics? Even assuming that the realist could show that an approximately true theory would be successful (which he cannot), how can he explain the success of a theory, like Newton's, which is – by his lights – not even close to the truth? The same goes in spades for most theories in the history of science. Because they have been based on what we now believe to be fundamentally mistaken theoretical models and structures, the realist cannot plausibly hope to explain the empirical success such theories enjoyed in terms of the truthlikeness of their constituent theoretical claims. So it appears that the realist is in a bind. Theories which are approximately true need not be successful; many theories which have been strikingly successful are evidently not approximately true. Under such circumstances, truthlikeness is a decidedly unpromising explanans for empirical success.

But the problems facing the realist go even deeper than this. There is a crucial ambiguity in the problem of scientific success, an ambiguity which highlights another weakness in the realist approach to that problem. When we ask why scientific theories work so well, we might be asking (and the realist response assumes as much) to be told what semantic features theories possess in virtue of which they have such an impressive range of true consequences. Alternatively, when we ask why science is successful, we might be asking an epistemic and methodological question about the selection procedures which scientists use for picking out theories with such impressive credentials. If, as I suspect, it is generally the latter which we are driving at, then the appropriate response to that problem will address itself to the probative and evaluative procedures which scientists use for identifying those theories which are likely to be reliable. And insofar as that is what is at stake, the realist response becomes even less availing than it already appears to be; for the realist's "explanation" of success (viz., theories work because they are true or nearly true) sheds no light whatever on how scientists come by these putatively true or truthlike theories. Because the realist makes no reference to the methods of investigation and warranting which scientists use for selecting their theories, he must leave that side of the question unaddressed. That side of the problem of success remains a mystery on realists principles, even if the realist can get his theory semantics in order.

I will conclude this section by offering several caveats. I have asserted here neither that realist epistemology is wholly irrelevant to the explanation of scientific success nor that realism's failure to explain the success of science is a disproof of realism. (Although that failure does raise serious questions about whether realism has any empirical content.) What I have insisted is that current realist approaches to this issue provide little more than pseudo-explanations of the success of science. Beyond that, I have suggested that realists have largely missed the point about the success of science, for they have failed to see that what is chiefly called for is an epistemic analysis of the methods of theory testing rather than an account of theory semantics.

RELATIVISM AND SUCCESS

There are numerous variants of relativism (cultural, historical, and epistemological among others). What most seem to have in common is a conviction that no method of inquiry can claim special or privileged status. Different cultures, different societies, different epochs will exhibit conflicting views about the appropriate ways of authenticating beliefs. Confronted by these differences, the relativist insists on remaining agnostic about the respective merits of different methodological and evaluative strategies for testing claims about the world. More than that, the relativist generally denies in principle that there can be any way of showing one doxastic or belief-forming policy to be superior to another, or one set of methods to be objectively preferable to another.[16]

The relativist circumvents the problem of success by writing it off his explanatory agenda. As he sees it, his task is the descriptive and explanatory one of explaining why agents believe what they do. He will thus quite happily offer us an explanation for why a particular scientist or group of scientists believes that their theories are successful. But what he is reluctant to confront head-on, in part because he mistakenly imagines it to be a purely normative and philosophical puzzle, is the question why certain theories or beliefs are, in fact, successful.

Indeed, many latter-day relativists explicitly repudiate any effort to acknowledge that certain systems of belief have been more successful than others. David Bloor, Barry Barnes, and a host of other sociologists of knowledge have argued for such agnosticism;[17] so too have such philosophers as Paul Feyerabend. My aim in this section is to explore the relations between relativism, so understood, and the problem of the success of science.

As I said earlier, this reluctance to grapple with the success of science is due, in part, to the relativists' failure to recognize that 'success' can be characterised in a thoroughly descriptive rather than an evaluative way. Equally, they have been understandably skeptical about approaches which assume that all human agents have the same cognitive goals. As the relativist sees it, terms like 'success' and 'failure' smack of cultural chauvinism because they seem to suggest that all human actors have the same aims. But if one takes seriously the arguments offered above, it becomes clear that the claim that a certain piece of science is successful relative to certain aims is not tantamount to the claim that all agents (or even all scientists) have the same aims. It is simply the empirical claim that the developments in science have in fact promoted certain aims. Since relativists do grant that there are some features of science which can be treated as data to be explained or accounted for, the relativist who refuses to countenance the success of science (so understood) must explain why the success of science is any less a fact about science than, say, that science is a social activity or that scientists use certain mechanisms for generating consensus.

But the relativist's uneasiness about the seemingly judgmental aspects of 'success' is only a part of the story. I submit that a second factor which encourages the relativist to finesse the issue of explaining why science is successful is his realization that his epistemology lacks the explanatory resources to give a plausible analysis of that success. Indeed, the fact that science is so successful constitutes a powerful anomaly to relativism; not because the success of science refutes relativism, but rather because it points up its descriptive and explanatory incompleteness as an empirical theory of the scientific enterprise.[18] Like some Victorians, who hoped venereal disease would go away if no one mentioned it, the relativist apparently thinks that aloof indifference to success is the preferred vehicle for wishing it away. Let me explain why I make this charge of incompleteness against relativism.

Consider one specific part of the success of science, its predictive ability. With respect to many sorts of phenomena, it is quite clear that science puts us in a position to anticipate what the world will do next with a rather higher reliability than can many systems of belief commonly regarded as nonscientific.[19] I submit that we have enormous support for the success thesis in that form. Let me particularize it still further in terms of a familiar example. In virtually every society which cultivates its own food, there is (what Habermas has called) a "technical interest" in anticipating when floods will occur, in judging their intensity, and in taking appropriate action to control the damage they wreak. Every agrarian and post-agrarian society has means for anticipating when and where major rivers will overflow their banks. Let us suppose, for the sake of argument, that modern Western scientific techniques yield predictions which are both more detailed and more accurate with respect to the phenomena of river flooding. If, as I believe, this greater accuracy could be convincingly established (even to those who were not products of Western culture), then we would be confronted with a situation where different cultures have a common technical interest in predicting a certain sort of phenomena and where scientific culture, by the standards of all concerned, yields more accurate predictions.[20]

What is the relativist to say about such a case? Well, what most of us non-relativists would say is that science has available certain methods of theory selection and theory testing which, over the long run, tend to pick out theories of high reliability. We might go on to point out precisely why one would expect methods of the sort scientists use to yield fairly reliable theories.[21] In the case in hand, we could contrast the methods of theory evaluation used in scientific predictions of rainfall and water run-off quantities with those rule-of-thumb methods utilized for generating predictions about flooding in non-scientific societies. I expect we would be able to show that our methods of theory selection are more robust than those of other cultures, thereby explaining why our theories about flooding were more efficacious in producing reliable predictions than the theories used by other societies.[22]

But, of course, such explanatory manoeuvres are not open to the relativist, for he denies that any methods can objectively be said to be better than any others.[23] Precisely because he makes that denial, he is in no position to cite the superior methods of science as the explanation of the greater predictive success of science compared to other forms of knowledge.[24] Since the relativist cannot explain the predictive superiority of science by invoking the greater rigor or robustness of its methods, what is he to say? Well, he might say that it is just a large cosmic coincidence that science is so successful; that the success of science reveals nothing about, and owes nothing to, the specific methods of inquiry used in science. But this is not to explain the success of science; it is, rather, to renounce any effort to account for that success.

I have suggested two causes for the relativist's reluctance to grapple with scientific success; one concerned the relativist's explanatory agenda (i.e., his uneasiness about "evaluative" concepts); the other, his limited explanatory resources. But I think that we must probe still further before we fully understand why many relativists are so reluctant to acknowledge the success of science as a datum to be explained. For more than a decade, relativist sociologists have been committed to the idea that the same sort of institutional analysis which they offer for other social structures and systems of belief (e.g., religion or the kinship system) can be applied indifferently to science. In their view, science is simply one among many institutions for the formation and perpetuation of beliefs. These new-wave sociologists have thereby sought to distance themselves from the older sociological tradition (e.g., associated with Robert Merton, among others) which tried to establish the cultural or sociological *uniqueness* of science as an institution. As soon as the relativist grants that science has been cognitively more successful than many other belief-building enterprises, then he can no longer argue for a monolithic or unitary account of cognitive practices. Put differently, the latter-day sociologist of knowledge wants to reduce the sociology of knowledge to the sociology of belief, and to conjoin that reduction with the thesis that the problem of explaining belief- or consensus-maintenance is to be handled in a unitary fashion across all institutions and cultures. To maintain this homogeneity thesis, the relativist must either deny that science has been successful, or insist that it has been no more successful than any other system for the generation of action-related beliefs (or, finally, hold that the success of science is fortuitous). In either case, the relativist's denial that science is *sui generis* disposes him to deny that the success of science is a datum requiring special explanation.

I have ventured into this lengthy digression about the causes that have evidently pushed the relativist in the direction of ignoring the problem of the success of science only because, when we find a group of thinkers denying what most of us take to be obvious, we have to cast about for some explanation of their apparently pathological behavior. But however far one goes in trying to understand why relativists might want to avoid grappling with the problem of the success of science, the fact remains that it is a phenomenon which they have left unexplained.

To that extent, relativism is radically incomplete as an explanatory theory about science.

If there is any plausibility in the arguments of these last two sections, we seem in fairly dire straits. Neither of the major epistemologies of our time seems to show much promise of handling one of the core intellectual issues of our time. The better part of valor might suggest that the problem of the success of science is simply intractable, a problem well beyond our limited explanatory capacities. It is, after all, conceivable, that – as Karl Popper once suggested[25] – why science works is just an insoluble problem which is best left well enough alone. But before we acquiesce too quickly in that view, and before the relativists and realists imagine that they are off the hook (for who can reasonably be expected to solve an insoluble mystery?), it is worth exploring briefly whether we really are in such a desperate position so far as explaining why science succeeds.

ACCOUNTING FOR THE SUCCESS OF SCIENCE

This is, of course, a tall order, and I have no intention of offering here a perfectly general solution to the problem. What I will do is to take one or two typical, if slightly idealized, cases of scientific success and offer a story about them, a story which will make it plausible why science works well in those circumstances. A different story would have to be told about other cases. But if the tale I have to tell is at all convincing, it will be easy to see how it could be adapted to a wide range of other situations.

But before I offer my narrative, I need to make some important disclaimers. It has often been assumed that the demand for an explanation for the success of science (i.e., an account of why science "works" so well) is really just a re-formulation of the hoary old problem of induction. And it is true that a solution to the riddle of induction, assuming one could be had, might well give us a solution to the problem of success. After all, if we could show under what circumstances it was reasonable to assume that unobserved instances of a generalization or theory will resemble observed ones, we would have shown the reasonableness of (at least some) inductive methods. I do not have a solution to the problem of induction; come to that, I do not regard it a particularly interesting problem in this form. The point I want to make here is that the problem of the success of science can be formulated in such a way that its solution does not require a prior solution to the problem of induction. We can see the independence of the two problems if we cast the problem of success in the following form: why is it that many of the theories of the natural sciences enable us to predict nature and to intervene in the natural order in ways we want to so much more frequently and more accurately than (say) the theories of the ancient Greeks permitted them to do? This is clearly a *comparative* version of the problem of success. Its solution does not require us, as the problem of induction apparently does, to show that our theories are always (or even usually) reliable guides to the course of nature. The problem of success requires us only to explain why certain sorts of theories, authenticated by certain sorts of probative procedures, tend to promote certain cognitive ends more effectively than other sorts of theories, grounded in other forms of legitimation, do. As I shall be construing the problem of success in this section, it is fundamentally the challenge of explaining why certain modes of knowledge authentication produce more reliable results than others do.[26] We can thus leave the problem of induction in its general form conveniently to one side.

So let us now turn to my pair of stories. For the first, let us imagine that I am having problems getting my car started on a cold morning. My mechanic hauls it into the garage and replaces the brake drums, returning the car to me the next day. In the meantime, the weather takes a decided turn for the better. I crank up the engine, and the car starts without difficulty. When the mechanic bills me for replacing my brakes, and I complain that he did not do what he was supposed to, he replies by pointing out that my car starts now, and that that was what I wanted all along. Moreover, he points out that *all* the cars in his shop that day suffering from ignition problems had their brakes replaced and invariably the problem was solved. In sum, he claims that his tinkering with the brakes cured my starting problem, and cites as evidence for his claim the acknowledged fact that my car – along with the others suffering similar problems in his shop – now starts without difficulty. In exasperation, I explain to him that the state of the brake drums could have nothing whatever to do with the operation of the starter. He replies that such happens to be *my* theory about how cars work, but that *he* has a different theory, according to which brake wear can be a cause of poor ignition. Being a reasonable fellow of sorts, he cites as evidence for *his* theory the fact that the car started smoothly once the brakes were replaced. What am I to do in this case? Well, the first thing I might do is to point out that there is a different explanation than his for the sudden improvement in my starter's performance (namely, the warmer weather). Because there is, and because he hopes to get paid, he must show me some empirical evidence which supports his explanation of the starter's improvement rather than my meteorological hypothesis. If, moreover, I can point to plenty of other cars whose starting performance has improved dramatically with warming weather when no one was let loose on their brakes, my case is won. (At least as far as I and the Better Business Bureau are concerned.) Now, what is going on here? In effect, my mechanic and I are comparing *different probative strategies for the evaluation of beliefs*. My mechanic is evidently quite willing to shape his beliefs according to a simple *post hoc ergo propter hoc* policy. By contrast, I am insisting that discriminating tests be designed in order to rule out some of the many incompatible hypotheses which his strategy supports. (After all, my hypothesis is, like his, supported on *post hoc* grounds.) Beyond that, I can point to improved starting performance in other automobiles, whose brakes were not replaced. I deny that anyone who thinks carefully about these two strategies for the evaluation of empirical claims can have any doubts about which one is more likely to produce reliable results. My mechanic's failure to impose any form of experimental controls on his causal claims is likely to lead him to make far less reliable predictions than I will. In short, my strategy will save me from several sorts of failure to which my mechanic friend will sometimes fall prey. This is not to say that hypotheses which pass my sorts of tests will never be mistaken, nor that theories which pass his tests will never lead to correct predictions. It is simply to say that my strategy will produce conjectures which break down less frequently and less quickly than his will, and that is precisely why we say that one theory is more successful than another.

Consider, as a second example, the testing of a new drug said to be efficacious in curing arthritis. If I want to test its effectiveness, I might begin by enlisting a group of physicians who would prescribe the drug for their arthritic patients. Suppose, in the first run of the test, the only evidence reported back to me is that 55% of those who took the drug reported a reduced level of pain 24 hours after the onslaught of an acute attack of arthritis. Well, what conclusion will I draw? I might, if I am hasty, pronounce the drug a qualified success. But if I am the least bit careful, I will draw no such conclusion whatever, for the test itself is very badly designed. For all I know, for instance, it might well be that 55% (or more!) of patients who take no medication whatever also report improvement after 24 hours.

So, in the second stage of the testing, we need to devise a more complex experiment. We might divide the patients into two groups, administering no drugs to one group and the drug being tested to the other. Suppose, after the experiments are performed, that it emerges that 55% of those given the drug reported improvement, while only 20% of those given no treatment reported improvement. Well, these results are rather more impressive, but again, we have to be careful about drawing any conclusions about therapeutic efficacy. We have introduced certain controls on the experiment, it is true, but we can still imagine all sorts of ways in which the reported results might be compatible with the fact that the drug is of no therapeutic value at all. Specifically, given what we know about the placebo effect and the psychosomatic character of pain, it may well be that patients given any pill, even a worthless one, will report an improvement – just because they expect medication to make them better and that expectation itself will sometimes have the desired effect. Because the control group was given no pill at all, the different results in the two cases might have nothing whatever to do with the specific character of the drug under investigation.

Realizing this, we re-design our test. For the third run of it, we give pills to *both* the control group and the test group, but only the administering physicians know that the control group receives sugar pills. Suppose the results of this experiment are as follows: the group given the real drug reports a 55% improvement in 24 hours while the group given the placebos reports a 30% improvement. Well, the evidence for therapeutic value is getting more impressive, but there are still causes for concern about the significance of the results. We have learned from many studies that those conducting experiments often have a way of transmitting their knowledge and their expectations to the human subjects on whom they are experimenting. There are all sorts of documented forms of conscious and unconscious suggestion that might be going on, even though the doctors are conscientiously trying to treat the two groups of patients identically. In short, the physicians might be conveying to their patients their knowledge of which pills are placebos, or the doctors might be interpreting their patients' comments so as to support their own expectations. If we want a strict test of the drug, we need to set up a situation where those giving out the pills and interviewing the patients have no idea whether the patients they are dealing with have been given the real drug or the placebo. With suitable precautions, such an experiment can be devised; indeed, this technique now represents a standard part of the repertoire for assessing the efficacy of therapies of many sorts, whether drugs or psychoanalysis.

Those familiar with experimental design will recognize in the chronology I have described the transition from an uncontrolled to a controlled experiment, from a controlled but unblinded experiment to a single-blind experiment, and finally from a single-blind to a double-blind experiment. For *each* transition, we can lay out some good reasons to expect the results of the later test to be more reliable than the results of the earlier one. Everyone who studies experimental methods knows the story to be offered in each case. Thus, in the first test (where we gave everyone the drug), we were using the simple method of agreement. Once we introduced the control group, we were using the joint method of agreement and difference. Anyone who thinks about these two methods will realize why theories which stand up to the latter sort of test are more likely to endure than theories which pass tests associated only with the method of agreement. *Some* of the mistakes into which the unaided method of agreement will lead us can be guarded against by using the two methods in conjunction. Even those unfamiliar with scientific methods can surely see why, if our concern is to find out whether the specific drug being tested is efficacious, procedures such as control groups and blinding will allow us better control over many extraneous variables and influences which can creep into experimental design.

What such examples vividly illustrate is that we need not engage in "high epistemology" to understand what is going on and why. The comparative reliability of various testing procedures can be explained without resorting to the realist's ambitious claims about the truthlikeness of scientific theories.[27] We already have in hand an informal logic for testing causal hypotheses which will rationalize and justify many of the methods of the natural sciences. (Every good textbook on experimental design goes much further toward explaining why science works than all the writings of scientific realists put together!) So far as I can see, scientific realism is just not needed to give a viable account of those methods. The "logic" of theory testing, imperfect as it is, puts us in a position to make some comparative judgments about the reliability of various methods of inquiry and, via those judgments, we can explain why theories which pass certain sorts of tests tend to endure longer than theories which pass other, less demanding sorts of tests. The explanation of the success of science, I submit, is no more mysterious and no more elusive than that.

This explanation of the success of science has the added virtue of being straightforwardly testable. It predicts, for instance, that where there are individuals or whole societies which shape their beliefs without the controls associated with science, those beliefs will be less reliable on the whole than the beliefs of a "scientific" culture. To be more specific, it predicts for instance that the medical practices of so-called primitive societies will tend to be less reliable and less efficacious than the practices of modern Western medicine (provided, that is, that physicians in those societies use less robust testing procedures for their theories of disease than their Western counterparts do – which may or may not be so). This is not to assert that non-scientific cultures can never discover "cures" which have eluded Western medicine, since weak heuristic and probative methods are sometimes capable of producing useful discoveries. The claim here would rather be that the frequency and reliability of such "discoveries" should be lower in societies which do not use controlled methods than in societies which do. I do not have the evidence at hand to confirm such predictions. My only point in making them is to show that the explanation of success offered here is distinctly non-vacuous.

If there are those, like the relativists, who refuse to accept this account, they must, for instance, counter the claim that a controlled experiment is an improvement on an uncontrolled one, and they must establish that a double-blind procedure is no improvement over a single-blind one. They must make plausible their claim that such experimental techniques are nothing more than socially-sanctioned conventions whose limited validity, such as it is, applies only to our culture and our time. They must show that we are just kidding ourselves in thinking that we have learned something in the last 300 years about how to put questions to nature.

CONCLUSION

Science is successful, to the extent that it is successful, because scientific theories result from a winnowing process which is arguably more robust and more discriminating than other techniques we have found for checking our empirical conjectures about the physical world. On a case-by-case basis, we can usually indicate why these methods and procedures are more likely to produce reliable results than certain other methods are.[28] Those procedures are not guaranteed to produce true theories; indeed, they generally do *not* produce true theories. But they do tend to produce theories which are more reliable than theories selected by the other belief-forming policies we are aware of. The methods of science are not necessarily the best possible methods of inquiry (for how could we conceivably show that?), nor are the theories they pick out likely to be completely reliable. But we lose nothing by conceding that the methods of science are imperfect and that the theories of science are probably false. Even in this less-than-perfect state, we have an instrument of inquiry which is arguably a better device for picking out reliable theories than any other instrument we have yet devised for that purpose. We can explain in great detail why that instrument works better than its extant rivals. Because we can, the success of science ceases to be quite the mystery which some philosophers and sociologists have made it out to be.[29]

NOTES

1. I am grateful to a variety of friends who commented on previous versions of this essay and helped clarify my thinking about several of its central themes. They include Alberto Coffa, Clark Glymour, Rick Creath, Tom Nickles, Peter Barker, Arthur Donovan, David Hull, Rachel Laudan, Adolf Grünbaum, Andrew Lugg, Robert Butts, Ilkka Niiniluoto, Nicholas Rescher, and Gary Gutting, as well as several of my immediate colleagues.
2. See letter xiv in Voltaire's *Lettres Philosophiques* (Rouen, 1734).
3. Careful readers will take exception to my juxtaposition of realism and relativism in this way, pointing out that they are really not so much "opposites", as they are orthogonal to one another. Strictly speaking, after all, those who deny realism tend to be instrumentalists or idealists (rather than relativists per se), while the natural opponents of relativism are what we might call "objectivists". Nonetheless, it is illuminating to play realist and relativist perspectives off against one another since (a) they are rival *epistemic* traditions which are genuine contraries of one another (i.e., they cannot both be correct), and (b) as I point out below, there is a widespread tendency to assume that weaknesses in either count as arguments for its rival.
4. To be more specific, the form of relativism which I shall be discussing chiefly involves the denial that any techniques or methods for warranting knowledge claims are "better" than any others. One might call this the thesis of 'methodological relativism'; it is different from, and much more ambitious than, the thesis (often called 'ontological relativism') to the effect that no ontological framework is "privileged".
5. See, for instance, B. Barnes and D. Bloor, "Relativism, Rationalism, and the Sociology of Knowledge", in M. Hollis and S. Luxes (ed.s), *Rationality and Relativism* (Oxford, 1982), 21–47.
6. As I shall show shortly, however, it is important not to draw too strong an analogy between judgments of rationality and judgments of success.
7. See, for example, Hilary Putnam's treatment of the success of science in his *Meaning and the Moral Sciences* (London, 1978).
8. I am grateful to Alberto Coffa (private correspondence) for persuading me of the urgency of avoiding pinning our characterizations of scientific success on the aims, real or avowed, of working scientists.

9. For a lengthy discussion of some of the changes which have taken place in the cognitive goals of scientists, see my *Science and Hypothesis* (Dordrecht, 1981) and *Science and Values* (Berkeley, 1984).

10. No claim about the relative success of science would be complete without reference to Paul Feyerabend's recent tirade against the thesis that science has been successful. In his *Science and a Free Society* (London, 1978), especially pp. 100 ff., Feyerabend asserts that the results of science (i.e., its successes) are no more impressive than the successes achieved by the cosmologies of many "primitive" societies. Feyerabend goes on to claim that science *appears* to be more successful than other systems of nature only because of the systematic suppression of other approaches in our culture: "*Today science prevails not because of its comparative merits, but because the show has been rigged in its favour*" (p. 102; italics in original). Earlier ways of studying nature "have disappeared or deteriorated not because science was better but because *the apostles of science were the more determined conquerors . . .* [who] *materially suppressed* the bearers of alternative cultures" (p. 102; Feyerabend's italics). As he warms to his topic, his claims become even more vitriolic: "The superiority of science is the result not of research, or argument, it is the result of political, institutional, and even military pressures." *Ibid.* As with most of Feyerabend's more provocative theses, this is more bluster than substance. Pointing (quite rightly) to the fact that pre-scientific cultures have made many very useful discoveries about how to manipulate nature, he concludes that the ideologies of those cultures are as successful empirically as the theories of science (or that they would have been if we had not systematically eradicated their advocates). This, clearly, is a monumental *non sequitur*. The judgment that science is more successful (in the sense spelled out above) than the nature philosophies of other cultures is, as Feyerabend is more than clever enough to realize, entirely compatible with the claim that non- or pre-scientific cultures have produced theories and methods which sometimes work well for their purposes. But since an acknowledgment that science has been more successful than those rivals would undercut Feyerabend's epistemic anarchism, he conveniently fails to alert his readers to the fact that the slide from the claim that pre-scientific theories have enjoyed some successes to the thesis that those theories have been as successful as, or more successful than, science is a monumental piece of bad reasoning.

11. H. Putnam, *Mathematics, Matter and Method* (Cambridge, 1975), p. 73.

12. See references to, and criticism of, the work of these authors in my "A Confutation of Convergent Realism," *Philosophy of Science*, 48 (1981), 19–49.

13. To say as much is probably to make life too easy for the realist. Since Duhem, it has been widely recognized that theories typically do not impinge on experience directly but only in conjunction with a wide variety of other auxiliary assumptions. Under such circumstances, it is entirely possible that a theory could be true *simpliciter*, and yet such that all the observed consequences we attribute to it (derived from conjunctions of that theory with auxiliary assumptions) could be false. (I owe this point to Jarrett Leplin.)

14. See my "A Confutation of Convergent Realism".

15. *Ibid.*

16. I shall resist the temptation to dwell on the fact that the relativist evidently *exempts* his own methods of theory evaluation from this general relativist critique. Since numerous authors have pointed to this apparently self-refuting feature of relativism, I shall not discuss it at length. (See, for instance, my "A Note on Collins's Blend of Relativism and Empiricism", *Social Studies of Science*, 12 (1982), pp. 131–132.)

17. Perhaps the most strenuously relativist of this group is Harry Collins, who seemingly denies that there is any sense in which we can say that science is successful in predicting the world. Because Collins believes that "the natural world has a small or non-existent role in the construction of scientific knowledge" (H. Collins, "Stages in the Empirical Program of Relativism", *Social Studies of Science*, 11 (1981), p. 3) and that "reality [does nothing] to circumscribe possible individual beliefs" (H. Collins and G. Cox, "Recovering Relativity: Did Prophecy Fail?", *Social Studies of Science*, 6 (1976), p. 437) he is presumably forced to deny that any meaning can be attached to the claim that any system of belief is any more successful than another.

18. In fact, I believe that the success of science does refute most of the extreme forms of relativism, but it would require an independent argument to show that, and its development would carry the narrative too far afield.

19. This is not to say that all parts of science are predictively reliable, nor even that science is always more reliable than other ways of second-guessing the future. The specific claim is a limited one: to wit, that certain scientific theories have a much better predictive track record than most of their extant, non-scientific counterparts.

20. In using this example, I do not mean to suggest that all the cognitive aims or interests associated with Western science correspond to identifiable technical interests which we can identify across a broad spectrum of cultures. I *do* mean to insist, however, that there are some interests which cut across boundaries of culture and society. It is an issue of great consequence, both intellectually and practically, whether the methods of warranting associated with science do or do not promote those aims.

21. Throughout this essay, I adopt three simplifying assumptions to make my task more manageable: (a) that there is a set of methods which we can identify as "scientific"; (b) that these methods are shared between proponents of rival theories; and (c) that these methods do not radically underdetermine theory choice. All these assumptions have been hotly contested by various relativists. I have elsewhere tried to defuse the force of the relativist's arguments on these issues. (I take on the first claim in "What Remains of the Scientific Method", forthcoming; the second in my *Science and Values* [Berkeley], and the third in my "Overestimating Underdetermination" forthcoming.) Readers will have to judge for themselves whether my arguments make it plausible for me to adopt here the simplifying assumptions indicated above.

22. I am not asserting categorically that Western hydrology would necessarily surpass all the rival, apparently "non-scientific" techniques for treating these phenomena. I do not know whether it would or not. I am, rather, showing how one might go about ascertaining which parts of science exhibit a degree of success which calls out for special explanation. (Obviously, should the folk wisdom of some societies produce

      theories which are consistently more successful than science, that would equally call for some form of special explanation.)

23. One sees precisely this assumption in the influential work of Mary Douglas. She argues, for instance, that "it is no more easy to defend... objective scientific truths than beliefs in gods and demons" (Mary Douglas, *Implicit Meanings* [London, 1975], p. xv). In effect, her analysis denies in principle that any methodological defense of the claims of science could be given which would show that those claims were better grounded than the beliefs of any non-scientific culture.

24. The relativist's cause is aided and abetted here from some unusual quarters, not least from the arch-rationalist, Imre Lakatos. He has claimed on numerous occasions that any point of view, however bizarre – if only it is provided with enough funds and talented advocates – can accumulate impressive empirical successes. (One should add, for the historical record, that Lakatos never provided any evidence for this assertion of his. One suspects he threw it in as a sop to his relativist friend Paul Feyerabend!)

25. Popper wrote: "No theory of knowledge should attempt to explain why we are successful in our attempts to explain things" *Objective Knowledge* (Oxford, 1972). The problem, of course, is that if epistemology cannot illuminate that problem, it is not clear what interesting tasks would remain for epistemology.

26. When I say that one theory is more reliable than another, I simply mean to refer to the fact that one theory is apt to be more useful, to be able to digest a larger and more disparate range of phenomena before it breaks down, than a theory which is less reliable.

27. The astute reader may note that neither of my examples of testing procedures involved the testing of that sort of deep-structure theory which is beloved by realists. But that omission reflects no limitation on the explanatory technique sketched out here. Basically, we test our most deep-structure theories, and credit them with success or failure, in precisely the same way that we test theories which are "closer" to "observation" (such as the two examples I discussed). Thus, if one wanted to explain the relative success of the atomic theory, one might venture to show that the battery of tests to which that theory had been subjected was more demanding than the sorts of tests which (say) hermetical theories of chemical structure had passed.

28. My discussion of the last few pages probably suggests that the justification of these methods is more straightforward and less problematic than it actually is. It would be less than candid not to note that there is some serious disagreement about exactly what rationale to give for some of the standard procedures of empirical control. But I would claim that the broad outlines of a rationale for such methods are clearly understood; that we know what technical and justificational problems confront us, and that we have some ideas about how to resolve them. In no case does it seem that such justificatory moves require us to go in the direction of scientific realism, i.e., of basing our explanation of the success of the methods of science on the thesis that the theories which science produces are true or nearly true.

29. Before I close, it is worth noticing how the approach to the problem of success sketched here exhibits the gratuitousness of the realist's would-be

solution to the problem. If we can explain why the methods of science are apt to produce theories which are more reliable than theories produced by other methods, then we need not commit ourselves, as the realist evidently must, to a dubious claim about the truth or truthlikeness of the theories of science. Going beyond that reliability to postulate that our theories correctly characterize the world via their deep-structural commitments is to assume both more, and less, than is necessary to explain why scientific theories work as well as they do.

6 Experimentation and Scientific Realism*

Ian Hacking

Experimental physics provides the strongest evidence for scientific realism. Entities that in principle cannot be observed are regularly manipulated to produce new phenomena and to investigate other aspects of nature. They are tools, instruments not for thinking but for doing.

The philosopher's standard "theoretical entity" is the electron. I will illustrate how electrons have become experimental entities, or experimenter's entities. In the early stages of our discovery of an entity, we may test hypotheses about it. Then it is merely a hypothetical entity. Much later, if we come to understand some of its causal powers and use it to build devices that achieve well-understood effects in other parts of nature, then it assumes quite a different status.

Discussions about scientific realism or antirealism usually talk about theories, explanation, and prediction. Debates at that level are necessarily inconclusive. Only at the level of experimental practice is scientific realism unavoidable – but this realism is not about theories and truth. The experimentalist need only be a realist about the entities used as tools.

A PLEA FOR EXPERIMENTS

No field in the philosophy of science is more systematically neglected than experiment. Our grade school teachers may have told us that scientific method is experimental method, but histories of science have become histories of theory. Experiments, the philosophers say, are of value only when they test theory. Experimental work, they imply, has no life of its own. So we lack even a terminology to describe the many varied roles of experiment. Nor has this one-sidedness done theory any good, for radically different types of theory are used to think

* Chapter 8 in J. Leplin (ed.), *Scientific Realism* (Berkeley: University of California Press, 1984), pp. 154–172.

about the same physical phenomenon (e.g., the magneto-optical effect). The philosophers of theory have not noticed this and so misreport even theoretical enquiry.

Different sciences at different times exhibit different relationships between "theory" and "experiment." One chief role of experiment is the creation of phenomena. Experimenters bring into being phenomena that do not naturally exist in a pure state. These phenomena are the touchstones of physics, the keys to nature, and the source of much modern technology. Many are what physicists after the 1870s began to call "effects": the photoelectric effect, the Compton effect, and so forth.[1] A recent high-energy extension of the creation of phenomena is the creation of "events," to use the jargon of the trade. Most of the phenomena, effects, and events created by the experimenter are like plutonium: they do not exist in nature except possibly on vanishingly rare occasions.[2]

In this paper I leave aside questions of methodology, history, taxonomy, and the purpose of experiment in natural science. I turn to the purely philosophical issue of scientific realism. Simply call it "realism" for short. There are two basic kinds: realism about entities and realism about theories. There is no agreement on the precise definition of either. Realism about theories says that we try to form true theories about the world, about the inner constitution of matter and about the outer reaches of space. This realism gets its bite from optimism: we think we can do well in this project and have already had partial success. Realism about entities – and I include processes, states, waves, currents, interactions, fields, black holes, and the like among entities – asserts the existence of at least some of the entities that are the stock in trade of physics.[3]

The two realisms may seem identical. If you believe a theory, do you not believe in the existence of the entities it speaks about? If you believe in some entities, must you not describe them in some theoretical way that you accept? This seeming identity is illusory. *The vast majority of experimental physicists are realists about entities but not about theories.* Some are, no doubt, realists about theories too, but that is less central to their concerns.

Experimenters are often realists about the entities that they investigate, but they do not have to be so. R. A. Millikan probably had few qualms about the reality of electrons when he set out to measure their charge. But he could have been skeptical about what he would find until he found it. He could even have remained skeptical. Perhaps there is a least unit of electric charge, but there is no particle or object with exactly that unit of charge. Experimenting on an entity does not commit you to believing that it exists. Only manipulating an entity, in order to experiment on something else, need do that.

Moreover, it is not even that you use electrons to experiment on something else that makes it impossible to doubt electrons. Understanding some causal properties of electrons, you guess how to build a very ingenious, complex device that enables you to line up the electrons the way you want, in order to see what will happen to something else. Once you have the right experimental idea, you know in advance roughly how to try to build the device, because you know that this is the way to get the electrons to behave in such and such a way. Electrons are no longer ways of organizing our thoughts or saving the phenomena that have been observed. They are now ways of creating phenomena in some other domain of nature. Electrons are tools.

There is an important experimental contrast between realism about entities and realism about theories. Suppose we say that the latter is belief that science aims at true theories. Few experimenters will deny that. Only philosophers doubt it. Aiming at the truth is, however, something about the indefinite future. Aiming a beam of electrons is using present electrons. Aiming a finely tuned laser at a particular atom in order to knock off a certain electron to produce an ion is aiming at present electrons. There is, in contrast, no present set of theories that one has to believe in. If realism about theories is a doctrine about the aims of science, it is a doctrine laden with certain kinds of values. If realism about entities is a matter of aiming electrons next week or aiming at other electrons the week after, it is a doctrine much more neutral between values. The way in which experimenters are scientific realists about entities is entirely different from ways in which they might be realists about theories.

This shows up when we turn from ideal theories to present ones. Various properties are confidently ascribed to electrons, but most of the confident properties are expressed in numerous different theories or models about which an experimenter can be rather agnostic. Even people in a team, who work on different parts of the same large experiment, may hold different and mutually incompatible accounts of electrons. That is because different parts of the experiment will make different uses of electrons. Models good for calculations on one aspect of electrons will be poor for others. Occasionally, a team actually has to select a member with a quite different theoretical perspective simply to get someone who can solve those experimental problems. You may choose someone with a foreign training, and whose talk is well-nigh incommensurable with yours, just to get people who can produce the effects you want.

But might there not be a common core of theory, the intersection of everybody in the group, which is the theory of the electron to which all the experimenters are realistically committed? I would say common lore, *not* common core. There are a lot of theories, models, approximations, pictures, formalisms, methods, and so forth involving electrons, but there is no reason to suppose that the intersection of these is a theory at all. Nor is there any reason to think that there is such a thing as "the most powerful nontrivial *theory* contained in the intersection of all the theories in which this or that member of a team has been trained to believe." Even if there are a lot of shared beliefs, there is no reason to suppose they form anything worth calling a theory. Naturally, teams tend to be formed from like-minded people at the same institute, so there is usually some real shared theoretical basis to their work. That is a sociological fact, not a foundation for scientific realism.

I recognize that many a scientific realism concerning theories is a doctrine not about the present but about what we might achieve, or possibly an ideal at which we aim. So to say that there is no present theory does not count against the optimistic aim. The point is that such scientific realism about theories has to adopt the Peircean principles of faith, hope, and charity. Scientific realism about entities needs no such virtues. It arises from what we can do at present. To understand this, we must look in some detail at what it is like to build a device that makes the electrons sit up and behave.

OUR DEBT TO HILARY PUTNAM

It was once the accepted wisdom that a word such as 'electron' gets its meaning from its place in a network of sentences that state theoretical laws. Hence arose the infamous problems of incommensurability and theory change. For if a theory is modified, how could a word such as "electron" go on meaning the same? How could different theories about electrons be compared, since the very word "electron" would differ in meaning from theory to theory?

Putnam saved us from such questions by inventing a referential model of meaning. He says that meaning is a vector, refreshingly like a dictionary entry. First comes the syntactic marker (part of speech); next the semantic marker (general category of thing signified by the word); then the stereotype (clichés about the natural kind, standard examples of its use, and present-day associations. The stereotype is subject to change as opinions about the kind are modified). Finally, there is the actual referent of the word, the very stuff, or thing, it denotes if it denotes anything. (Evidently dictionaries cannot include this in their entry, but pictorial dictionaries do their best by inserting illustrations whenever possible.)[4]

Putnam thought we can often guess at entities that we do not literally point to. Our initial guesses may be jejune or inept, and not every naming of an invisible thing or stuff pans out. But when it does, and we frame better and better ideas, then Putnam says that, although the stereotype changes, we refer to the same kind of thing or stuff all along. We and Dalton alike spoke about the same stuff when we spoke of (inorganic) acids. J.J. Thomson, H.A. Lorentz, Bohr, and Millikan were, with their different theories and observations, speculating about the same kind of thing, the electron.

There is plenty of unimportant vagueness about when an entity has been successfully "dubbed," as Putnam puts it. "Electron" is the name suggested by G. Johnstone Stoney in 1891 as the name for a natural unit of electricity. He had drawn attention to this unit in 1874. The name was then applied to the subatomic particles of negative charge, which J.J. Thomson, in 1897, showed cathode rays consist of. Was Johnstone Stoney referring to the electron? Putnam's account does not require an unequivocal answer. Standard physics books say that Thomson discovered the electron. For once I might back theory and say that Lorentz beat him to it. Thomson called his electrons "corpuscles", the subatomic particles of electric charge. Evidently, the name does not matter much. Thomson's most notable achievement was to measure the mass of the electron. He did this by a rough (though quite good) guess at e, and by making an excellent determination of e/m, showing that m is about $1/1800$ the mass of the hydrogen atom. Hence it is natural to say that Lorentz merely postulated the existence of a particle of negative charge, while Thomson, determining its mass, showed that there is some such real stuff beaming off a hot cathode.

The stereotype of the electron has regularly changed, and we have at least two largely incompatible stereotypes, the electron as cloud and the electron as particle. One fundamental enrichment of the idea came in the 1920s. Electrons, it was found, have angular momentum, or "spin." Experimental work by O. Stern and W. Gerlach first indicated this, and then S. Goudsmit and G.E. Uhlenbeck provided the theoretical understanding of it in 1925. Whatever we think, Johnstone Stoney, Lorentz, Bohr, Thomson, and Goudsmit were all finding out more about the same kind of thing, the electron.

We need not accept the fine points of Putnam's account of reference in order to thank him for giving us a new way to talk about meaning. Serious discussion of inferred entities need no longer lock us into pseudo-problems of incommensurability and theory change. Twenty-five years ago the experimenter who believed that electrons exist, without giving much credence to any set of laws about electrons, would have been dismissed as philosophically incoherent. Now we realize it was the philosophy that was wrong, not the experimenter. My own relationship to Putnam's account of meaning is like the experimenter's relationship to a theory. I do not literally believe Putnam, but I am happy to employ his account as an alternative to the unpalatable account in fashion some time ago.

Putnam's philosophy is always in flux. His account of reference was intended to bolster scientific realism. But now, at the time of this writing (July 1981), he rejects any "metaphysical realism" but allows "internal realism."[5] The internal realist acts, in practical affairs, as if the entities occurring in his working theories did in fact exist. However, the direction of Putnam's metaphysical antirealism is no longer scientific. It is not peculiarly about natural science. It is about chairs and livers too. He thinks that the world does not naturally break up into our classifications. He calls himself a transcendental idealist. I call him a transcendental nominalist. I use the word "nominalist" in the old-fashioned way, not meaning opposition to "abstract entities" like sets, but meaning the doctrine that there is no nonmental classification in nature that exists over and above our own human system of naming.

There might be two kinds of internal realist, the instrumentalist about science and the scientific realist. The former is, in practical affairs where he uses his present scheme of concepts, a realist about livers and chairs but thinks that electrons are only mental constructs. The latter thinks that livers, chairs, and electrons are probably all in the same boat, that is, real at least within the present system of classification. I take Putnam to be an internal scientific realist rather than an internal instrumentalist. The fact that either doctrine is compatible with transcendental nominalism and internal realism shows that our question of scientific realism is almost entirely independent of Putnam's internal realism.

INTERFERING

Francis Bacon, the first and almost last philosopher of experiments, knew it well: the experimenter sets out "to twist the lion's tail." Experimentation is interference in the course of nature; "nature under constraint and vexed; that is to say, when by art and the hand of man she is forced out of her natural state, and squeezed and moulded."[6] The experimenter is convinced of the reality of entities, some of whose causal properties are sufficiently well understood that they can be used to interfere *elsewhere* in nature. One is impressed by entities that one can use to test conjectures about other, more hypothetical entities. In my example, one is sure of the electrons that are used to investigate weak neutral currents and neutral bosons. This should not be news, for why else are we (nonskeptics) sure of the reality of even macroscopic objects, but because of what we do with them, what we do to them, and what they do to us?

Interference and interaction are the stuff of reality. This is true, for example, at the borderline of observability. Too often philosophers imagine that microscopes carry conviction because they help us see better. But that is only part of the story. On the contrary, what counts is what we can do to a specimen under a microscope, and what we can see ourselves doing. We stain the specimen, slice it, inject it, irradiate it, fix it. We examine it using different kinds of microscopes that employ optical systems that rely on almost totally unrelated facts about light. Microscopes carry conviction because of the great array of interactions and interferences that are possible. When we see something that turns out to be unstable under such play, we call it an artifact and say it is not real.[7]

Likewise, as we move down in scale to the truly unseeable, it is our power to use unobservable entities that makes us believe they are there. Yet, I blush over these words "see" and "observe". Philosophers and physicists often use these words in different ways. Philosophers tend to treat opacity to visible light as the touchstone of reality, so that anything that cannot be touched or seen with the naked eye is called a theoretical or inferred entity. Physicists, in contrast, cheerfully talk of observing the very entities that philosophers say are not observable. For example, the fermions are those fundamental constituents of matter such as electron neutrinos and deuterons and, perhaps, the notorious quarks. All are standard philosophers' "unobservable" entities. C.Y. Prescott, the initiator of the experiment described below, said in a recent lecture, that "of these fermions, only the t quark is yet unseen. The failure to observe $t\bar{t}$ states in e^+e^- annihilation at PETRA remains a puzzle."[8] Thus, the physicist distinguishes among the philosophers' "unobservable" entities, noting which have been observed and which not. Dudley Shapere has just published a valuable study of this fact.[9] In his example, neutrinos are used to see the interior of a star. He has ample quotations such as "neutrinos present the only way of directly observing" the very hot core of a star.

John Dewey would have said that fascination with seeing-with-the-naked-eye is part of the spectator theory of knowledge that has bedeviled philosophy from earliest times. But I do not think Plato or Locke or anyone before the nineteenth century was as obsessed with the sheer opacity of objects as we have been since. My own obsession with a technology that manipulates objects is, of course, a twentieth-century counterpart to positivism and phenomenology. Its proper rebuttal is not a restriction to a narrower domain of reality, namely, to what can be positivistically seen with the eye, but an extension to other modes by which people can extend their consciousness.

MAKING

Even if experimenters are realists about entities, it does not follow that they are right. Perhaps it is a matter of psychology: maybe the very skills that make for a great experimenter go with a certain cast of mind which objectifies whatever it thinks about. Yet this will not do. The experimenter cheerfully regards neutral bosons as merely hypothetical entities, while electrons are real. What is the difference?

There are an enormous number of ways in which to make instruments that rely on the causal properties of electrons in order to produce desired effects of unsurpassed precision. I shall illustrate this. The argument – it could be called the "experimental argument for realism" – is not that we infer the reality of electrons from our success. We do not make the instruments and then infer the reality of the electrons, as when we test a hypothesis, and then believe it because it passed the test. That gets the time-order wrong. By now we design apparatus relying on a modest number of home truths about electrons, in order to produce some other phenomenon that we wish to investigate.

That may sound as if we believe in the electrons because we predict how our apparatus will behave. That too is misleading. We have a number of general ideas about how to prepare polarized electrons, say. We spend a lot of time building prototypes that do not work. We get rid of innumerable bugs. Often we have to give up and try another approach. Debugging is not a matter of theoretically explaining or predicting what is going wrong. It is partly a matter of getting rid of "noise" in the apparatus. "Noise" often means all the events that are not understood by any theory. The instrument must be able to isolate, physically, the properties of the entities that we wish to use, and damp down all the other effects that might get in our way. *We are completely convinced of the reality of electrons when we regularly set to build – and often enough succeed in building – new kinds of devices that use various well understood causal properties of electrons to interfere in other more hypothetical parts of nature.*

It is not possible to grasp this without an example. Familiar historical examples have usually become encrusted by false theory-oriented philosophy or history, so I will take something new. This is a polarizing electron gun whose acronym is PEGGY II. In 1978, it was used in a fundamental experiment that attracted attention even in *The New York Times*. In the next section I describe the point of making PEGGY II. To do that, I have to tell some new physics. You may omit reading this and read only the engineering section that follows. Yet it must be of interest to know the rather easy-to-understand significance of the main experimental results, namely, that parity is not conserved in scattering of polarized electrons from deuterium, and that, more generally, parity is violated in weak neutral-current interactions.[10]

PARITY AND WEAK NEUTRAL CURRENTS

There are four fundamental forces in nature, not necessarily distinct. Gravity and electromagnetism are familiar. Then there are the strong and weak forces (the fulfillment of Newton's program, in the *Optics*, which taught that all nature would be understood by the interaction of particles with various forces that were effective in attraction or repulsion over various different distances, i.e., with different rates of extinction).

Strong forces are 100 times stronger than electromagnetism but act only over a minuscule distance, at most the diameter of a proton. Strong forces act on "hadrons," which include protons, neutrons, and more recent particles, but not electrons or any other members of the class of particles called "leptons."

The weak forces are only 1/10,000 times as strong as electromagnetism, and act over a distance 100 times greater than strong forces. But they act on both hadrons and leptons, including electrons. The most familiar example of a weak force may be radioactivity.

The theory that motivates such speculation is quantum electrodynamics. It is incredibly successful, yielding many predictions better than one part in a million, truly a miracle in experimental physics. It applies over distances ranging from diameters of the earth to 1/100 the diameter of the proton. This theory supposes that all the forces are "carried" by some sort of particle: photons do the job in electromagnetism. We hypothesize "gravitons" for gravity.

In the case of interactions involving weak forces, there are charged currents. We postulate that particles called "bosons" carry these weak forces.[11] For charged currents, the bosons may be either positive or negative. In the 1970s, there arose the possibility that there could be weak "neutral" currents in which no charge is carried or exchanged. By sheer analogy with the vindicated parts of quantum electrodynamics, neutral bosons were postulated as the carriers in weak neutral interactions.

The most famous discovery of recent high-energy physics is the failure of the conservation of parity. Contrary to the expectations of many physicists and philosophers, including Kant,[12] nature makes an absolute distinction between right-handedness and left-handedness. Apparently, this happens only in weak interactions.

What we mean by right- or left-handed in nature has an element of convention. I remarked that electrons have spin. Imagine your right hand wrapped around a spinning particle with the fingers pointing in the direction of spin. Then your thumb is said to point in the direction of the spin vector. If such particles are traveling in a beam, consider the relation between the spin vector and the beam. If all the particles have their spin vector in the same direction as the beam, they have right-handed (linear) polarization, while if the spin vector is opposite to the beam direction, they have left-handed (linear) polarization.

The original discovery of parity violation showed that one kind of product of a particle decay, a so-called muon neutrino, exists only in left-handed polarization and never in right-handed polarization.

Parity violations have been found for weak *charged* interactions. What about weak *neutral* currents? The remarkable Weinberg–Salam model for the four kinds of force was proposed independently by Stephen Weinberg in 1967 and A. Salam in 1968. It implies a minute violation of parity in weak neutral interactions. Given that the model is sheer speculation, its success has been amazing, even awe-inspiring. So it seemed worthwhile to try out the predicted failure of parity for weak neutral interactions. That would teach us more about those weak forces that act over so minute a distance.

The prediction is: slightly more left-handed polarized electrons hitting certain targets will scatter, than right-handed electrons. Slightly more! The difference in relative frequency of the two kinds of scattering is 1 part in 10,000, comparable to a difference in probability between 0.50005 and 0.49995. Suppose one used the standard equipment available at the Stanford Linear Accelerator Center in the early 1970s, generating 120 pulses per second, each pulse providing one electron event. Then you would have to run the entire SLAC beam for twenty-seven years in order to detect so small a difference in relative frequency. Considering that one uses the same beam for lots of experiments simultaneously, by letting different experiments use different pulses, and considering that no equipment remains stable for even a month, let alone twenty-seven years, such an experiment is impossible. You need enormously more electrons coming off in each pulse. We need between 1000 and 10,000 more electrons per pulse than was once possible. The first attempt used an instrument now called PEGGY I. It had, in essence, a high-class version of J.J. Thomson's hot cathode. Some lithium was heated and electrons were boiled off. PEGGY II uses quite different principles.

PEGGY II

The basic idea began when C.Y. Prescott noticed (by chance!) an article in an optics magazine about a crystalline substance called gallium arsenide. GaAs has a curious property; when it is struck by circularly polarized light of the right frequencies, it emits lots of linearly polarized electrons. There is a good, rough and ready quantum understanding of why this happens, and why half the emitted electrons will be polarized, three-fourths of these polarized in one direction and one-fourth polarized in the other.

PEGGY II uses this fact, plus the fact that GaAs emits lots of electrons owing to features of its crystal structure. Then comes some engineering – it takes work to liberate an electron from a surface. We know that painting a surface with the right stuff helps. In this case, a thin layer of cesium and oxygen is applied to the crystal. Moreover, the less air pressure around the crystal, the more electrons will escape for a given amount of work. So the bombardment takes place in a good vacuum at the temperature of liquid nitrogen.

We need the right source of light. A laser with bursts of red light (7100 Ångstroms) is trained on the crystal. The light first goes through an ordinary polarizer, a very old-fashioned prism of calcite, or Iceland spar[13] – this gives linearly polarized light. We want circularly polarized light to hit the crystal, so the polarized laser beam now goes through a cunning device called a Pockel's cell, which electrically turns linearly polarized photons into circularly polarized ones. Being electric, it acts as a very fast switch. The direction of circular polarization depends on the direction of current in the cell. Hence, the direction of polarization can be varied randomly. This is important, for we are trying to detect a minute asymmetry between right- and left-handed polarization. Randomizing helps us guard against any systematic "drift" in the equipment.[14] The randomization is generated by a radioactive decay device, and a computer records the direction of polarization for each pulse.

A circularly polarized pulse hits the GaAs crystal, resulting in a pulse of linearly polarized electrons. A beam of such pulses is maneuvered by magnets into the accelerator for the next bit of the experiment. It passes through a device that checks on a proportion of polarization along the way. The remainder of the experiment requires other devices and detectors of comparable ingenuity, but let us stop at PEGGY II.

BUGS

Short descriptions make it all sound too easy; therefore, let us pause to reflect on debugging. Many of the bugs are never understood. They are eliminated by trial and error. Let me illustrate three different kinds of bugs: (1) the essential technical limitations that, in the end, have to be factored into the analysis of error; (2) simpler mechanical defects you never think of until they are forced on you, and (3) hunches about what might go wrong.

Here are three examples of bugs:

1. Laser beams are not as constant as science fiction teaches, and there is always an irremediable amount of "jitter" in the beam over any stretch of time.

2. At a more humdrum level, the electrons from the GaAs crystal are back-scattered and go back along the same channel as the laser beam used to hit the crystal. Most of them are then deflected magnetically. But some get reflected from the laser apparatus and get back into the system. So you have to eliminate these new ambient electrons. This is done by crude mechanical means, making them focus just off the crystal and, thus, wander away.

3. Good experimenters guard against the absurd. Suppose that dust particles on an experimental surface lie down flat when a polarized

pulse hits it, and then stand on their heads when hit by a pulse polarized in the opposite direction. Might that have a systematic effect, given that we are detecting a minute asymmetry? One of the team thought of this in the middle of the night and came down next morning frantically using antidust spray. They kept that up for a month, just in case.[15]

RESULTS

Some 10^{11} events were needed to obtain a result that could be recognized above systematic and statistical error. Although the idea of systematic error presents interesting conceptual problems, it seems to be unknown to philosophers. There were systematic uncertainties in the detection of right- and left-handed polarization, there was some jitter, and there were other problems about the parameters of the two kinds of beam. These errors were analyzed and linearly added to the statistical error. To a student of statistical inference, this is real seat-of-the-pants analysis with no rationale whatsoever. Be that as it may, thanks to PEGGY II the number of events was big enough to give a result that convinced the entire physics community.[16] Left-handed polarized electrons were scattered from deuterium slightly more frequently than right-handed electrons. This was the first convincing example of parity-violation in a weak neutral current interaction.

COMMENT

The making of PEGGY II was fairly nontheoretical. Nobody worked out in advance the polarizing properties of GaAs – that was found by a chance encounter with an unrelated experimental investigation. Although elementary quantum theory of crystals explains the polarization effect, it does not explain the properties of the actual crystal used. No one has got a real crystal to polarize more than 37 percent of the electrons, although in principle 50 percent should be polarized.

Likewise, although we have a general picture of why layers of cesium and oxygen will "produce negative electron affinity," that is, make it easier for electrons to escape, we have no quantitative understanding of why this increases efficiency to a score of 37 percent.

Nor was there any guarantee that the bits and pieces would fit together. To give an even more current illustration, future experimental work, briefly described later in this paper, makes us want even more electrons per pulse than PEGGY II can give. When the aforementioned parity experiment was reported in *The New York Times*, a group at Bell Laboratories read the newspaper and saw what was going on. They had been constructing a crystal lattice for totally unrelated purposes. It uses layers of GaAs and a related aluminum compound. The structure of this lattice leads one to expect that virtually all the electrons emitted would be polarized. As a consequence, we might be able to double the efficiency of PEGGY II. But, at present, that nice idea has problems. The new lattice should also be coated in work-reducing paint. The cesium–oxygen compound is applied at high temperature. Hence the aluminum tends to ooze into the neighboring layer of GaAs, and the pretty artificial lattice becomes a bit uneven, limiting its fine polarized-electron-emitting properties.[17] So perhaps this will never work. Prescott is simultaneously reviving a souped up new thermionic cathode to try to get more electrons. Theory would not have told us that PEGGY II would beat out thermionic PEGGY I. Nor can it tell if some thermionic PEGGY III will beat out PEGGY II.

Note also that the Bell people did not need to know a lot of weak neutral current theory to send along their sample lattice. They just read *The New York Times.*

MORAL

Once upon a time, it made good sense to doubt that there were electrons. Even after Thomson had measured the mass of his corpuscles, and Millikan their charge, doubt could have made sense. We needed to be sure that Millikan was measuring the same entity as Thomson. Thus, more theoretical elaboration was needed, and the idea had to be fed into many other phenomena. Solid state physics, the atom, and superconductivity all had to play their part.

Once upon a time, the best reason for thinking that there are electrons might have been success in explanation. Lorentz explained the Faraday effect with his electron theory. But the ability to explain carries little warrant of truth. Even from the time of J.J. Thomson, it was the measurements that weighed in, more than the explanations. Explanations, however, did help. Some people might have had to believe in electrons because the postulation of their existence could explain a wide variety of phenomena. Luckily, we no longer have to pretend to infer from explanatory success (i.e., from what makes our minds feel good).

Prescott and the team from the SLAC do not explain phenomena with electrons. They know how to use them. Nobody in his right mind thinks that electrons "really" are just little spinning orbs about which you could, with a small enough hand, wrap your fingers and find the direction of spin along your thumb. There is, instead, a family of causal properties in terms of which gifted experimenters describe and deploy electrons in order to investigate something else, for example, weak neutral currents and neutral bosons. We know an enormous amount about the behavior of electrons. It is equally important to know what does *not* matter to electrons. Thus, we know that bending a polarized electron beam in magnetic coils does not affect polarization in any significant way. We have hunches, too strong to ignore although too trivial to test independently: for example, dust might dance under changes of directions of polarization. Those hunches are based on a hard-won sense of the kinds of things electrons are. (It does not matter at all to this hunch whether electrons are clouds or waves or particles.)

WHEN HYPOTHETICAL ENTITIES BECOME REAL

Note the complete contrast between electrons and neutral bosons. Nobody can yet manipulate a bunch of neutral bosons, if there are any. Even weak neutral currents are only just emerging from the mists of hypothesis. By 1980, a sufficient range of convincing experiments had made them the object of investigation. When might they lose their hypothetical status and become commonplace reality like electrons? – when we use them to investigate something else.

I mentioned the desire to make a better electron gun than PEGGY II. Why? Because we now "know" that parity is violated in weak neutral interactions. Perhaps by an even more grotesque statistical analysis than that involved in the parity experiment, we can isolate just the weak interactions. For example, we have a lot of interactions, including electromagnetic ones, which we can censor in various ways. If we could also statistically pick out a class of weak interactions, as precisely those where parity is not conserved, then we would possibly be on the road to quite deep investigations of matter and antimatter. To do the statistics, however, one needs even more electrons per pulse than PEGGY II could hope to generate. If such a project were to succeed, we should then be beginning to use weak neutral currents as a manipulable tool for looking at something else. The next step toward a realism about such currents would have been made.

The message is general and could be extracted from almost any branch of physics. I mentioned earlier how Dudley Shapere has recently used "observation" of the sun's hot core to illustrate how physicists employ the concept of observation. They collect neutrinos from the sun in an enormous disused underground mine that has been filled with old cleaning fluid (i.e., carbon tetrachloride). We would know a lot about the inside of the sun if we knew how many solar neutrinos arrive on the earth. So these are captured in the cleaning fluid. A few neutrinos will form a new radioactive nucleus (the number that do this can be counted). Although, in this study, the extent of neutrino manipulation is much less than electron manipulation in the PEGGY II experiment, we are nevertheless plainly using neutrinos to investigate something else. Yet not many years ago, neutrinos were about as hypothetical as an entity could get. After 1946 it was realized that when mesons disintegrate giving off, among other things, highly energized electrons, one needed an extra non-ionizing particle to conserve momentum and energy. At that time this postulated "neutrino" was thoroughly hypothetical, but now it is routinely used to examine other things.

CHANGING TIMES

Although realisms and antirealisms are part of the philosophy of science well back into Greek prehistory, our present versions mostly descend from debates at the end of the nineteenth century about atomism. Antirealism about atoms was partly a matter of physics; the energeticists thought energy was at the bottom of everything, not tiny bits of matter. It also was connected with the positivism of Comte, Mach, K. Pearson, and even J.S. Mill. Mill's young associate Alexander Bain states the point in a characteristic way, apt for 1870:

> Some hypotheses consist of assumptions as to the minute structure and operation of bodies. From the nature of the case these assumptions can never be proved by direct means. Their merit is their suitability to express phenomena. They are Representative Fictions.[18]

"All assertions as to the ultimate structure of the particles of matter," continues Bain, "are and ever must be hypothetical...The kinetic theory of heat serves an important intellectual function." But we cannot hold it to be a true description of the world. It is a representative fiction.

Bain was surely right a century ago, when assumptions about the minute structure of matter could not be proved. The only proof could be indirect, namely, that hypotheses seemed to provide some explanation and helped make good predictions. Such inferences, however, need never produce conviction in the philosopher inclined to instrumentalism or some other brand of idealism.

Indeed, the situation is quite similar to seventeenth-century epistemology. At that time, knowledge was thought of as correct representation. But then one could never get outside the representations to be sure that they corresponded to the world. Every test of a representation is just another representation. "Nothing is so much like an idea as an idea," said Bishop Berkeley. To attempt to argue to scientific realism at the level of theory, testing, explanation, predictive success, convergence of theories, and so forth is to be locked into a world of representations. No wonder that scientific antirealism is so permanently in the race. It is a variant on "the spectator theory of knowledge."

Scientists, as opposed to philosophers, did, in general, become realists about atoms by 1910. Despite the changing climate, some antirealist variety of instrumentalism or fictionalism remained a strong philosophical alternative in 1910 and in 1930. That is what the history of philosophy teaches us. The lesson is: think about practice, not theory. Antirealism about atoms was very sensible when Bain wrote a century ago. Antirealism about *any* submicroscopic entities was a sound doctrine in those days. Things are different now. The "direct" proof of electrons and the like is our ability to manipulate them using well-understood low-level causal properties. Of course, I do not claim that reality is constituted by human manipulability. Millikan's ability to determine the charge of the electron did something of great importance for the idea of electrons, more, I think, than the Lorentz theory of the electron. Determining the charge of something makes one believe in it far more than postulating it to explain something else. Millikan got the charge on the electron; but better still, Uhlenbeck and Goudsmit in 1925 assigned angular momentum to electrons, brilliantly solving a lot of problems. Electrons have spin, ever after. The clincher is when we can put a spin on the electrons, and thereby get them to scatter in slightly different proportions.

Surely, there are innumerable entities and processes that humans will never know about. Perhaps there are many in principle we can never know about, since reality is bigger than us. The best kinds of evidence for the reality of a postulated or inferred entity is that we can begin to measure it or otherwise understand its causal powers. The best evidence, in turn, that we have this kind of understanding is that we can set out, from scratch, to build machines that will work fairly reliably, taking advantage of this or that causal nexus. Hence, engineering, not theorizing, is the best proof of scientific realism about entities. My attack on scientific antirealism is analogous to Marx's onslaught on the idealism of his day. Both say that the point is not to understand the world but to change it. Perhaps there are some entities which in theory we can know about only through theory (black holes). Then our evidence is like that furnished by Lorentz. Perhaps there are entities which we shall only measure and never use. The experimental argument for realism does not say that only experimenter's objects exist.

I must now confess a certain skepticism, about, say, black holes. I suspect there might be another representation of the universe, equally consistent with phenomena, in which black holes are precluded. I inherit from Leibniz a certain distaste for occult powers. Recall how he inveighed against Newtonian gravity as occult. It took two centuries to show he was right. Newton's ether was also excellently occult – it taught us lots: Maxwell did his electromagnetic waves in ether, H. Hertz confirmed the ether by demonstrating the existence of radio waves. Albert A. Michelson figured out a way to interact with the ether. He thought his experiment confirmed G.G. Stoke's ether drag theory, but, in the end, it was one of many things that made ether give up the ghost. A skeptic such as myself has a slender induction: long-lived theoretical entities which do not end up being manipulated commonly turn out to have been wonderful mistakes.

NOTES

1. C.W.F. Everitt suggests that the first time the word "effect" is used this way in English is in connection with the Peltier effect, in James Clerk Maxwell's 1873 *Electricity and Magnetism*, par. 249, p. 301. My interest in experiment was kindled by conversation with Everitt some years ago, and I have learned much in working with him on our joint (unpublished) paper, "Theory or Experiment, Which Comes First?"

2. Ian Hacking, "Spekulation, Berechnung und die Erschaffnung der Phänomenen," in *Versuchungen: Aufsätze zur Philosophie, Paul Feyerabends*, no. 2, ed. P. Duerr (Frankfort, 1981), pp. 126–158.

3. Nancy Cartwright makes a similar distinction in her book, *How the Laws of Physics Lie* (Oxford: Oxford University Press, 1983). She approaches realism from the top, distinguishing theoretical laws (which do not state the

facts) from phenomenological laws (which do). She believes in some "theoretical" entities and rejects much theory on the basis of a subtle analysis of modeling in physics. I proceed in the opposite direction, from experimental practice. Both approaches share an interest in real life physics as opposed to philosophical fantasy science. My own approach owes an enormous amount to Cartwright's parallel developments, which have often preceded my own. My use of the two kinds of realism is a case in point.

4. Hilary Putnam, "How Not to Talk About Meaning," "The Meaning of 'Meaning,'" and other papers in *Mind, Language and Reality,* Philosophical Papers, vol. 2 (Cambridge: Cambridge University Press, 1975).

5. These terms occur in, e.g., Hilary Putnam, *Meaning and the Moral Sciences* (London: Routledge & Kegan Paul, 1978), pp. 123–130.

6. Francis Bacon, *The Great Instauration, in The Philosophical Works of Francis Bacon,* trans. Ellis and Spedding, ed. J.M. Robertson (London, 1905), p. 252.

7. Ian Hacking, "Do We See Through a Microscope?" *Pacific Philosophical Quarterly* 62 (1981), pp. 305–322.

8. C.Y. Prescott, "Prospects for Polarized Electrons at High Energies," SLAC-PUB-2630, Stanford Linear Accelerator, October 1980, p. 5.

9. "The Concept of Observation in Science and Philosophy," *Philosophy of Science* 49 (1982): 485–526. See also K. S. Shrader-Frechette, "Quark Quantum Numbers and the Problem of Microphysical Observation," *Synthèse* 50 (1982): 125–146, and ensuing discussion in that issue of the journal.

10. I thank Melissa Franklin, of the Stanford Linear Accelerator, for introducing me to PEGGY II and telling me how it works. She also arranged discussion with members of the PEGGY II group, some of whom are mentioned below. The report of experiment E-122 described here is "Parity Non-conservation in Inelastic Electron Scattering," C.Y. Prescott et al., in *Physics Letters.* I have relied heavily on the in-house journal, the *SLAC Beam Line,* report no. 8, October 1978, "Parity Violation in Polarized Electron Scattering." This was prepared by the in-house science writer Bill Kirk.

11. The odd-sounding bosons are named after the Indian physicist S.N. Bose (1894–1974), also remembered in the name "Bose-Einstein statistics" (which bosons satisfy).

12. But excluding Leibniz, who "knew" there had to be some real, natural difference between right- and left-handedness.

13. Iceland spar is an elegant example of how experimental phenomena persist even while theories about them undergo revolutions. Mariners brought calcite from Iceland to Scandinavia. Erasmus Bartholinus experimented with it and wrote it up in 1609. When you look through these beautiful crystals you see double, thanks to the so-called ordinary and extraordinary rays. Calcite is a natural polarizer. It was our entry to polarized light which for three hundred years was the chief route to improved theoretical and experimental understanding of light and then electromagnetism. The use of calcite in PEGGY II is a happy reminder of a great tradition.

14. It also turns GaAs, a 3/4 to 1/4 left-hand/right-hand polarizer, into a 50–50 polarizer.

15. I owe these examples to conversation with Roger Miller of SLAC.
16. The concept of a "convincing experiment" is fundamental. Peter Gallison has done important work on this idea, studying European and American experiments on weak neutral currents conducted during the 1970s.
17. I owe this information to Charles Sinclair of SLAC.
18. Alexander Bain, *Logic, Deductive and Inductive* (London and New York, 1870), p. 362.

7 The Paradox of Scientific Subjectivity*

Evelyn Fox Keller

O Telescope, instrument of much knowledge, more precious than any sceptre! Is not he who holds thee in his hand made king and lord of the works of God? Truly,

"All that is overhead, the mighty orbs
With all their motions, thou dost subjugate
To man's intelligence."
(Kepler, *A Conversation with Galileo's Sidereal Messenger*, 86.)

CLASSICAL PERSPECTIVE AND MODERN SCIENCE

It has often been said that the revolution in consciousness that gave rise to modern science was felt first, not in sixteenth- or seventeenth-century experimental philosophy, but in fifteenth-century representational art. Filippo Brunelleschi, the inventor of perspective drawing, was also "the first professional engineer," and much has been made of this fact – perhaps especially by Giorgio de Santillana (1959). Rolling Brunelleschi's innovations into one general procedure, de Santillana characterizes that procedure as an "early counterpart or rather first rehearsal" of "the 'social breakthrough' that the new science of Galileo effected through the telescope" (44–45). The history that entwines perspective, engineering, and the emerging science over the two centuries between Brunelleschi and Galileo is surely more complex than de Santillana intuited, yet the novelty of Brunelleschi's perspective does seem clearly to have been linked with new possibilities of both planning and intervention (see, e.g., Kuhn, 1990) and with new notions of descriptive knowledge. From its origins, classical perspective provided at least a

* Essay in A. Megill (ed.), *Rethinking Objectivity* (Durham: Duke University Press, 1994), pp. 313–331.

metaphor for faithful knowledge of the natural world acquired by rule-bound observation and documentation.

But if classical perspective is to be taken as a metaphor for the origins of modern science, it is a metaphor that cuts two ways at once. As such, it holds within it an image or early analog of the paradox of subjectivity bequeathed to us from the earliest days of modern science. On the one hand, the practice of perspective explicitly inscribes the point from which an observation is made, and accordingly makes evident the need to recognize the kinds of difference a change in viewpoint makes. On the other hand, by providing an image so lifelike "that you thought you saw the proper truth" (Antonio Manetti, quoted in de Santillana, 35), it invites the claim that faithful obedience to specified procedures that are in accordance with Nature's geometrical laws will result in an image for which Nature, not the individual observer, is responsible – as Leon Battista Alberti put it, an image "depicted by Nature." Thus, out of its very contingency, perspective extracts a new kind of veridicality: it locates in the vantage point of a particular somewhere at least the tacit promise of a view from nowhere.

In classical perspective, the observer is simultaneously named by his location and made anonymous by his adherence to specified rules (any individual subject will do). Bodily attached to his viewpoint as a surveyor is to his theodolite, he is at the same time released from himself and invested instead in his technique – that very technique which authorizes him as a vehicle of secure knowledge. What is more, he transmits this privilege to the viewer of the resulting image. A depicted scene reaches out to its viewer, inviting him (or her) to occupy the original vantage point, to identify his or her experience with that of the draftsman/observer. This invitation is didactically explicit where there is one central vanishing point, signalled by the converging orthogonal lines of a square pavement or a receding arcade. Such a vanishing point mirrors the viewer's position, transfixing him or her on a visual beam, as if image and viewer had made compelling eye-contact. Yet what the vanishing point embodies is utter disappearance; it is where objects in the pictorial space vanish on the horizon. The picture does not offer the viewer a place within the depicted scene, but rather marks his own absence from it – an absence that is the very condition of his privilege as observer. Perspectival realism, in other words, both depends on and guarantees the presence of an actual physical subject with a particular "point of view" (Rotman, 19) – a presence always outside the painting, yet connoted by an isomorphic absence *in* the painting.

Taken as a metaphor for the identity and location of the subject/ author of scientific representations, one could almost say that the contradiction embedded in this dual semiotic of classical perspective was *the* problem that modern science needed to solve: In order to generate a representation of the world in its entirety – as Husserl put it, "more comprehensive, more reliable, and in every respect more perfect than that offered by the information received by experience" (cited in Bryson, 4), one which could include the observer in the representation of the world of which he is part, which could bring him into the canvas from outside, the task of modern science was to eradicate his presence as an external observer, and fill the lacuna created by his internal absence.

The history of this task over the centuries that followed bears close resemblance to what Lorraine Daston calls the "history of objectivity." It is a history by no means linear, but rather multilayered and entangled, accompanied by complex resistances and anxieties – and, perhaps especially, by radical changes in the very meaning of the term "objective." As Daston points out, the term "objective" had a very different – effectively opposite – meaning in the seventeenth century from what it has today; it referred neither to a state of mind nor to a mode of perception, but to the objects of thought and perception, to what Hobbes called the "effects of nature." Only in the nineteenth century did the term "objective" acquire the current meaning of aperspectival – a "view from nowhere," knowledge without a knower. Indeed, far from supposing an absence of a "point of view," the possibility of veridical representation in seventeenth-century science, more like that of classical perspective than nineteenth-century conceptions, depended critically on the very particular point of view of the observer – on his special skills and instruments, as well as on the credibility of his testimony conferred by social status.[1] But perhaps a conjunction between veridicality and a view from a particular somewhere could more easily be maintained in the seventeenth century than in the nineteenth, underwritten as it was in the earlier period by the tacit understanding that there was one vantage point, namely God's view, that was simultaneously absolutely special and absolutely knowing.

Furthermore, not all perspectival privilege in seventeenth-century natural philosophy depended on being in the right place at the right time, on having the right instruments and the right character. Descartes, e.g., assumed another kind of epistemic privilege, conferred not by ocular evidence, but by the discipline of a disembodied rationality. Yet what appears in retrospect as so manifestly a contradiction – between the vantage point of disembodied thought and that of an embodied (or empirical) observer – then seemed more like different routes· to truths that began and ended with God. Clearly, a proper history of scientific representation (or of objectivity) needs to trace these disparate strands along with their changing interrelations. It needs also, as Daston and others remind us, to embed these strands in the even more complex history of changing practices in the conduct and organization of the natural sciences that were themselves intimately linked to larger cultural transformations. Still, it is nonetheless possible to trace a distinctly linear arc in this highly nonlinear history, a story line in the history of scientific subjectivity that is rooted in the very logic of scientific representation. Such a story line closely parallels the semiotic history traced by Brian Rotman (1987); it too is a history of erasure, of the progressive disembodiment and dislocation of the scientific observer and author that ultimately became sufficiently complete to permit the comprehensive and apparently subjectless representation of the world that emerges today, in the late twentieth century. Indeed, I suggest that it is only in the late twentieth century that this story comes to an end, and it is precisely its ending that constitutes the major focus of this paper. First, however, I want briefly to indicate how some of the familiar landmarks of the history of science might be located in such a linear narrative.

STORY LINE

Descartes provides a canonical starting point. Conventionally at least, the *cogito* promises a virtually instant solution to the problem of representation. Liberated from its bodily locus, the knowing subject needs no physical vantage point in order to be near to God, to stand over and above the object of H/his knowledge. From such a transcendent vantage point, simultaneously everywhere and nowhere, it seemed possible to see the entire universe, including not only the moons of Jupiter that Galileo had espied through his telescope, but also

the digesting of food, the beating of the heart and arteries, nourishment and growth, respiration, waking, and sleeping, the reception of light, sounds, odors, tastes, warmth, ... the impression of the corresponding ideas upon a common sensorium and on the imagination; the retention or imprint of these ideas in the Memory; the internal movements of the Appetites and Passions; and finally, the external motions of all the members of the body. (Descartes, *Oeuvres* [1953], 873).

Descartes had posited a longer and more powerful telescope, through which one ought be able, in principle, to see even Galileo's eye. By severing mental from ocular vision, he effectively split the observing subject. The sensible and experiencing self (he who touches, sees, walks, feels, and even remembers), being material and hence existing in space, could now at least in principle be included on the canvas; only the knowing, thinking self, by definition lacking extension, could not appear. The cogito is not only the ghost in the machine, but a holy ghost – made invisible by its absence from space but omnipotent by its union with God; it eludes description, self-consciousness, and, of course, constraint. The mind's I/eye can itself be neither known, thought, nor governed. Like the representation of the world it points to, posited in a state of absolute being, it just *is*, ontologically both anterior to the inquiring gaze and beyond or outside its scope.

For men like Robert Boyle, however, the enduring location and physicality of the sensible and experiencing self implied a critical problem of residual personal – and hence subjective – authority. Even if some truths could be fixed by reason, by mathematical or logical demonstration, particular propositions about the physical world could not, and the operation of reason in natural philosophy depended on these. Propositions about the physical world, being contingent rather than necessary, had to be inferred from observation, from the empirical facts of the matter, i.e., from direct scrutiny of the actual workings of the "great pregnant automaton" that Boyle called Nature. In other words, Boyle concurred with Descartes in his picture of the world as machine, but the actual behavior of the machine, "the nature of the spring, that gets all [the parts] amoving" (4, 358) – these, he argued, were ascertainable only from observational experience. And if the facts elicited by such observation are to be clearly distinguishable from mere opinion, if they are to be indisputable, they cannot be permitted to depend on the private viewing of any particular individual who happens to be doing the looking.

The contingency of physical events, Boyle thought, required grounding scientific knowledge not in the privileged vantage point of pure thought but in the more humble stance of observation; universality required that the locus of that activity be dispersed. Shapin (1984) and Shapin and Schaffer (1987) have argued that the essence of Robert Boyle's experimental program lay in his management of just this shift: in particular, in his definition of the rules by which the personal authority of a knowing, seeing "I" could be transformed into a depersonalized and "public" authority.

Shapin and Schaffer contend that Boyle engineered the solidity of the facts of the "great system of things corporeal" by locating their genesis in devices (e.g., the air pump) that were themselves mechanical, and, in parallel, by grounding the perception of these facts in what Shapin calls "a multiplication of the witnessing experience" (1984, 488). Such a multiplication was to be accomplished trivially by moving the experimental demonstration to the public rooms of the Royal Society, and more ambitiously by creating the possibility of an indefinite number of "virtual witnesses" through the development of a particular literary technique – the technique of the scientific report.[2] The form of such reports was designed, Boyle pointed out, so that "the person I addressed them to might, without mistake, and with as little trouble as possible, be able to repeat such unusual experiments..." (quoted in Shapin, 1984, 490).

In actuality, even with the very simple technologies employed by Robert Boyle, such replication proved to be exceedingly difficult. But this merely reinforced the need to "set down divers things with [such] minute circumstances" that would create the sense that the reader had been there" (490). As Shapin puts it, "the technology of virtual witnessing involves the production in a reader's mind of such an image of an experimental scene as obviates the necessity for either its direct witness or its replication" (491). Boyle was a self-conscious master of rhetoric. To persuade the reader of the credibility of the author, and at the same time to effect the conviction of virtual presence on the part of the reader, he advocated a plain, unadorned mode of presentation, "a naked way of writing" that avoided both rhetorical flourish and personal reference, that included the reporting of an occasional failed experiment, and that made use of carefully designed naturalistic visual images which, when combined with the copious details of the verbal text, would create the conviction of verisimilitude. These narratives were to be "standing records" that would enable readers to participate vicariously in the experience of discovery, "*to have as distinct an idea of it*, as may suffice to ground their reflexions and speculations upon" without needing themselves to reiterate the experiment (493). In short, they were to enable the reader to see with the author/observer's eye without having actually to stand where the author/observer had stood.

In Boyle's combined literary and mechanical technology, the residual subjectivity inhering in the act of observation – along with the activities of constructing the experiment, of defining the vantage point which permits the observation to make sense, and, of course, of the process of making sense – could effectively be displaced by an imagined (or projected) subjectivity that was not born in actual experience but precisely in the act of reading. It was a subjectivity, like that of the reader of the eighteenth-century novel, already pre-scripted on the written page (see Gallagher, 1993). But a crucial difference separated the reading of a scientific report from the reading of a novel: the ultimate target of projection (and hence, also of introjection) for the reader of the scientific report was not the fictionalized emotional, intentional, alter ego found in the novel, but a machine – a device for receiving and recording experience that was itself incapable of either sentiment or purpose. (To quote Shapin and Shaffer, 'it is not I who says this, it is the machine' [1987:77]). The virtual witness resembled a truly perfect observer, an alter ego that modelled itself after a mechanical detector.

But this description surely imposes too much hindsight on Boyle's mechanical and literary technology, both of which were, in fact, still quite rudimentary. Most important, the air pump, like Galileo's telescope, was not yet an easily reproducible device, and indeed the very realism and particularity of detail of Boyle's report was in part designed to compensate for the lack of reproducibility of the experiment. Thus the credibility of Boyle's scientific report still depended crucially on the status, skill, and integrity of the original observer. In this sense, Boyle was himself a more ambiguous figure than Shapin and Schaffer suggest, for the persuasiveness of his scientific report was still very much dependent on the particular position (both social and material) of the original observer, still dependent, in short, on a first person pronoun. A more complete decentering of the original observer awaited at least two crucial additional developments: first, the standardization of his material technology enabled by the mass production of machines in the late eighteenth century, and, second, the quantification of the literary technology of scientific reporting that came with the rise of statistics in the nineteenth century (see Porter, 1986; also, Megill, 1994).[3]

The history I am so blithely skimming over is of course so complex and thickly layered that it perforce requires many different kinds of readings. Ted Porter, e.g., has suggested that the history of standardization – simultaneously of content and viewer – be read as a *political* history in which modern science provides the locus for the development of a "technology of distrust" appropriate not only to a dispersed and impersonal scientific practice, but also to the needs of the "modern suspicious democratic political order" (1990, or Megill, 1994). For my purposes, however, the relevant point is that once air pumps and telescopes became freely available as standardized instruments, neither the author of the text nor the original observer *needed* any longer to be identified. By the second half of the nineteenth century, the first person pronoun narrator of the scientific text could be effectively replaced by abstract 'scientist' – then a newly coined word – who could speak for everyman but was no-man. Noman in the double sense of being not any particular man and also as a site for the being of not-man within each and every particular observer. Between the seventeenth and nineteenth centuries, a hollow place had been carved out in the mind of every actual or virtual witness into which a machine could vicariously be placed – the lacuna of classical perspective now filled in on the canvas, but emerging, instead, in the mind of the viewer.

The result of this semiotic progression was an enduring and final erasure: with the disappearance of all consciousness of the representation qua representation, both the presence of the knowing, doing subject behind representation and its corresponding absence *in* representation have been thoroughly occluded. The scientific subject arising in place of the embodied crafter, interpreter, and reporter of experiments is a classic instance of what Brian Rotman calls a "metasubject": invisible, autonomous, virtual – floating above the situated, dependent, and very real work that scientists actually engage in the complex production of the scientific corpus that can itself neither be seen on the canvas, nor be noted for its absence. But if the original subject cannot be signified in the new representational scheme, the metasubject can: Its sign is the conspicuously replicable and representable mechanical detector – as it were, the machine in the ghost: a surrogate that stands in not for the presence of the knowing, doing subject, but for its absence.

In *Signifying Nothing: The Semiotics of Zero*, Rotman recounts the progressive loss of the anteriority of things to signs throughout European culture, a story of the growing distance of metasubject from subject in art, in mathematics, and in finance. Each of his examples charts the same dilemma (or paradox) of subjectivity, namely, that while the viewing, acting, and doing subject – he who paints, counts, or trades – remains as necessary as always for the actual production of art, mathematics, or financial transactions, the new logic of representation allows only the metasubject to be signified, denying the subject behind the metasubject its signification even as an absence. The same dynamic, I suggest, can also be seen in science. Indeed, I am suggesting that it is in the natural sciences that we see the dilemma of subjectivity in its most critical form. The man-made (or, more accurately, men-made) nature of scientific knowledge cannot be represented in the texts of science, for precisely what has been constructed is the illusion that this knowledge is *not* man-made – not crafted, articulated or constructed, but discovered (Shapin, 510), in a word, simply true. It is this erasure, this representational logic, that underlies – indeed, that guarantees – the representation of a single, unified spatio-temporal reality of which the human subject could ultimately itself become part, represented in the only coinage available: as a machine among machines.

One face of the history of modern science is thus a history of semiotic repositioning that seems finally, in the twentieth century, to enable us to realize Descartes' early dream of depicting, in humans, the processes of motion and emotion, of growth and reproduction, of waking and sleeping, even of perception and memory – of presenting the viewer of the natural world as "an object or spectacle before his own vision." Not surprisingly, Descartes' actual portrait of the human machine has been transformed over the intervening centuries beyond recognition; perhaps the only surprise is how stable, through all this transformation, his legacy of scientific subjectivity has been. Despite its radical transfiguration, the human subject that modern biology promises to reveal turns out after all to be in insignificant ways, still, a Cartesian object – merely knowable, never itself capable of knowing. Even where molecular determinants can be identified for memory and vision, for sleep and dreams, for depression and perhaps even for love, the agentic subjects responsible for this representation – still tacitly coded as the thinking searcher for the mechanical basis of thought – remain apart, off a canvas that has become coextensive with the world, outside a domain that denies it has an outside. After centuries of erasure, however invisible and unthinkable, the knowing subject remains resistantly anterior. The question is where, in an age of atheism, the anteriority that had earlier been ceded to God but can no longer be, is now to be located? It is to this question, surely one of the most critical questions that faces us in the late twentieth century, that I now turn.

THE SUBJECT OF/IN TWENTIETH-CENTURY SCIENCE

In 1929, J.D. Bernal, assessing our overall progress in this long struggle (an assessment, as it happens, resurrected and endorsed by Freeman Dyson in 1972), ventured an answer. Bernal called his treatise "The

World, the Flesh and the Devil: An Enquiry into the Future of the Three Enemies of the Rational Soul." As Bernal saw it, the greatest impediment that remains in the struggle of the rational soul against the inorganic forces of the world and the organic structure of our flesh is the endurance of the irrational forces of desire and fear, i.e., the Devil. He wrote,

> [W]e can abandon the world and subdue the flesh only if we first expel the devil, and the devil, for all that he has lost individuality, is still as powerful as ever...: he is inside ourselves, we cannot see him. Our capacities, our desires, our inner confusions are almost impossible to understand or cope with in the present... (48)

The goal awaiting us is to rout desire from its hiding place and bring it into line with our objective aims, "using and rendering innocuous the power of the id and leading to a life where a full adult sexuality would be balanced with objective activity" (57). We need to bring "feeling, or at any rate, feeling-tones... under conscious control... induced to favour the performance of a particular kind of operation" (67). As if aware of the absence of a subject in these pronouncements, Bernal, a Marxist, sought to relieve at least most of humanity from the dilemma of a still unrepresented subjectivity. He speculated about a future "splitting of the human race – the one section developing a fully-balanced humanity, and the other groping unsteadily beyond it" (60). The latter, which Bernal himself described as "an aristocracy of scientific intelligence" (73), would have

> a dual function: to keep the world going as an efficient food and comfort machine, and to worry out the secrets of nature for themselves... A happy prosperous humanity enjoying their bodies, exercising the arts, patronizing the religions, may well be content to leave the machine, by which their desires are satisfied, in other and more efficient hands. Psychological and physiological discoveries will give the ruling powers the means of directing the masses in harmless occupations and of maintaining a perfect docility under the appearance of perfect freedom. But this cannot happen unless the ruling powers are the scientists themselves. (74)

Perhaps, Bernal goes on to speculate (in a fashion close to Dyson's own heart), the colonization of space will offer a particularly convenient "solution." In that case,

Mankind – the old mankind – would be left in undisputed possession of the earth, to be regarded by the inhabitants of the celestial
spheres with a curious reverence. The world might, in fact, be transformed into a human zoo, a zoo so intelligently managed that its inhabitants are not aware that they are there merely for the purposes
of observation and management. (79–80)

Bernal's diplacement of the scientific observer/manager to the celestial
sphere is of course no solution at all to the problem of representing
that observer's residual (albeit now fully rationalized) subjectivity, but
it is at least evocative. The very phrase, "celestial spheres," antiquated
as it is, recalls an idea of proximity to God that Bernal would surely
not have wanted to claim, but could nonetheless rely upon. Perhaps
more to his point, it also evokes an ideal of ethereality and disembodiment that, for a scientist committed to monism, is curiously reminiscent of Cartesian dualism.

Dyson, in his tribute to Bernal, remarks that "The decisive change
that has enabled us to see farther in 1972 than we could in 1929 is the advent of molecular biology." And indeed, molecular biology has provided
us with the tools of identifying molecular determinants of perception,
emotion, and perhaps even of thought. Today, in the aperspectival genius
of molecular biology, we seem to have come closer yet to Descartes'
dream than even Bernal imagined. Only a few short decades after James
Watson and Francis Crick's momentous discovery of "the secret of
life," those whose business it is to "worry out the secrets of nature" can finally reveal, as Watson has put it, "what it really means to be human."

In the discourse of molecular biology, there is no talk of a scientific
aristocracy explicitly set apart from a malleable but contented humanity. Here, the knowing subject asserts itself not by a residual identification with God, but by projection onto the canvas, through
identification with those molecules, at once master architect and master builder, that both encode the blueprint for life and direct its production. (It is, Crick says, precisely through an identification with such
molecules as these that we can restore our sense of unity with nature.)
Yet side by side with this willful identification with DNA, it is still possible to detect an anterior knowing subject – not removed to the celestial spheres, but right here on earth, behind the DNA, behind even (or
perhaps especially) the authorial "we." One way in which this anteriority is exhibited is by an unwitting yet routine syntactical distancing of
the authorial and subject pronoun from the represented object, be it
"nature," "people," or "society."

To see this, consider the use of pronouns in some of Crick's more informal asides. Take, e.g., his remarks about "Nature's own analogue computer" in his 1966 book, *Of Molecules and Men:* "[T]he system itself... works so fantastically fast. Also, she knows the rules more precisely than we do. But we still hope, if not to beat her at her game, at least to understand her..." (1966, 12). My point here is to note that the subject "we" in this sentence is not itself part of, but rather on a par with, and hopefully, even better than Nature's own computer.

A similar distancing, only now from "people" or "society," can be detected in a series of comments Crick offered on the subject of "Eugenics and Genetics" in 1963. I quote three of these:

> (1) "Do people have the right to have children at all?... I think that if we can get across to people the idea that their children are not entirely their own business... it would be an enormous step forward." (275)

Here we have three subjects: The "I" who speculates, the "we" who know, and finally, those to whom "we" must communicate our knowledge, namely "people."[4]

In a related remark, a similar syntactical distancing is expressed once again, but this time – perhaps because of the explicit message conveyed – further enhanced by invoking the even more abstract pronoun "one":

> (2) "The question... as to whether there is a drive for women to have children and whether [my proposal of licencing] would lead to disturbances is very relevant. I would add, however, that there are techniques by which one can inconspicuously apply social pressure and thus reduce such disturbances." (284)

And for a final and truly embarrassing example that reads like a throwback to an earlier part of this century,

> (3) "... we are likely to achieve a considerable improvement [in the human stock through genetics]... that is by simply taking the people with the qualities we like and letting them have more children." (295)

CONSTRUCTING A NEW SUBJECT

I began this essay with a discussion of the basic (and general) paradox of authorial representation that once found so graphic an image in the vanishing point of classical perspective; I then attempted to trace the construction of a progressively more abstract and more dispersed scientific subject as attempts to alleviate the particular difficulties the paradox of authorship presented for the natural sciences – until the scientist becomes (in the nineteenth century) a mere cipher, a depersonalized reporter of the recordings of a mechanical detector. Finally, I have tried to show how, well into the twentieth century, we can still see the endurance of a semblance of the original paradox, although these days perhaps evident only in the informal discourse of actual scientists, in the locutory "we" or "one" in such off-the-cuff, unwitting remarks as these of Crick's. Though the paradox is still present, its days are palpably numbered. As molecular neuro-biology extends its frontier ever deeper into the brain, promising to converge even on the problem of consciousness (Crick's own favorite), the scientist himself comes to be drawn ever deeper into the machine he has created. At such a point, the circle of scientific knowledge threatens to close in on itself, and the anchoring that has till now been provided by a residually anterior knowing subject threatens to disappear altogether. Clearly, a new and radically different kind of subjectivity is called for. Indeed, we can already discern within the scientific project itself the outlines of the form this new, postmodern, sensibility has, even as we speak, already begun to take.

In the late twentieth century, as Francis Crick boldly searches for the structure of consciousness in molecular arrangements, another kind of science, and with it, another discourse, has developed to meet him at least halfway. Designed to simulate the arts of mathematical reasoning, computer technology is, by definition, a science of artificial intelligence. As such, it works simultaneously to bridge the gaps between Descartes and Boyle, between the mathematical and the experimental, and between the mental and the physical.[5] And out of this quintessentially hybrid science of computers has emerged a discourse that has as little need for the hypothetical "I" or "we" of molecular biology as the discourse of classical physics earlier had for God. In its place it offers the promise of a mind that can indeed exist without the body, as Descartes once believed, but a mind that makes a mockery of all of Descartes' hopes, a mind appropriated by its own alter ego. This mind neither lacks extension nor meets God; it is spatial and banally corporeal: a bank of self-regulating knobs and switches that needs no outside agent to maintain it. Danny Hillis, designer of *The Connection Machine*, anticipates the development of computers that would yield "not so much an artificial intelligence, but rather a human intelligence sustained within an artificial mind" (1988, 189). Hillis acknowledges:

Of course, I understand that this is just a dream, and I will admit that I am propelled more by hope than by the probability of success. But if this artificial mind can sustain itself and grow of its own accord, then for the first time human thought will live free of bones and flesh, giving this child of mind an earthly immortality denied to us. (1988, 189)

Descartes too imagined that human thought could live "free of flesh and bones" – but when he wrote, "I am truly distinct from my body, and... I can exist without it" (Sixth Meditation), he never dreamt that this marvelously autonomous "I" could itself be constituted of crass and earthly matter.[6]

Only one twist is still missing from this apparently post-Cartesian vision: Hillis's artificial mind may be able to sustain itself, but for it to live and grow, it must be able to reproduce itself. As it happens, almost forty years ago, just one year before the announcement of Watson and Crick, John von Neumann shows us how this problem too, could in principle be solved.[7] Because of von Neumann, the technical culture of human robotics is emboldened to produce an auxiliary dream to Hillis's. For one of its more provocative expressions, I quote from the introduction to Hans Moravec's recent work, *Mind Children:*

Unleashed from the plodding pace of biological evolution, the children of our minds will be free to grow to confront immense and fundamental challenges in the larger universe. We humans will benefit for a time from their labors, but sooner or later, like natural children, they will seek their own fortunes while we, their aged parents, silently fade away. Very little need be lost in this passing of the torch – it will be in our artificial offspring's power, and to their benefit, to remember almost everything about us, even, perhaps, the detailed workings of individual human minds. (1988, 1)

In Moravec's vision, the human subjects who have begotten these machines have become entirely superfluous. Like the God who was once but is no longer thought necessary to account for our existence, our authorship too has become redundant – no longer required even for the intellectual and technical prowess that made such machines possible in the first place. Moravec and Hillis are the new magicians, presenting us in this ultimate vanishing act with a prosthetic subject that is fully autonomous. They paint a portrait, and a world, from which all residual signs of anteriority are erased – a representation that is complete and unto itself; that needs neither the eye of God, the eye of the artist, nor the eye of an observer. It begets itself, it makes itself, it sees itself. If this new kind of subjectivity – unmoored and floating free – resembles that of poststructuralist theory, might even be taken as providing ironic grounding for such theory – it should be noted that the subject Hillis and Moravec describe is less a departure from than the culmination of over three hundred years of representational logic that has anchored the entire tradition of modernity. It is the end point of a semiotic system in which, as Rotman puts it,

> the signs of the system become creative and autonomous. The things that are ultimately "real," that is numbers, visual scenes, and goods, [and now we add machines], are precisely what the system allows to be represented as such. The system becomes both the source of reality, it articulates what is real, and provides the means of "describing" this reality as if it were some domain external and prior to itself. . . (28)

CONCLUSION

Still, it may be possible, even at this late date, to articulate some resistance against this sensibility on behalf of the flesh and blood humans who must live out its consequences and even on behalf of those who have nurtured it and labored to embody it – i.e., to remind ourselves of something like the facts of the matter. If so, they would have to be these: whatever lies in store for the future of robotics and artificial intelligence, the imagings of Moravec and Hillis point at least as much to the past as to the future. Not only is it in fact not possible to construct such machines now, but the expectation that it ever will be possible in the future needs to be recognized as a particular, and very potent, fantasy – a fantasy that, like the machines that enliven it, grows out of, and speaks for, a particular and very potent historical development. The replacement of God's I/eye by a thinking and knowing machine may indeed mark a new way of speaking and thinking – perhaps neither more nor less veridical than the old – but it is one that emerges logically and with a certain inexorability out of half a millenium of history, in which human subjects have increasingly sought particular kinds of representation in the interests of constructing particular kinds of products. We do, in fact, now know how to build parallel processors and computer viruses, just as we know how to construct nucleotide polymers, and soon we will undoubtedly be able to synthesize real viruses, and more. But these products of computer science and molecular biology do not arise autonomously. Rather, they emerge from the arts and artifacts of human industry, expressing human needs and desires. They are effected by particular material and social practices that are themselves both enabled by and enabling of a tradition of representation that has sought for more than three hundred years the erasure of all evidence of the human agency behind both the practices and the representations, and even behind all allure of mechanical surrogacy that has provided such essential fuel for this search. In this sense, the vision that Hillis and Moravec offer us today clearly speaks to an ultramodern rather than postmodern sensibility. To insist that humanly meaningful texts (scientific or otherwise) cannot be authored by God, is to resist erasure of the dependence of meaning on cultural and material history. It is to resist in the name of another sensibility, on behalf of the human beings who act, think, and speak in the terms made available to them by the culture, and who, with their actions, thinking, and speaking, have not only created these terms, but also act out the meanings they inscribe.

NOTES

I am indebted to readers of earlier drafts of this article for their many helpful suggestions and criticisms, especially to Jehane Kuhn, Sam Schweber, Lorraine Daston, Peter Dear, Ted Porter, and Allan Megill.

1. Galileo's account of the moons of Jupiter provides a case in point. As de Santillana implies, Galileo is an exemplar of the classical attitude in natural science. "[B]y the help of a telescope devised by me, through God's grace first enlightening my mind" (Kepler, 9) he acquired a unique vantage point, enabling him to see what had till then been visible only to God. But as Peter Dear has noted (1992), when Kepler received word of Galileo's sightings, he did not need to peer through Galileo's telescope himself in order to embrace his report; he had enough confidence in Galileo's optical instruments, as well as in his character, his style, and his social position: "I may perhaps seem rash in accepting your claims so readily with no support from my own experience. But why should I not believe a most learned mathematician, whose very style attests the soundness of his judgment? . . . Shall I disparage him, a gentleman of Florence, for the things he has seen? . . . Shall he with his equipment of optical instruments be disparaged by me, who must use my naked eyes because I lack these aids?" (Kepler, 12–13; quoted in Dear, 1992)

2. In his analysis of "Rhetoric and Authority in the Early Royal Society" (1985), Peter Dear attributes the development of the technique of the scientific report not so much to Boyle as to the collective stylistic practices of the early Royal Society. He writes, "In such reports, the operator or observer (the two were equivalent) was central to the episode recounted – and episodes they were. Located, explicitly or implicitly, at a precise point in space and time, the observer's reported experience of a singular phenomenon constituted his authority." (199)

3. It is worth noting that the inversion that Daston (and others) observe in the meaning of objectivity – from objects to minds (in her terms, from ontological to perspectival and mechanical objectivity [see Daston, 1990; Daston and Galison, 1992]) – was accompanied by a parallel inversion in the reference of machine. In the seventeenth and early eighteenth centuries, the paradigmatic machine (Boyle's air pump notwithstanding) was the clock; its metaphoric reference, the universe of objects. By the nineteenth century, the more relevant machine for natural science had become a device for recording, a measurement apparatus; its metaphoric reference was less likely to be the world than the observer.

4. It should be noted, at least parenthetically, that the easy slide between the first two of these is surely as significant as the distance marked between "I/we" and "they": Indeed, the use of the first person singular pronoun is highly unusual, employed here merely in the service of a small hedge. The first person plural is a different kind of hedge – employed simultaneously to stand in for the singular, to evade personal responsibility, and to invoke the anonymous authority of a seamless community not, of course, to be confused with "people."

5. See an interesting discussion by Peter Galison on the development of computer simulation as a *"tertium quid,"* cutting across conventional divisions between "pure" and "applied" mathematics, between theoretical and experimental physics, and between the artificial and the real (Galison, 1991).

6. In a sense, of course, neither does Hillis. His response to my argument is worth quoting, if only to underscore the coexistence in his own vision of modern (even Cartesian) and postmodern conceptions of the subject: "I was touched that you understood exactly what I was saying, even if you did not seem to emphathize with it entirely... I sympathize with your argument that we have historically tried to cast away the observer, piece by piece. But I interpret this very differently than you. To me, with each step we are not denying the essential "I", but we are rather trying to find it. We say 'what is important about me is not my position in space nor my position in society, nor even is it the substance of my body.' This is not an attempt to whittle our self down to nothing, but rather to expose its essential core. Each time we shed some unessential symbol of our own self-importance we gain new freedom." (Personal [electronic mail] correspondence, Nov. 29, 1990.) The dualism still lurking in Hillis's vision perhaps finds its clearest expression in the proximity and interdependence of the two principal byproducts of AI currently under development – namely, the twin technologies of virtual reality and robotics.

7. The explicit connections between the ambitions to escape the body (where "human thought will live free of bones and flesh") and the scientific and

technological drive to decode the mysteries of biological reproduction (the ultimate "secret of life") have yet to be fully drawn. But a beginning (focusing especially on the symbolic role of "woman" in scientific discourse) can be found in Keller (1990a; 1990b). And for another perspective on many of the same issues, see the work of Donna Haraway, especially Haraway, 1991.

REFERENCES

Alpers, Svetlana, 1987. "The Mapping Impulse in Dutch Art," in Woodward, D. (ed.), *Art and Cartography: Six Historical Essays* (Chicago: The University of Chicago Press).

Bernal, J.D., 1929. *The World, the Flesh and the Devil: An Enquiry into the Future of the Three Enemies of the Rational Soul* (New York: E.P. Dutton & Co.).

Boyle, R., 1744. *The Works of Robert Boyle*, (ed.) Thomas Birch, 5 vols (London: A. Miller).

Bryson, Norman, 1983. *Vision and Painting* (New Haven: Yale University Press).

Carlos, Edward Stafford, 1880. *The Sidereal Messenger of Galileo Galilei* (London: Dawsons of Pall Mall, reprinted 1960).

Crick, Francis, 1963. 'Eugenics and Genetics,' in Wolstenholme, G. (ed.), *Man and His Future* (Boston: Little, Brown, and Co.).

Crick, Francis, 1966. *Of Molecules and Men* (Seattle: University of Washington Press).

Daston, Lorraine, 1990. "The Objectivity of Interchangeable Observers; 1830–1930," unpublished ms. Presented at the History of Science Society Meeting, Seattle (October 28).

—— "Baconian Facts, Academic Civility, and the Prehistory of Objectivity," in Megill (ed.) (1994), *op. cit.*

Daston, Lorraine and Peter Galison, 1992. "The Image of Objectivity" *Representations* 40 (Fall), 81–128.

Dear, Peter, 1985. "*Totius in verba*: Rhetoric and Authority in the Early Royal Society," *ISIS* 76: 145–61.

—— 1992. "From Truth to Disinterestness in the Seventeenth Century," *Social Studies of Science* 22 (November 1992), 619–31.

de Santillana, Giorgio, 1959. "The Role of Art in the Scientific Renaissance," in Clagett, M. (ed.), *Critical Problems in the History of Science* (Madison: University of Wisconsin Press).

Descartes, René, 1664. "Traité de l'homme" in *Oeuvres*, Pléiade, (ed.) [1953].

Dyson, Freeman, 1972. "The World, The Flesh and the Devil," Third Bernal Lecture, Birkbeck College, London, May 16 (unpublished).

Galison Peter, 1991. "Artificial Reality," presented to the Conference on Disunity and Contextualism, Stanford University (March 31–April 1); published in (Chicago: The University of Chicago Press). *Image and Logic: The Material Culture of Microphysics.*

Gallagher, Catherine, 1993. *British Women Writers and the Literary Marketplace: 1670–1820* (Berkeley: University of California Press).

Haraway, Donna, 1991. *Simians, Cyborgs and Women: The Reinvention of Nature* (New York: Routledge).

Hillis, Daniel, 1988. in Graubard, S.R., (ed.), *The Artificial Intelligence Debate* (Cambridge, MA: MIT Press).

Keller, Evelyn Fox, 1990a. "From Secrets of Life to Secrets of Death," in *Body/Politics: Women in the Discourse of Science*, M. Jacobis, E.F. Keller, S. Shuttleworth (eds.) (New York: Routledge: 1990).

——"Secrets of God, Nietzsche, and Life," *History of the Human Sciences* 3, no. 2, pp. 229–42.

Kepler, Johannes, 1965. *A Conversation with Galileo's Sidereal Messenger*, trans Rosen, Edward, Johnson Reprint Corp.

Kuhn, Jehane R., 1990. "Measured Appearances," *Journal of the Warburg and Courtauld Institutes* 53, pp. 114–34.

Megill, A. (ed.) 1994. *Rethinking Objectivity* (Durham: Duke University Press).

Moravec, Hans, 1988. *Mind Children* (Cambridge, MA: Harvard University Press).

Porter, Theodore M., 1986. *The Rise of Statistical Thinking: 1820–1900* (Princeton: Princeton University Press).

——"Objectivity as Standardization: The Rhetoric of Impersonality in Measurement, Statistics, and Cost-Benefit Analysis," in Megill (ed.) (1994), *op. cit.*.

——1990. "Quantification and the Accounting Ideal in Science," presented at the History of Science Annual Meeting, Seattle (October 28).

Rotman, Brian, 1987. *Signifying Nothing: The Semiotics of Zero* (London: MacMillan).

Shapin, Steven, 1984. "Pump and Circumstance: Robert Boyle's Literary Technology," *Social Studies of Science* 14, pp. 481–520.

Shapin, Steven and Simon Schaffer, 1987. *Leviathan and the Air Pump: Hobbes, Boyle and the Experimental Life*, including a Translation of Thomas Hobbes, *Dialogus physicus de natura aeris* by Simon Schaffer (Princeton: Princeton University Press).

Part III
The Nature of Scientific Change

8 Theories of Scientific Change*

Robert Richards

The writing of science history may itself be regarded as a scientific enterprise, involving evidence, hypotheses, theories, and models. I wish here to investigate several historiographic models and their variants. While these undoubtedly do not exhaust the store available to imaginative historians of science, they nonetheless represent, I believe, those that have played the significant roles in the development of the discipline, either as models that have long functioned in historical writing or as models more recently proposed in metahistorical works.

The models described in the first part of this [chapter] represent major assumptions that have guided the construction of histories of science since the Renaissance. They thus embody directive ideas concerning the character of science, its advance, and the nature of scientific knowing. Since the models are idealizations, they do not always precisely reflect the structures of particular written histories. Yet they can serve to elucidate those controlling assumptions that have shaped our understanding of science and its history.

The second part of this [chapter] will attend to the class of models that appears the most powerful for capturing the actual movement of science: evolutionary models. I will briefly examine two instances of this class, the models of Popper and Toulmin, and consider their deficiencies. I will then develop a natural selection variant that, I believe, escapes their liabilities...

FIVE MODELS IN THE HISTORIOGRAPHY OF SCIENCE

The Static Model

Many historians and scientists of the late Renaissance and early Enlightenment shared R. Bostocke's conviction, as expressed in his *The*

---

* Appendix I, "The Natural Selection Model and Other Models in the History of Science," in R.J. Richards, *Darwin and the Emergence of Evolutionary Theories of Mind and Behavior* (Chicago: The University of Chicago Press, 1987), pp. 559–591.

difference Between the Auncient Phisicke and the Latter Phisicke (1585), that God had infused certain men (such as Adam or Moses) with scientific knowledge, which was passed on to successive generations intact.[1] Even Newton, in his historical musings, employed a static model, maintaining that his *Principia* was a recovery of wisdom known to the ancients.[2]

Use of a static model in history of science accorded with the Renaissance presumption that ancient thought embodied the highest standards of knowledge and style. But another consideration also promoted the acceptance of the model. This may be found in Olaus Borrichius's *De ortu et progressu chemiae dissertatio* (1668), a standard textbook of the history of chemistry during the late seventeenth and early eighteenth centuries.[3] In accord with the tradition, Borrichius credited Tubalcain, a descendant of Cain and a figure he identified with Vulcan, as having received from God the divine knowledge of chemistry. The Cartesian argument he used to fortify his baroque sentiments displays an important justification for use of the static model. He reasoned that "the priests of Tubalcain would have been unable to discover, shape, and form the metals of iron and copper except that their *ratio* was prior known; that the natures of these minerals might be investigated and that they might be cooked, purged, and segregated could not occur except that knowledge of this were divinely inborn. Once this knowledge is had, however, these techniques follow for any skillful people."[4] Borrichius, tinctured with the Cartesian spirit, knew that chemical knowledge and science in general must be innate, at least in their fundamentals; for unilluminated natural induction could never of itself lead to such scientific achievements as his age had witnessed. And if the essential features of a science had this kind of origin, then from its first discoverer such knowledge could only be passed on or rediscovered again by succeeding generations. This model of the origin and course of science can be detected in transmogrified form in Thomas Kuhn's Gestalt model (described below), which assumes that in a moment of insight the transformed vision of an inspired genius may establish the framework and fundamental premises of a science, the details of which may be left to the normal plodding of disciples.

The Growth Model

After the late Renaissance, historians of science began to discard the static model, replacing it with one still in use today. By the eighteenth century, the growth model clearly prevailed, as Freind's *The History of Physick from the Time of Galen to the Beginning of the Sixteenth Century* (1725)[5] and Watson's essay "On the Rise and Progress of Chemistry" (1793)[6] testify. Indeed, Freind's history may be read as a sustained argument against the Renaissance tendency to overprize the ancients and to suppose that the essential concepts and principles of science lay with them, only to be ornamented by succeeding generations. Freind proposed to show that the knowledge of medicine did not begin and end with Hippocrates and Galen. Instead, as a careful study of the writings of subsequent physicians demonstrated, "Physick was still making progress 'till the Year 600".[7] (He charted the gradual advance of the science since the beginning of the medieval period in the second volume of his history.) As a consequence of the particular model he had chosen, that of gradual, cumulative growth, Freind could recommend reading in the history of medicine as "the surest way to fit a man for the Practice of this Art."[8] This is a piece of advice annulled by historians advocating other models.

Watson's essay highlights an assumption of the growth model that was to have particular importance in later controversies, namely, that science in its conceptual development is relatively isolated from other human occupations, even from the technology that fostered it. Watson felt assured of this independence, since he understood science to have a rational integrity not found in the less 'liberal and philosophical' pursuits.[9] This presumption also bound together the various parts of that monument to the growth model, the *Encyclopédie* of Diderot and d'Alembert. In the *Discours Préliminaire* to the *Encyclopédie,* d'Alembert projected the prescriptions of the growth model back even to prerecorded thought, suggesting that primitive sensory awareness might gradually have established the foundational principles of scientific advance.[10]

In the nineteenth century, William Whewell supplied the most elaborate employment and justification of the growth model in his *History of the Inductive Sciences* (1837).[11] He rejected the idea that discontinuous intellectual upheaval marked the development of the various sciences: "On the contrary, they consist in a long-continued advance; a series of changes; a repeated progress from one principle to another, different and often apparently contradictory."[12] If the progress of science occurred by contradictory ideas replacing one another, there would not, of course, be organic growth, but revolutionary saltation. That is why Whewell urged his reader to remember that the contradictions were only apparent:

The principles which constituted the triumph of the preceding stages of the science may appear to be subverted and ejected by the later discoveries, but in fact they are (so far as they were true) taken up in the subsequent doctrines and included in them. They continue to be an essential part of the science. The earlier truths are not expelled but absorbed, not contradicted but extended; and the history of each science, which may thus appear like a succession of revolutions, is, in reality, a series of developments.[13]

The central assumptions embodied in the growth model are compendiously present in the work of George Sarton, the doyen of historians of science in the middle of this century. His several observations on the nature of science therefore afford a convenient summary of the implications of the model. The primary feature of the model is its affirmation of the unalterable and clearly discernible progress of science toward the fullness of truth, progress that can be only momentarily delayed by retarding forces. "The history of science," Sarton declared, "is an account of definite progress, the only progress clearly and unmistakably discernible in human evolution. Of course, this does not mean that scientific progress is never interrupted; there are moments of stagnation and even regression here or there; but the general sweep across the times and across the countries is progressive and measurable."[14]

The steady advance of science, accomplished by the rationally exact methods of quantification and experimentation, and "its astounding consistency (in spite of occasional, partial, temporary contradictions due to our ignorance) prove at one and the same time the unity of knowledge and the unity of nature."[15] Since the unity and continuity of knowledge, which are grounded in the unity and intelligibility of nature, are not, in Sarton's estimation, enjoyed in other human pursuits, these latter are unable conceptually to affect the course of science. Moreover, the clear evidence of history gives no support, he thought, to attempts at the sociologizing of scientific knowledge. The internal progress of science has a force beyond the vicissitudes of men's passions and the subtle pressures of social life. To be sure, science does not grow in a social vacuum: men need food, they are called to war; money for equipment is required. But, Sarton avowed, the man of science remains ultimately untouched in his theoretical endeavor by the ideologies or conditions of society: "Nobody can completely control his spirit; he may be helped or inhibited, but his scientific ideas are not determined by social factors."[16] Insofar as the history of science is independent of the cultural life of the larger community, it can serve as a standard of truth and error in those other domains: "the history of science describes man's exploration of the universe, his discovery of existing relations in time and space, his defense of whatever truth has been attained, his fight against errors and superstitions. Hence, it is full of lessons which one could not expect from political history, wherein human passions have introduced too much arbitrariness."[17]

The Revolutionary Model

A brief examination of the history of the term *revolution* suggests that its application to scientific thought is not necessarily derived from analogies with political overthrow. The *Oxford English Dictionary* indicates that its use to describe dramatic changes in thought antedates by a considerable period its use to designate political upheavals. By the late eighteenth century, the term was widely employed to denote important transformations in the course of science. When Kant referred to particular "revolutionary" events in the history of science, he employed the word in the manner of contemporary historians: to describe a profound shift in thinking, after which there is relatively smooth scientific progress to the present time. For Kant, as well as for most recent historians using the model, revolution in a science is a one-time affair. In the preface to the second edition of the *Kritik der reinen Vernunft*, Kant depicted the intellectual revolution undergone by the mathematical and physical sciences, before which we had no science proper and after which we had unimpeded advance into the modern period. Mathematics had to grope during the Egyptian era, but with the Greeks came the revolution that set it on its present course. Natural science had to wait a bit longer for its revolution, as Kant explained: "It took natural science much longer before it entered on to the road of science; for it is only about a century and a half since the proposal of the ingenious Bacon of Verulam partly fostered its discovery and, since some were already on its trail, partly gave encouragement. But this can be explained only as a suddenly occurring revolution in the mode of thought (*eine schnell vorgegangene Revolution der Denkart*)."[18]

While the use of the term *revolution* to describe radical changes in thought is older than its use in the specifically political context, the political analogy is often implied and does seem justified. Political revolutionaries have particular enemies with whom they wage their ideological and bloody battles; the scientific revolutionaries of the sixteenth and seventeenth centuries also had their foes: Aristotle, Ptolemy, Galen, and the Scholastics. Political revolutionaries aim at overturning an undesirable system and replacing it with one that will perdure and serve as a base for further progress; their scientific counterparts harbor similar goals. Significant political revolutions are not usually spontaneous; they have their doctrinal basis formed in the work of men who may be long dead before the revolution. Historiographers of scientific revolutions also acknowledge necessary foundations: the groundwork of modern physics laid, for example, by the Merton school of mathematical physics or the Paduan Aristotelians of the early Renaissance. The historical importance of a political revolution lies more in the fruit of the new ideas and systems that the revolution inaugurates – fruit that may take time in ripening. Those writing of the scientific revolution fomented by Copernicus, Kepler, Galileo, Harvey, and Descartes construe the ideas of these scientists as establishing the foundations for thoroughly modern science, even though their specific conceptions may no longer be acceptable.

The most influential historian to employ the revolutionary model was Alexandre Koyré, whose views set out its essential features. To the history of science Koyré brought the philosopher's eye for metaphysical assumptions and the intellectual historian's concern for doctrinal context. In his view, the scientific revolution of the sixteenth and seventeenth centuries outwardly expressed a more fundamental turn of mind, a "spiritual revolution" having two basic features. There was, first of all, the Platonically motivated dismissal of the qualitative space of Aristotle and the Scholastics and its replacement with abstract geometrical space. Galileo's contribution to the scientific revolution was precisely his insistence on mathematical reasoning rather than sense experience as the foundation for scientific success.[19] But this alteration in thought about the universe was only a phase, though the crucial one, of a more pervasive revolution, one that brought about, according to Koyré,

> the destruction of the Cosmos, that is, the disappearance, from philosophically and scientifically valid concepts, of the conception of the world as a finite, closed, and hierarchically ordered whole... - and its replacement by an indefinite and even infinite universe which is bound together by the identity of its fundamental components on the same level of being. This, in turn, implies the discarding by scientific thought of all the considerations based upon value concepts, such as perfection, harmony, meaning, and aim, and finally the utter devalorization of being, the divorce of the world of value and the world of facts.[20]

Insofar as the revolution had banished from explanatory rule such concepts as "perfection, harmony, meaning, and aim," historians of science under the banner of Koyré have felt justified in dismissing from serious consideration neo-Platonic mysteries and Paracelsian occultism, which were contemporary with what has become known as the new science. Perhaps paradoxically, Koyré himself was quite willing to consider the influence of such spirits on the new science, though he denied the influence was specifically scientific.[21]

Models in historiography, as well as in science, provide more than a mere heuristic for investigation. They focus attention, exclude possibilities, and reveal hidden connections. Whether as covert assumptions or as consciously accepted devices, models intervene (inevitably, I believe) between the historian and his or her subject. Yet the sensitive historian is not often led far astray by the odd magnifications a model might produce; a distorted perspective can be corrected by the feel of hard facts that he or she continues to accumulate. Moreover, the use of a model and the application of its embedded hypotheses require an artful intelligence, one that individualizes the crafted product. Thus historians who generally employ the revolutionary model may offer different perspectives on the same issues. Alistair Crombie, Rupert Hall, and Charles Gillispie, for example, in some contrast to Koyré, locate the revolution in scientific thought in the application of mathematics to mechanics and in the resultant construction of formal systems for the construal of nature.[22] Hall believes that the instruments and techniques developed by craftsmen have provided a stimulus and auxiliary to the new sciences.[23] But Koyré virtually ignores the crafts, since the science of Galileo and Descartes "is made not by engineers or craftsmen, but by men who seldom built or made anything more real than a theory."[24] Hall regards pre-seventeenth-century investigations of nature as essentially discontinuous with science after that period.[25] Crombie, who devotes considerable attention to the medieval development of the foundations of modern science, believes that "a more accurate view of seventeenth-century science is to regard it as the second phase of an intellectual movement in the West that began when philosophers of the thirteenth century read and digested in Latin translation the great scientific authors of classical Greece and Islam."[26] Gillispie, too, acknowledges the debt of Renaissance science to greek mathematical rationalism.[27]

Yet those who generally employ the revolutionary model agree – and this constitutes the essential feature of the model – that a revolution in thought, a decisive overthrow of distinctly ancient modes of conception, is necessary to set a discipline on the smooth course of modern science. Hall clearly highlights the core of the model. For him the medieval period did have its quasi-science; and though that enterprise set the stage for the appearance of modern science, yet the mathematical methods of the latter were radically different from the methods of its predecessor: "Rational science, then, by whose methods alone the phenomena of nature may be rightly understood, and by whose application alone they may be controlled, is the creation of the seventeenth and eighteenth centuries."[28] It is the method of rational science that guarantees its further progress – without fear of taking fundamentally wrong paths. Whatever revisions in science have come since the revolution are revisions in content only, not in structure.[29]

The Gestalt Model

In recent years ideas from particular currents within the social and psychological sciences have joined those springing from conceptual studies in the history of science, especially those studies whose epistemological channels run to neo-Kantianism. From this confluence has emerged what might be called a Gestalt model of science. Among those most influential in employing this model are Norwood Russell Hanson, Thomas Kuhn, and Michel Foucault.

Both Hanson and Kuhn explicitly use devices drawn from Gestalt psychology and the psychology of perception. The Necker cube, the goblet-faces display, pictures of creatures looking alternately like birds or antelope, and similar puzzles illustrate for them the ways in which context, past experience, and familiar assumptions control our perceptual and conceptual experiences of things. In the scientific domain, as construed by Hanson, it is the well-entrenched theory that determines the perception of facts: "Physical theories provide patterns within which data appear intelligible. They constitute a 'conceptual Gestalt.' A theory is not pieced together from observed phenomena, it is rather what makes it possible to observe phenomena as being of a certain sort, and as related to other phenomena."[30] Likewise in Kuhn's judgment: "Assimilating a new sort of fact demands more than additive adjustment of theory, and until that adjustment is completed – until the scientist has learned to see nature in a different way – the new fact is not quite a scientific fact at all."[31]

If facts and their organizing theories are mutually implicative and constitute a perceptual – conceptual whole – a "paradigm," to use the by-now debased coin – and if "the switch of Gestalt ... is a useful elementary prototype for what occurs in full-scale paradigm shift," then the model of scientific advance through the gradual increment of new facts and ideas proves inadequate for the historian's needs. "The transition from a paradigm in crisis to a new one from which a new tradition of normal science can emerge," argues Kuhn, "is far from a cumulative process, one achieved by an articulation or extension of the old paradigm. Rather, it is a reconstruction that changes some of the field's most elementary theoretical generalizations as well as many of its paradigm methods and applications."[32]

The lens of the Gestalt model can be focused more narrowly on the immediate scientific community, or dilated to situate the scientific community within a broader cultural context, as Foucault attempted in *The Order of Things: An Archaeology of the Human Sciences.* He intended in this work to explore the *"positive unconscious* of knowledge: a level that eludes the consciousness of the scientist and yet is part of scientific discourse."[33] He held that there were different epochs in Western history in which the sciences and related disciplines were bound together in the general cultural reticulum by unconscious principles of order. These principles yielded an "entire system of grids which analysed the sequences of representations (a thin temporal series unfolding in men's minds), arresting movement, fragmenting it, spreading it out and redistributing it in a permanent table."[34] Such ordering structures functioned, in Foucault's estimation, to determine both the domain of problems existing for the sciences and the methods of their resolution. His inquiry revealed to him three distinct epistemological ages – the Renaissance, the classical period (seventeenth and eighteenth centuries), and the modern period – each radically discontinuous from the previous one, so that terms of description used by one (e.g., "man," "society," "language," "nation") would have fundamentally different meanings when used by others. In the transition from one epoch to the next, it was "not that reason made any progress: it was simply that the mode of being of things, of the order that divided them up before presenting them to understanding, was profoundly altered."[35] As a consequence of the shift in the patterns of representation – the switch of the Gestalt – man as we now construe him in the human sciences came into existence only at the beginning of the nineteenth century. This is the paradoxical thesis of Foucault's work.

The Gestalt model makes two principal demands: first, that the historian should attempt sympathetically to assimilate and reconstruct the context of scientific discourse of a given period, and in this way to determine the social, psychological, and historical influences that controlled the ways scientists patterned their theoretical concepts and perceived through them the facts constituting the domain of scientific inquiry; and second, that the historian should regard history of science not as an internal and smooth flow of observations and theoretical generalizations across the ages but as the sudden shift of different world views, linked only by the extrinsic contingencies of time and place.

The Gestalt model as employed by Hanson, Kuhn, and Foucault bears similarities to the revolutionary model – and, of course, Kuhn's express aim is to describe the structure of revolutions in science. But the differences between the revolutionary model as commonly used and the Gestalt model are marked. Those employing a revolutionary model discover in the course of a particular science a signal awakening of thought, an overturning of what the model characterizes as a decidedly archaic mode of thinking, and the establishment of a lasting foundation for future progress by, as Hall puts it, "accretion."[36] Since the logic of the revolutionary model hinges on the dichotomy between ancient and modern methods of scientific thought, historians of this persuasion usually assume that revolution in a science is a one-time affair. The Gestaltists, however, emphasize multiple "scientific revolutions," no one of which secures a position that is any more scientific or more stable than others that have preceded it. The revolutionists believe that revolutions happen for good reasons, reasons that sustain the future growth of a science. The Gestaltists, as is consonant with the source of their model, tend to stress psychological and sociological factors in scientific change. In their view, scientific change is rarely the result of good reasons; indeed, reasons have weight only against a background of commonly accepted theory. The revolutionists view science as a search for truth about the world. The Gestaltists argue that there is no truth about the world; truth is a function of the coherence of the theoretical arrangement which holds at any one time; there are no independent, theory-free standards against which a hypothesis might be measured to assess its truth.[37] The revolutionists are apt to regard postrevolutionary science as better than or more true than prerevolutionary science. The Gestaltists believe that the perceptual-conceptual paradigm adopted by a given community of scientists is incommensurable with those assumed by their predecessors: in the Gestalt switch of the goblet-faces display, the goblet is no better or truer than the faces.[38]

The Gestalt model encourages the historian to interpret scientific ideas as parts of a larger complex of meanings; it emphasizes the mutual determination of these elements. The hermeneuticist of the scientific Gestalt begins with a node of experience or a paradigmatic idea and moves laterally, interpreting one symbol of the pattern in terms of the others, ultimately including socially and culturally entwined meanings. Another recent model, however, suggests that the interpretive relation is vertical and unidirectional, and that scientific patterns of thought merely reflect deeper and more covert social or psychological structures.

Social–Psychological Model

From the ancient through the modern periods, scientists have frequently justified their theories by appeal to the more general doctrines – metaphysical, religious, or social – to which those theories have been related. Samuel Clarke defended Newton's science, since it would "confirm, establish, and vindicate against all objections those great and fundamental truths of natural religion."[39] But it was only at the beginning of our century, after transformations in the social and psychological sciences (by Marxism, Durkheimian social anthropology, Freudianism, and similar conceptual movements) that historians seriously attempted to organize their narratives under the assumption that scientific programs might be fueled by social interests and psychological needs. What united both socially oriented and psychologically disposed historians was the conviction that apparently extrinsic conceptual structures, whether embedded in social relationships or in psychological complexes, might covertly determine the generation, formulation, and acceptance of scientific ideas. Moreover, though Freudians have insisted on the primacy of sedimented attitudes, they usually have admitted that these originated in certain real or imagined social situations. Similarly, Marxist historians have recognized that the effects of class stratification are mediated by subtle patterns of individual belief. Because of these common features, social and psychological models may be considered as forming one class of historiographic models.

Social–psychological models can be divided into those prescribing weak determination and those prescribing strong determination of scientific development. The weak version of the model is the central organizing device of J.D. Bernal's four-volume *Science in History*. The model guided Bernal in mapping an enlarged field of investigation. "Science," he proposed, "may be taken as an institution; as a method; as a cumulative tradition of knowledge; as a major factor in the maintenance and development of production; and as one of the most powerful influences molding beliefs and attitudes to the universe and man."[40] Such a generous conception compelled him to trace the social and psychological patterns in the terrain of science. For example, he initially explained Darwin's hypothesis of natural selection as a reformulation of Malthusian economics in other terms – that is, as a biological construction of a "theory built to justify capitalist exploitation."[41] For Bernal, the source of scientific thought, the institutions of science, its methods, the economic forces driving it, and its impact on society were all fit subjects for social-psychological analysis. Yet science as a "cumulative tradition of knowledge" was not.

Bernal could not bring himself to extend his Marxist vision to the heart of science. He confessed that the cumulative nature of science distinguished it from such other human pursuits as law, religion, and art. Though science, like these other enterprises, grows in a field of social relations and class interests, its claims, unlike theirs, can be checked directly "by reference to verifiable and repeatable observations in the material world."[42] The weak model thus protects the internal logical and justificatory structure of science from the hands of the sociologist and the psychologist. That is, it does so when the science goes right.

When it goes wrong, the historian has a sure sign that extrinsic social or psychological factors have intruded. For instance, Erik Nordenskiöld, in his influential *History of Biology*, felt constrained to invoke the weak model in his account of the unwarranted (as he believed) acceptance of Darwinian theory by scientists in the latter half of the nineteenth century:

> From the beginning Darwin's theory was an obvious ally to liberalism; it was at once a means of elevating the doctrine of free competition, which had been one of the most vital cornerstones of the movement of progress, to the rank of natural law, and similarly the leading principle of liberalism, progress, was confirmed by the new theory... It was no wonder, then, that the liberal-minded were enthusiastic; Darwinism must be true, nothing else was possible.[43]

It is ironic that one powerful tradition in the sociology of science, led by Robert Merton[44] and Joseph Ben-David,[45] endorses the weak model as the only one appropriate for respecting the cognitive content of science. Ben-David, for example, admits that socially conditioned biases and ideology "might have played some role in the blind alleys entered by science." In those darkened corners, sociology can prove illuminating. But the main scientific roads are "determined by the conceptual state of science and by individual creativity – and these follow their own laws, accepting neither command nor bribe."[46] Sociologists in this tradition confine their empirical analyses to questions of institutional organization, the spread of scientific knowledge, social controls · on the focus of scientific interest, and public attitudes toward science. They regard the cognitive content of science, however, as the reserve, not of the sociologists, but of those intellectual historians whose concerns are principally logical and methodological.

Yet even with the support of the dominant tradition in the sociology of science, can the weak model be justified? Those employing it usually fail to supply a convincing reason why social or psychological analyses might explain error in science but not truth. The persuasiveness of this model is further diminished when one considers that, in a strict sense, most past science is "erroneous," at least by contemporary standards. Hence if the logicist assumptions of the weak model are consistently heeded, the content of virtually all past science ought to be amenable to social and psychological interpretation. Nor should contemporary science be exempt, since there is no reason to suspect that it has achieved final truth.

The logic of the preceding line of reasoning appears to have persuaded, implicitly at least, those using a strong version of the model. For example, Margaret Jacob has detected social interests at the root of seventeenth-century mechanical philosophy. In her *The Newtonians and the English Revolution*, she argues that the Newtonians, those traditional harbingers of contemporary science, constructed matter as passive (i.e., having no occult powers) not because reason and evidence required it but because their latitudinarian and religious ideology demanded it.[47] The Edinburgh sociologist of science David Bloor supports Jacob's use of the strong version of the model. In similar fashion, he maintains that the seventeenth-century scientist Robert Boyle, his colleagues, and his opponents "were arranging the fundamental laws and classifications of their natural knowledge in a way that artfully aligned them with their social goals." The lesson Bloor drew from this drama of Restoration science was that quite generally in the history of science, "the classification of things reproduces the classification of men."[48]

The strong version of the model, then, asserts that the structure of scientific knowledge is determined not by nature but by social patterns or psychological complexes. The model stipulates that logic and the appeal to natural facts are on the surface and that what really matters in comprehending the work of scientists are dominance struggles with the father – as in the case of Mitzman's reconstruction of Weber's social science[49] – or the social practices of a society – as in the case of Bloor's account of Greek mathematics.[50]

The strong version of the social-psychological model, despite initial implausibility, does focus the historian's sight on a cardinal feature of scientific development: that science depends on norms – norms suggesting what is appropriate both to investigate and to accept. Norms, however, are dictated not directly by nature but by the decisions of men. The logic of scientific argument cannot coerce, except insofar as men feel moved to abide by its rules and adopt its premises. Ultimately, the acceptance of metarules and first premises appears to be a function of social enculturation, of psychological conditioning, and perhaps of biological disposition. For as Aristotle pointed out, only the fool tries to demonstrate the principles upon which all his arguments are based.

Nonetheless, the strong version of the social-psychological model seems too strong. It is liable to a *tu quoque* response. Why, after all, should we be convinced by the account of a historian who uses the strong version, if that account itself merely reflects his inferiority complex or his Calvinistic upbringing? The destruction of scientific rationality also undermines the plausibility of historical argument. To restrain the destructive relativism of both the social-psychological model and the Gestalt model, while preserving the edge of their insights, is one of the chief tasks for which evolutionary models have been constructed.

EVOLUTIONARY MODELS OF SCIENTIFIC DEVELOPMENT

The use of evolutionary theory in explanations of cultural phenomena can easily be traced back to the mid-nineteenth century. John Lubbock, Walter Bagehot, Lewis Henry Morgan, Edward Tylor, Herbert Spencer, and a host of others applied evolutionary concepts to societal institutions in an effort to account for the descent from primitive culture.[51] More recently, the specialized use of evolutionary notions, aping its biological counterpart, has proceeded from the macroconsideration of culture to the microconsideration of the development of ideas, particularly scientific ideas. Gerald Holton, for instance, makes detailed use of the evolution analogy in his *Thematic Origins of Modern Science*;[52] and in reconsidering his theory of paradigms, Kuhn has suggested that the appropriate approach to science history is evolutionary.[53] But Holton and Kuhn wield evolutionary constructs only as vague analogies. Others believe they promise more. Evolutionary theory, duly generalized, provides, it is argued, the very explanation of scientific growth. Not only are ideas conceived, but like Darwin's finches they also evolve. In the nineteenth century, George Romanes, Conwy Lloyd Morgan, William James, and James Mark Baldwin all proposed that Darwinian theory explained the development of ideas. More recently Karl Popper, Stephen Toulmin, and Donald Campbell have advanced a strict epistemological Darwinism. In what follows, I will briefly examine the proposals of Popper and Toulmin, indicate the deficiencies of their models, and then elaborate a natural selection version that comes close in spirit to the fertile ideas of Campbell.

The Models of Popper and Toulmin

In *The Logic of Scientific Discovery*, Popper describes the scientific community's selection of theories not as a process by which a given theory is justified by the evidence but as one by which a theory survives because its competitors are less fit. Thus he argues that the preference for one theory over another "is certainly not due to anything like an experimental justification of the statements composing the theory; it is not due to a logical reduction of the theory to experience. We choose the theory which best holds its own in competition with other theories; the one which, by natural selection, proves itself the fittest to survive."[54]

In Popper's judgment, our scientific and pedestrian quests for knowledge always begin not with pure observation but with a problem that has arisen because some expectation has not been met. In confronting the problem, the cognizer makes unrestrained conjectures about possible solutions, much as nature makes chance attempts at solving particular survival problems.[55] These conjectures are then tested against empirical evidence and rational criticism. The rational progress of science, therefore, consists in replacing unfit theories with those that have solved more problems. These latter, according to Popper, should imply more empirical statements that have been confirmed than their predecessors.[56] This condition enables us to describe successor theories as closer to the truth and consequently more progressive. I will not expand further on Popper's conception, since Lakatos has already done this with concision, fashioning from it a model of scientific research programs, which I will discuss below.

The evolutionary model permits Popper to avoid the presumption that theories are demonstrated by experience; it also allows him to dismiss the view that theories and creative ideas arise from any sort of logical induction from observation. Thus the older and newer problems of induction are skirted. Popper believes that the model directs one to interpret scientific discovery as fundamentally an accidental occurrence, a chance mutation of ideas. He consequently fails to emphasize that the intellectual environment not only selects ideas but restricts the kinds of ideas that may be initially entertained by a scientist. Attention to the environment of scientific ideas, however, is precisely what Toulmin requires for an adequate account of scientific growth.

Toulmin's thesis is that scientific disciplines are like evolving biological populations, that is, like species. Each discipline has certain methods, general aims, and explanatory ideas that provide its coherence over time, its specific identity, while its more rapidly changing content is constituted of loosely related conceptions and theories, "each with its own separate history, structure, and implications."[57] To comprehend the evolution of a science so structured requires that one attend to the cultural environment promoting the introduction of new ideas, as well as to the selection processes by which some few of these ideas are perpetuated.

The content of a discipline, according to Toulmin's scheme, adapts to two different (though merging) environmental circumstances: the intellectual problems the discipline confronts and the social situations of its practitioners. Novel ideas emerge as scientists attempt rationally to resolve the conceptual difficulties with which their science deals; but often those new sports will also be influenced by institutional demands and social interests. Therefore, in explaining the appearance of innovative ideas within an evolving science, one must consider both *reasons* and *causes*. After such variations are generated, however, one must turn to the processes, rationally and socially causal, by which the variations are selected and preserved.

The processes that shape the growth of a discipline – selection processes – also occur within particular intellectual and social settings. The intellectual milieu consists of the immediate problems and entrenched concepts of the science and its neighbors. Within this environment, rational appraisal by the scientific community tests the mettle of new ideas. The survivors are incorporated into the advancing discipline. The social and professional conditions of the discipline also work to cull ideas, sanctioning some and eliminating others. Both of these selection processes – selection against intellectual standards and against social demands – may act either in complementary fashion or in opposition. But both must be heeded, in Toulmin's judgment, if one is to understand the actual history of a science. The historian will look only to intellectual conditions, however, when pursing a rational account of the development of a particular science. When investigating, say, the causes accelerating or retarding scientific growth, he or she will turn to the social and professional institutions of that science.[58]

It is the mark of the recent past in the historiography of science that it is the rational continuity of science that appears to require explanation. Toulmin has proposed his evolutionary model to meet this need. In his view, the continuity of disciplines, like the continuity of biological species, involves transmission of previously selected traits to new generations. In science this process is, according to Toulmin, one of enculturation: junior members of a discipline serve an apprenticeship in which they learn by tutored doing, by exercising certain "intellectual techniques, procedures, skills, and methods of representation, which are employed in 'giving explanations' of events and phenomena within the scope of the science concerned."[59] What principally gets inherited, he believes, is not a disembodied set of mental concepts but particular constellations of explanatory procedures, techniques, and practices that give muscle to the explicating representations and methodological goals of the science. Through the active participation in an ongoing scientific community, the novice inherits two kinds of instantiated concepts. The first comprises the specific substantive ideas and theories, the special explanations and techniques that solve recognized problems at any one period in the evolution of a discipline. The second kind of inheritance remains continuous over much longer periods and changes only slowly. It consists of the explanatory ideals, the general aims, and the ultimate goals that distinguish the disciplines from one another. It is within this more general inherited tradition that large-scale conceptual changes in substantive theory occur "by the accumulation of smaller modifications, each of which has been selectively perpetuated in some local and immediate problem situation."[60] But such changes should not suggest, as they do for those adopting the Gestalt model, that there are not good reasons for the shifts. The basic structure against which reasons can be measured is the continuity of explanatory aims and ideals which a discipline manifests through long periods of its history.

Toulmin further attempts to ensure that his model will allow rational criteria to operate in science by adjusting it with the postulate of "coupled evolution." The neo-Darwinian theory of organic evolution requires that variability within a species be independent ("decoupled" as Toulmin puts it) of natural selection. According to his interpretation of the modern synthesis, there is no preselection or direction given classes of variations. But coupled evolution, which he regards as another species of the larger genus of evolutionary processes, postulates that variation and selection "may involve related sets of factors, so that the novel variants entering the relevant pool are already preselected for characteristics bearing directly on the requirements for selective perpetuation."[61]

What Toulmin has suggested by his postulate of coupled evolution is not, however, a Darwinian sort of mechanism, in which production of variations is blind or random, but a Lamarckian one, in which conscious acts preshape the material in anticipation of the exigencies of survival. Accordingly the cardinal feature of the Darwinian perspective, competitive struggle against environmental demands, is largely obviated. Natural selection has no pivotal role in Toulmin's scheme.

But Toulmin need not have abandoned the device of natural selection so quickly. For the neo-Darwinian synthesis does recommend clearly acceptable senses in which individuals within a species might be described as preselected or preadapted to an altered environment: when, for example, heterozygote superiority leads to the retention of alleles that would be fit in different circumstances; or when linkage holds in a population certain alleles that would enhance adaptation to changed surroundings; or when alleles at certain loci have fixed rates of mutation. Such mechanisms for storing variation act as constraints on selection, making specific kinds of adaptive responses to a given situation more likely. It is of course true that the variations stored and the methods of their preservation are products of previous selection over many generations. In any case, classes of variations characteristic of elephants are not likely to occur in the species *Rattus rattus*. The genetic background of a species will restrict, and in that sense preselect, the kinds of variations that are immediately possible. In a moment I will indicate what this feature of biological evolution suggests for understanding conceptual evolution.

The Natural Selection Model

Popper's version of the evolutionary model of science emphasizes that theories succeed one another something like species: that theory is selected which solves more problems than its competitors. Toulmin's version complements Popper's by focusing on the cultural environments in terms of which new ideas appear and are incorporated into an evolving discipline. But Toulmin relinquishes a formal Darwinian device in an attempt to capture the way problem solutions originally emerge. As just indicated, abandonment of a natural selection mechanism is not necessary in order to model the birth of new scientific theories. In this section I want to build upon the Popperian and Toulminian variations and thereby refine a natural selection model for historiographic use. I will do this in two stages: first, by further specifying exactly what it is that evolves in scientific change; and second, by adding a psychosocial theory of idea production and selection, one similar to that proposed by Campbell.

According to Toulmin's model, the specieslike entity that evolves is the intellectual discipline. But this, I think, is the wrong analogue. Intellectual disciplines are, after all, composed of heterogeneous theories, methods, and techniques, while a species is a population of interbreeding individuals that bear genetic and phenotypic resemblance. Disciplines, moreover, are organized formally into subdisciplines and overlapping and competing specialties and are interlaced with invisible networks of communication.[62] Disciplines seem more like evolving ecological niches, consisting of symbiotic, parasitic, and competing species. The proper analogue of a species is, I believe, the conceptual system, which may be a system of theoretical concepts, methodological prescriptions, or general aims. The gene pool constituting such a species is, as it were, the theory's individual ideas, which are united into genotypes or genomic individuals by the bonds of logical compatibility and implication and the ties of empirical relevance. These connecting principles may themselves, of course, be functions of higher-order regulatory ideas. Biological genotypes vary by reason of their components, the genes, and the specific linkage relations organizing them; these genotypes display different phenotypes according, both as they have slightly different components and componential relationships and as they react to altered environments. Analogously, the cognitive representation of a scientific theory – its phenotypic expression in terms of the model here proposed – will vary from scientist to scientist by reason of the slightly different ideas constituting it, their relations, and the changing intellectual and social environment that supports it. So, for instance, Darwin and Wallace both advanced *specifically* the same evolutionary theory, though the components of their respective representation were not exactly the same, and the intellectual problems to which they applied their views and for which they sought resolutions also differed in some respects. Yet we still want to say that Darwin and Wallace developed the "same" – specifically the same – theory of evolution by natural selection. Constructing the model in this way also allows us to appreciate that, like the boundaries between species, the boundaries separating theories may be indefinite and shifting.

If a historiographic model of scientific development proposes that conceptual systems, like biological species, evolve against a problem environment, then that model, to tighten the Darwinian analogy, should include a mechanism accounting for adaptive change in scientific thought. During the last quarter century, Donald Campbell has worked out a psychological theory of idea production and selection that meets this demand.[63] His mechanism of "blind variation and selective retention" not only illuminates a fundamental feature of creative thinking in science (and in other cognitive pursuits) but, as an unintended consequence, also explains why some ideas seem to come (as Toulmin believes) preadapted to their intellectual tasks. Let me first sketch the essential aspects of Campbell's natural selection mechanism and then add some refinements.

In the Darwinian scheme, species become adapted to solve the problems of their environment through chance variations and selective perpetuation. Campbell supposes that the creative thinker exhibits counterpart cognitive mechanisms; these mechanisms blindly generate possible solutions to intellectual problems, select the best-adapted thought trials, and reproduce consequently acquired knowledge on the appropriate occasions. A distinctive postulate of this model is that cognitive variations are produced blindly, which is to say that initial thought trials are not justified by induction from the environment, or by previous trials, or by "the eventual fit or structured order that is to be explained."[64] The production of thought variations by the scientist – or the creative thinker in any realm – is therefore precisely analogous to chance mutations and recombinations in organic evolution.

The bones of Campbell's conception can be fleshed out in ways that make it fit for the historiographic model I have in mind. The following additional postulates serve this function.

1. The generation and selection of scientific ideas, both as the hypotheses that guide a scientist's work and as the relatively sedimented doctrine of the scientific community, should be understood as the result of a feedback mechanism. Such a mechanism, which we may consider only formally without worrying about its physiological realization, will generate ideas in a biased rather than in a purely random fashion. For without some restraints on generation, a scientist might produce an infinity of ideas with virtually no probability of hitting on a solution to even the simplest problem. But of course even mutations and recombinations of genes do not occur completely at random. The constraints on idea production are determined by the vagaries of education and in-

tellectual connections, the social milieu, psychological dispositions, previously settled theory, and recently selected ideas. This postulate therefore suggests that, though ideas may come serendipitously, their generation is not unregulated but can be comprehended by the historian. Thus, for example, when Darwin began musing on the nature of a mechanism to explain species change, he did so in a conceptual environment formed partly of ideas stimulated by his *Beagle* voyage and partly of ideas acquired from his grandfather, from Lamarck, and from a host of authors he read between 1836 and 1838. These ideas not only determined the various problems against which successful hypotheses were selected, but they also initially fixed the restraints on the generation of trial solutions. It is within a certain (albeit vaguely defined and shifting) conceptual space that chance variations are displayed. And it is because of such constraints that even a scientist's rejected hypotheses can make sense to the historian.

2. To think scientifically is to direct the mind to the solution of problems posed by the intellectual environment. Novel ideas are not produced in an environment where perceptual or theoretical situations are settled. As Popper (and Dewey before him) has argued, for thinking to occur there must be a troubled, unsettled cognitional matrix; the perceived environment must be changing. Conversely, alterations in the intellectual situation that are not perceived or that are ignored must lead to the arrest of scientific thought and the eventual extinction of a scientific system.

3. Ideas and ultimately well articulated theories are originally generated and selected within the conceptual domain of the individual scientist. Only after an idea system has been introduced to the scientific community (or communities, since scientists usually belong to several interlocking social networks) does public scrutiny result. The broader conceptual environment established by the community may present somewhat different problem situations and standards of competitive survival. To the extent, however, that the problem environments of the individual and the community coincide, individually selected ideas or theories will be fit for life in the community. If the historian neglects (as Toulmin does) to consider the processes of idea generation and evaluation at the individual level, then scientific ideas will appear to come mysteriously preadapted to their public environment.

4. Finally, if this model is to be used in construing the acquisition of knowledge in science, then one must suppose that selection components operate in accord with certain essential criteria: logical consistency, semantic coherence, standards of verifiability and falsifiability,

and observational relevance. These criteria may function only implicitly, but they form a necessary subset of criteria governing the development of scientific thought throughout its history. Without such norms, we would not be dealing with the selection of *scientific* ideas. The criteria thus aid historians of science in distinguishing their subject from other cognitive occupations. It should be stressed, however, that these selection criteria are themselves the result of previous idea generation and continuous selection, processes by means of which science has descended from protoscience – just as the mammals have descended from the reptiles. The complete set of selection criteria define what in a given historical period constitutes the standard of scientific acceptability. The above-specified criteria are only elements of this more comprehensive set.

Natural Selection Model vs. Scientific Research Programs

Since the natural selection model of science (NSM for short) is a model, it implicitly represents a theory about science, about its structure, growth, and rationality. The model and its imbedded theory portray scientific conceptual systems as quasi-organisms that compete for survival; and it proposes that the system which best solves the problems of its cultural environment will survive, gradually displacing its competitors. In order to assess the model's viability, we might compare it with another powerful model, which appears to offer it the keenest competition – Lakatos's model of scientific research programs (SRP).[65] Lakatos designed SRP to serve both as a standard for appraising the scientific and rational status of contemporary conceptual systems and as a historiographic device for constructing explanatory accounts of science's growth. Because of this explicit intention and the model's rigorous formulation, SRP furnishes an exceptional standard by which to evaluate NSM.

Lakatos formulated his model expressly for the purpose of interpreting the history of science as rationally progressive. He contrasts his conception with the Kuhnian model, which he regards, correctly I believe, as forbidding judgments of general scientific progress across problem shifts and as supplying no criteria for distinguishing scientific rationality from doctrinaire opinion.[66] Yet like Kuhn, Lakatos chooses a larger unit of analysis than the solitary idea or theory, since he recognizes the historical and epistemological fact that ideas cannot be evaluated in isolation from the auxiliary concepts which specify normal conditions, relevant evidence, and theoretical pertinence. He takes this larger conceptual scheme, the SRP, as the entity to be judged as progressing (or degenerating), as competitive with other programs, and as the basis for estimating the rationality of a particular scientific enterprise.

As Lakatos characterizes its structure, SRP has a "hard core" of central principles and a "belt of surrounding auxiliary hypotheses" that continues to change during the life of a program.[67] Newton's program, for example, had a stable center consisting of his three laws of dynamics and principle of attraction; it also had a belt of hypotheses composed of assumptions about the gravitational center of large bodies, the viscosity of different resisting media, the paths of planets, the distance of the fixed stars, and a host of other boundary conditions. If a program is to be pursued, the hard core embodying its defining ideas must be protected, especially in the early stages of growth, from noxious facts and the harmful competition of rival programs. The program's "negative heuristic," then, bids falsification attempts be deflected to the auxiliary hypotheses. It is the protective girdle of hypotheses that is challenged by the facts and adjusted to escape the force of contrary evidence. The "positive heuristic" of the program complements the negative imperative by proposing means of advancing the empirical content of the program through development of the auxiliary hypotheses. The positive heuristic discharges this function principally by suggesting replacement hypotheses when evidence or internal logic require that and by setting the plan which the program will stubbornly follow in the face of anomalies and the claims of rival programs.

Lakatos offers his model as a refinement of Popper's. It nevertheless differs from Popper's selection model on an important point. Popper sometimes suggests that theories can be falsified directly, by infection from toxic facts, and that such falsified theories are (or scientific honor demands that they should be) rendered immediately extinct.[68] Lakatos, in contrast, recognizes that theories may accumulate anomalies but that scientists properly adjust their auxiliary hypotheses to avoid them, even, if possible, to turn them into dramatic corroborations of the program. Darwin, for example, was initially stumped by the seemingly inexplicable adaptations of neuter insects – they left no progeny to inherit favorable variations. But when after several years he finally developed his mechanism of community selection, what originally threatened to falsify his theory became the strongest evidence for it. Such manipulations, however, can be abused.

To prevent ad hoc alterations from turning rational science into empirically immune pseudoscience, Lakatos stipulates that such adjustments should be capacious enough to extend the empirical content of the theory beyond the refuting cases, so that such extension yields the prediction of new facts: "A given fact is explained scientifically only if a new fact is also explained with it."[69] Predictive extension is for Lakatos the mark of a scientifically authentic research program. If a program accounts for the empirical content of rivals but also generates further predictions, then the program is "theoretically progressive" and thus scientific. If the predictive excess is corroborated, then the program is also "empirically progressive"; otherwise, it is "degenerating." But if a program confronts incompatible facts that its constituent theories cannot neutralize, while theories of a rival program can give them account, and if that rival also generates further predictions, some of which are corroborated, then the original program is "falsified." Programs that persist in the face of a rival's success can only be judged "pseudo-scientific."[70]

Lakatos has constructed his model as a device for appraising research programs, both recent and historically remote ones. He believes that appropriate standards are required if the historian is to do his job properly. SRP allows the historian to distinguish the internal history of science, which expresses the rational growth of objective knowledge, from "empirical (sociopsychological) 'external history.'"[71] Since the principal meaning for "science" is "accomplished objective knowledge," the internal history of science captures all that is essential to it. SRP allows investigators to select out of the morass of historical clutter precisely their special subject, the internal logical development of theories. After all,

> most theories of the growth of knowledge are theories of the growth of disembodied knowledge: whether an experiment is crucial or not, whether a hypothesis is highly probable in the light of the available evidence or not, whether a problem shift is progressive or not, is not dependent in the slightest on the scientists' beliefs, personalities, or authority. These subjective factors are of no interest for any internal history.[72]

Indeed, with the help of SRP, the historian of science should construct an ideal history, the normative fabula that the logic of a given research program demands. Lakatos offers Niels Bohr's program as illustrative:

> Bohr, in 1913, may not have even thought of the possibility of electron spin. He had more than enough on his hands without the spin. Nevertheless, the historian, describing with hindsight the Bohrian

program, should include electron spin in it, since electron spin fits naturally in the original outline of the program. Bohr might have referred to it in 1913. Why Bohr did not do so is an interesting problem which deserves to be indicated in a footnote.[73]

Despite his comic exaggeration, Lakatos does not intend that the historian should literally write a bilevel history, one narrative in the text and another in the footnotes. But his model does require a history structured with a logically distinctive internal core which is to control the significance assigned to external social and psychological events. This core, in his view, should express not only the research program that some scientist actually established but also an enlarged program that contains features implicitly derivable from the original.

In Lakatos's application, SRP exudes a peculiar Platonic odor, which I think most historians would find offensive. For the model appears to demand not a historian but a Laplacean demon who could extract from a program all that was logically or compatibly contained therein. Insofar as SRP is used by human historians, it seems to urge them to read history backwards, to find in earlier, inchoate concepts the results of more recent research. In Lakatos's hands, SRP would obscure the vision of historians wishing to detect the emergence of scientific ideas from previously developed ideas, community expectations, and personal aims.

EDITOR'S NOTE

Richards completes this discussion with a defense of his own "natural selection" model against what he regards as the most serious contender, "scientific research programs." Although of interest to professional historians and philosophers of science, the overview of historiographic models has been sufficently summarized and this concluding section has been omitted.

NOTES

1. R. Bostocke, *The Difference Between the Auncient Phisicke and the Latter Phisicke* (1585), in Allen Debus, "An Elizabethan History of Medical Chemistry," *Annals of Science* 18 (1962): 1–29.
2. J.E. McGuire and P. Rattansi, "Newton and the 'Pipes of Pan,'" *Notes and Records of the Royal Society of London* 21 (1966): 108–43.
3. Olaus Borrichius, *De ortu et progressu chemiac dissertatio* (1668), in *Bibliotheca chemica curiosa*, ed. J. Manget (Geneva: Chouet, 1702).
4. Ibid., p. 1.
5. John Freind, *The History of Physick from the Time of Galen to the Beginning of the Sixteenth Century* (London: Walthe, 1725).
6. R. Watson, "On the Rise and Progress of Chemistry," in vol. 1 of his *Chemical Essays*, 6th ed. (London: Evans, 1793).
7. Friend, *History of Physick*, 1: 298.
8. Ibid., p. 9.
9. Watson, "On the Rise and Progress of Chemistry," p. 30.
10. Jean d'Alembert, *Discours Préliminaire*, in vol. I of *Encyclopédie on dictionnaire raisonne des sciences, des arts et des métiers*, 2d ed., ed. Dennis Diderot and Jean D'Alembert (Paris: Lucques, 1758–1771).
11. William Whewell, *History of the Inductive Sciences* (London: Parker, 1837).
12. Ibid. 1: 9.
13. Ibid., p. 10.
14. George Sarton, *Sarton on the History of Science: Essays by George Sarton* (Cambridge: Harvard University Press, 1962).
15. Ibid., p. 15.
16. Ibid., p. 13.
17. Ibid., p. 21.
18. Immanuel Kant, *Kritik der reinen Vernunft* (1787), vol. 2 of *Immanuel Kant Werke in sechs Bänden*, ed. W. Weischedel (Wiesbaden: Insel, 1956), p. 23 (B xii).
19. Alexandre Koyré, *Metaphysics and Measurement: Essays in Scientific Revolution* (Cambridge: Harvard University Press, 1968).
20. Alexandre Koyré, *From the Closed World to the Infinite Universe* (Baltimore: Johns Hopkins University Press, 1957), p. 2.
21. The role of occult influences on the development of science is highly controverted. The dispute may be followed in the following discussions: Francis Yates, *Giordanu Bruno and the Hermetic Tradition* (Chicago: University of Chicago Press, 1964); P. Rattansi, "Some Evaluations of Reason in Sixteenth- and Seventeenth-Century Natural Philosophy,' in *Changing Perspectives in the History of Science*, ed. M. Teich and R. Young (London: Heinemann, 1973); and Mary Hesse "Reasons and Evaluation in the History of Science," in *Changing Perspectives in the History of Science*. The various parties are brought together in Roger Stuewer, ed., *Historical and Philosophical Perspectives on Science*, vol. 5 of *Minnesota Studies in the Philosophy of Science* (Minneapolis: University of Minnesota Press, 1970).
22. Alistair Crombie, *Medieval and Early Modern Science*, 2d ed. (Cambridge: Harvard University Press, 1961), 2 : 125; A. Rupert Hall, *The Scientific*

Revolution, 1500–1800 (Boston: Beacon, 1966), pp. 370–71; Charles Gillispie, *The Edge of Objectivity* (Princeton: Princeton University Press, 1960), pp. 8–16.

23. Hall, *The Scientific Revolution,* pp. 217–43.
24. Koyré, *Metaphysics and Measurement,* p. 17.
25. Hall, *The Scientific Revolution,* p. 370.
26. Crombie, *Medieval and Early Modern Science* 2: 110.
27. Gillispie, *The Edge of Objectivity,* pp. 8–16.
28. Hall, *The Scientific Revolution,* p. xii.
29. Ibid., p. xiii.
30. Norwood Russell Hanson, *Patterns of Discovery* (Cambridge: Cambridge University Press, 1970), p. 90.
31. Thomas Kuhn, *The Structure of Scientific Revolutions,* 2d ed. (Chicago: University of Chicago Press, 1970), p. 53.
32. Ibid., pp. 84–85.
33. Michel Foucault, *The Order of Things: An Archaeology of the Human Sciences* (New York: Vintage, [1966] 1973), p. xi.
34. Ibid., pp. 303–4.
35. Ibid., p. xii.
36. Hall, *The Scientific Revolution,* pp. xiii–xiv.
37. Hanson, *Patterns of Discovery,* p. 15.
38. Kuhn, *Structure of Scientific Revolutions,* pp. 170–71.
39. Samuel Clarke, *The Leibniz–Clarke Correspondence,* ed. H. Alexander (Manchester: Manchester University Press, [1717] 1956), p. 6.
40. J.D. Bernal, *Science in History,* 3d ed. (Cambridge, Mass.: M.I.T. Press, 1971), 1: 31.
41. Ibid. 2: 644.
42. Ibid. 1: 43–44.
43. Erik Nordenskiöld, *The History of Biology,* 2d ed. (New York: Tudor, 1936), p. 477.
44. Robert Merton, *The Sociology of Science* (Chicago: University of Chicago Press, 1973).
45. Joseph Ben-David, *The Scientist's Role in Society* (Englewood Cliffs, N.J.: Prentice-Hall, 1971).
46. Ibid., pp. 11–12.
47. Margaret Jacob, *The Newtonians and the English Revolution* (Ithaca, N.Y.: Cornell University Press, 1976).
48. David Bloor, "Klassifikation und Wissenssoziologie: Durkheim und Mauss neu betrachtet," *Kölner Zeitschrift für Soziologie und Sozialpsychologie* 22 (1980): 20–51.
49. Arthur Mitzman, *The Iron Cage: An Historical Interpretation of Max Weber* (New York: Grosset & Dunlap, 1971).
50. David Bloor, *Knowledge and Social Imagery* (London: Routledge & Kegan Paul, 1976), pp. 95–116.
51. See John Burrow, *Evolution and Society* (Cambridge: Cambridge University Press, 1966); and George Stocking, *Race, Culture, and Evolution,* 2d ed. (Chicago: University of Chicago Press, 1981).
52. Gerald Holton, *Thematic Origins of Scientific Thought: Kepler to Einstein* (Cambridge: Harvard University Press, 1973).

53. Thomas Kuhn, "Reflections on My Critics," in *Criticism and the Growth of Knowledge*, ed. Imre Lakatos and Alan Musgrave (Cambridge: Cambridge University Press, 1970), p. 264.

54. Karl Popper, *The Logic of Scientific Discovery*, 2d ed. (New York: Harper & Row, 1968), p. 108.

55. Karl Popper, *Objective Knowledge* (Oxford: Oxford University Press, 1972), p. 145.

56. Karl Popper, *Conjectures and Refutations: The Growth of Scientific Knowledge*, 2d ed. (New York: Harper & Row, 1968), pp. 215–50.

57. Stephen Toulmin, *Human Understanding* (Oxford: Oxford University Press, 1972), p. 130.

58. Ibid., pp. 307–13.

59. Ibid., p. 159.

60. Ibid., p. 130.

61. Ibid., p. 337.

62. See Diana Crane, *Invisible College* (Chicago: University of Chicago Press, 1972).

63. Donald Campbell has developed his theory in a series of papers: "Methodological Suggestions from a Comparative Psychology of Knowledge Processes," *Inquiry* 2 (1959): 152–82; "Blind Variation and Selective Retention in Creative Thought as in Other Knowledge Processes," *Psychological Review* 67 (1960): 380–400; "Blind Variation and Selective Retention in Socio-Cultural Evolution," in *Social Change in Developing Areas*, ed. H. Barringer, G. Blanksten, and R. Mack (Cambridge, Mass.: Schenkman, 1965); 'Evolutionary Epistemology,' in *The Philosophy of Karl Popper*, ed. Paul Schilpp (La Salle, Ill.: Open Court, 1974); "Unjustified Variation and Selective Retention in Scientific Discovery," in *Studies in the Philosophy of Biology*, ed. Francisco Ayala and Theodosius Dobzhansky (London: Macmillan, 1974); "Discussion Comment on "The Natural Selection Model of Conceptual Evolution,'" *Philosophy of Science* 44 (1977): 502–507.

64. Campbell, 'Unjustified Variation and Selective Retention in Scientific Discovery,' p. 150.

65. See Imre Lakatos, "Falsification and the Methodology of Scientific Research Programmes" and "History of Science and Its Rational Reconstructions," in *The Methodology of Scientific Research Programmes: Philosophical Papers of Imre Lakatos*, vol. 1, ed. John Worrall and Gregory Currie (Cambridge: Cambridge University Press, 1978).

66. Lakatos, "Falsification and the Methodology of Scientific Research Programmes," pp. 8–10.

67. Ibid., pp. 48–52.

68. Popper, *The Logic of Scientific Discovery*, pp. 86–87.

69. Lakatos, "Falsification and the Methodology of Scientific Research Programmes," p. 34.

70. Ibid., pp. 32–35.

71. Lakatos, "History of Science and Its Rational Reconstructions," p. 102.

72. Ibid., p. 118.

73. Ibid., p. 119.

9 The Road Since *Structure**

Thomas S. Kuhn

On this occasion, and in this place, I feel that I ought, and am probably
expected, to look back at the things which have happened to the philosophy of science since I first began to take an interest in it over half a
century ago. But I am both too much an outsider and too much a
protagonist to undertake that assignment. Rather than attempt to situate the present state of philosophy of science with respect to its past –
a subject on which I've little authority – I shall try to situate my present
state in philosophy of science with respect to its own past – a subject
on which, however imperfect, I'm probably the best authority there is.

As a number of you know, I'm at work on a book, and what I mean to
attempt here is an exceedingly brief and dogmatic sketch of its main
themes. I think of my project as a return, now underway for a decade,
to the philosophical problems left over from the *Structure of Scientific
Revolutions*. But it might better be described more generally, as a study
of the problems raised by the transition to what's sometimes called the
historical and sometimes (at least by Clark Glymour, speaking to me)
just the "soft" philosophy of science. That's a transition for which I get
far more credit, and also more blame, than I have coming to me. I
was, if you will, present at the creation, and it wasn't very crowded.
But others were present too: Paul Feyerabend and Russ Hanson, in particular, as well as Mary Hesse, Michael Polanyi, Stephen Toulmin, and
a few more besides. Whatever a *Zeitgeist* is, we provided a striking illustration of its role in intellectual affairs.

Returning to my projected book, you will not be surprised to hear that
the main targets at which it aims are such issues as rationality, relativism
and, most particularly, realism and truth. But they're not primarily
what the book is about, what occupies most space in it. That role is taken
instead by incommensurability. No other aspect of *Structure* has concerned me so deeply in the thirty years since the book was written, and I
emerge from those years feeling more strongly than ever that incommen-

* *PSA 1990*, vol. 2 (1991), pp. 3–13. This paper was delivered by Kuhn as the Presidential
Address at the 1990 Biennial Meeting of the Philosophy of Science Association (Minneapolis, Minnesota).

surability has to be an essential component of any historical, developmental, or evolutionary view of scientific knowledge. Properly understood – something I've by no means always managed myself – incommensurability is far from being the threat to rational evaluation of truth claims that it has frequently seemed. Rather, it's what is needed, within a developmental perspective, to restore some badly needed bite to the whole notion of cognitive evaluation. It is needed, that is, to defend notions like truth and knowledge from, for example, the excesses of post-modernist movements like the strong program. Clearly, I can't hope to make all that out here: it's a project for a book. But I shall try, however sketchily, to describe the main elements of the position the book develops. I begin by saying something about what I now take incommensurability to be, and then attempt to sketch its relationship to questions of relativism, truth, and realism. In the book, the issue of rationality will figure, too, but there is no space here even to sketch its role.

Incommensurability is a notion that for me emerged from attempts to understand apparently nonsensical passages encountered in old scientific texts. Ordinarily they had been taken as evidence of the author's confused or mistaken beliefs. My experiences led me to suggest, instead, that those passages were being misread: the appearance of nonsense could be removed by recovering older meanings for some of the terms involved, meanings different from those subsequently current. During the years since, I've often spoken metaphorically of the process by which later meanings had been produced from earlier ones as a process of language change. And, more recently, I've spoken also of the historian's recovery of older meanings as a process of language learning rather like that undergone by the fictional anthropologist whom Quine misdescribes as a radical translator (Kuhn, 1983a). The ability to learn a language does not, I've emphasized, guarantee the ability to translate into or out of it.

By now, however, the language metaphor seems to me far too inclusive. To the extent that I'm concerned with language and with meanings at all – an issue to which I'll shortly return – it is with the meanings of a restricted class of terms. Roughly speaking, they are taxonomic terms or kind terms, a widespread category that includes natural kinds, artifactual kinds, social kinds, and probably others. In English the class is coextensive, or nearly so, with the terms that by themselves or within appropriate phrases can take the indefinite article. These are primarily the count nouns together with the mass nouns, words which combine with count nouns in phrases that take the indefinite article. Some terms require still further tests hinging, for example, on permissible suffixes.

Terms of this sort have two essential properties. First, as already indicated, they are marked or labelled as kind terms by virtue of lexical characteristics like taking the indefinite article. Being a kind term is thus part of what the word means, part of what one must have in the head to use the word properly. Second – a limitation I sometimes refer to as the no-overlap principle – no two kind terms, no two terms with the kind label, may overlap in their referents unless they are related as species to genus. There are no dogs that are also cats, no gold rings that are also silver rings, and so on: that's what makes dogs, cats, silver, and gold each a kind. Therefore, if the members of a language community encounter a dog that's also a cat (or, more realistically, a creature like the duck-billed platypus), they cannot just enrich the set of category terms but must instead redesign a part of the taxonomy. *Pace* the causal theorists of reference, "water" did not always refer to H_2O (Kuhn, 1987; 1990, pp. 309–14).

Notice now that a lexical taxonomy of some sort must be in place before description of the world can begin. Shared taxonomic categories, at least in an area under discussion, are prerequisite to unproblematic communication, including the communication required for the evaluation of truth claims. If different speech communities have taxonomies that differ in some local area, then members of one of them can (and occasionally will) make statements that, though fully meaningful within that speech community, cannot in principle be articulated by members of the other. To bridge the gap between communities would require adding to one lexicon a kind-term that overlaps, shares a referent, with one that is already in place. It is that situation which the no-overlap principle precludes.

Incommensurability thus becomes a sort of untranslatability, localized to one or another area in which two lexical taxonomies differ. The differences which produce it are not any old differences, but ones that violate either the no-overlap condition, the kind-label condition, or else a restriction on hierarchical relations that I cannot spell out here. Violations of those sorts do not bar intercommunity understanding. Members of one community can acquire the taxonomy employed by members of another, as the historian does in learning to understand old texts. But the process which permits understanding produces bilinguals, not translators, and bilingualism has a cost, which will be particularly important to what follows. The bilingual must always remember within which community discourse is occurring. The use of one taxonomy to make statements to someone who uses the other places communication at risk.

Let me formulate these points in one more way, and then make a last remark about them. Given a lexical taxonomy, or what I'll mostly now call simply a lexicon, there are all sorts of different statements that can be made, and all sorts of theories that can be developed. Standard techniques will lead to some of these being accepted as true, others rejected as false. But there are also statements which could be made, theories which could be developed, within some other taxonomy but which cannot be made with this one and vice versa. The first volume of Lyons' *Semantics* (1977, pp. 237–8) contains a wonderfully simple example, which some of you will know: the impossibility of translating the English statement, "the cat sat on the mat", into French, because of the incommensurability between the French and English taxonomies for floor coverings. In each particular case for which the English statement is true, one can find a co-referential French statement, some using "tapis", others "paillasson," still others "carpette," and so on. But there is no single French statement which refers to all and only the situations in which the English statement is true. In that sense, the English statement cannot be made in French. In a similar vein, I've elsewhere pointed out (Kuhn, 1987, p. 8) that the content of the Copernican statement, "planets travel around the sun", cannot be expressed in a statement that invokes the celestial taxonomy of the Ptolemaic statement, "planets travel around the earth". The difference between the two statements is not simply one of fact. The term "planet" appears as a kind term in both, and the two kinds overlap in membership without either's containing all the celestial bodies contained in the other. All of which is to say that there are episodes in scientific development which involve fundamental change in some taxonomic categories and which therefore confront later observers with problems like those the ethnologist encounters when trying to break into another culture.

A final remark will close this sketch of my current views on incommensurability. I have described those views as concerned with words and with *lexical* taxonomy, and I shall continue in that mode: the sorts of knowledge I deal with come in explicit verbal or related symbolic forms. But it may clarify what I have in mind to suggest that I might more appropriately speak of concepts than of words. What I have been calling a lexical taxonomy might, that is, better be called a conceptual scheme, where the "very notion" of a conceptual scheme is not that of a set of beliefs but of a particular operating mode of a mental module prerequisite to having beliefs, a mode that at once supplies and bounds the set of beliefs it is possible to conceive. Some such taxonomic module I take to be pre-linguistic and possessed by animals. Presumably it evolved originally for the sensory, most obviously for the visual, system. In the book I shall give reasons for supposing that it developed from a still more fundamental mechanism which enables individual living organisms to reidentify other substances by tracing their spatio–temporal trajectories.

I shall be coming back to incommensurability, but let me for now set it aside in order to sketch the developmental framework within which it functions. Since I must again move quickly and often cryptically, I begin by anticipating the direction in which I am headed. Basically, I shall be trying to sketch the form which I think any viable evolutionary epistemology has to take. I shall, that is, be returning to the evolutionary analogy introduced in the very last pages of the first edition of *Structure*, attempting both to clarify it and to push it further. During the thirty years since I first made that evolutionary move, theories of the evolution both of species and of knowledge have, of course, been transformed in ways I am only beginning to discover. I still have much to learn, but to date the fit seems extremely good.

I start from points familiar to many of you. When I first got involved, a generation ago, with the enterprise now often called historical philosophy of science, I and most of my coworkers thought history functioned as a source of empirical evidence. That evidence we found in historical case studies, which forced us to pay close attention to science as it really was. Now I think we overemphasized the empirical aspect of our enterprise (an evolutionary epistemology need not be a naturalized one). What has for me emerged as essential is not so much the details of historical cases as the perspective or the ideology that attention to historical cases brings with it. The historian, that is, always picks up a process already underway, its beginnings lost in earlier time. Beliefs are already in place; they provide the basis for the ongoing research whose results will in some cases change them; research in their absence is unimaginable though there has nevertheless been a long tradition of imagining it. For the historian, in short, no Archimedean platform is available for the pursuit of science other than the historically situated one already in place. If you approach science as an historian must, little observation of its actual practice is required to reach conclusions of this sort.

Such conclusions have by now been pretty generally accepted: I scarcely know a foundationalist any more. But for me, this way of abandoning foundationalism has a further consequence which, though widely discussed, is by no means widely or fully accepted. The discussions I have in mind usually proceed under the rubric of the rationality or relativity of truth claims, but these labels misdirect attention. Though both rationality and relativism are somehow implicated, what is fundamentally at stake is rather the correspondence theory of truth, the notion that the goal, when evaluating scientific laws or theories, is to determine whether or not they correspond to an external, mind-independent world. It is that notion, whether in an absolute or probabilistic form, that I'm persuaded must vanish together with foundationalism. What replaces it will still require a strong conception of truth, but not, except in the most trivial sense, correspondence truth.

Let me at least suggest what the argument involves. On the developmental view, scientific knowledge claims are necessarily evaluated from a moving, historically-situated, Archimedean platform. What requires evaluation cannot be an individual proposition embodying a knowledge claim in isolation: embracing a new knowledge claim typically requires adjustment of other beliefs as well. Nor is it the entire body of knowledge claims that would result if that proposition were accepted. Rather, what's to be evaluated is the desirability of a particular change-of-belief, a change which would alter the existing body of knowledge claims so as to incorporate, with minimum disruption, the new claim as well. Judgements of this sort are necessarily comparative: which of two bodies of knowledge – the original or the proposed alternative – is *better* for doing whatever it is that scientists do. And that is the case whether what scientists do is solve puzzles (my view), improve empirical adequacy (Bas van Fraassen's), or increase the dominance of the ruling elite (in parody, the strong program's). I do, of course, have my own preference among these alternatives, and it makes a difference (Kuhn, 1983b). But no choice between them is relevant to what's presently at stake.

In comparative judgements of the kind just sketched, shared beliefs are left in place: they serve as the given for purposes of the current evaluation; they provide a replacement for the traditional Archimedean platform. The fact that they may – indeed probably will – later be at risk in some other evaluation is simply irrelevant. Nothing about the rationality of the outcome of the current evaluation depends upon their, in fact, being true or false. They are simply in place, part of the historical situation within which this evaluation is made. But if the actual truth value of the shared presumptions required for the evaluation is irrelevant, then the question of the truth or falsity of the changes made or rejected on the basis of that evaluation cannot arise either. A number of classic problems in philosophy of science – most obviously Duhemian holism – turn out on this view to be due not to the nature of scientific knowledge but to a misperception of what justification of belief is all about. Justification does not aim at a goal external to the historical situation but simply, in that situation, at improving the tools available for the job at hand.

To this point I have been trying to firm-up and extend the parallel between scientific and biological development suggested at the end of the first edition of *Structure*: scientific development must be seen as a process driven from behind, not pulled from ahead – as evolution from, rather than evolution towards. In making that suggestion, as elsewhere in the book, the parallel I had in mind was diachronic, involving the relation between older and more recent scientific beliefs about the same or overlapping ranges of natural phenomena. Now I want to suggest a second, less widely perceived parallel between Darwinian evolution and the evolution of knowledge, one that cuts a synchronic slice across the sciences rather than a diachronic slice containing one of them. Though I have in the past occasionally spoken of the incommensurability between the theories of contemporary scientific specialities, I've only in the last few years begun to see its significance to the parallels between biological evolution and scientific development. Those parallels have also been persuasively emphasized recently in a splendid article by Mario Biagioli of UCLA (1990). To both of us they seem extremely important, though we emphasize them for somewhat different reasons.

To indicate what is involved I must revert briefly to my old distinction between normal and revolutionary development. In *Structure* it was the distinction between those developments that simply add to knowledge, and those which require giving up part of what's been believed before. In the new book it will emerge as the distinction between developments which do and developments which do not require local taxonomic change. (The alteration permits a significantly more nuanced description of what goes on during revolutionary change than I've been able to provide before.) During this second sort of change, something else occurs that in *Structure* got mentioned only in passing. After a revolution there are usually (perhaps always) more cognitive specialties or fields of knowledge than there were before. Either a new branch has split off from the parent trunk, as scientific specialties have repeatedly split off in the past from philosophy and from medicine. Or else a new specialty has been born at an area of apparent overlap between two preexisting specialties, as occurred, for example, in the cases of physical chemistry and molecular biology. At the time of its occurrence this second sort of split is often hailed as a reunification of the sciences, as was the case in the episodes just mentioned. As time goes on, however, one notices that the new shoot seldom or never gets assimilated to either of its parents. Instead, it becomes one more separate specialty, gradually acquiring its own new specialists' journals, a new professional society, and often also new university chairs, laboratories, and even departments. Over time a diagram of the evolution of scientific fields, specialties, and sub-specialties comes to look strikingly like a layman's diagram for a biological evolutionary tree. Each of these fields has a distinct lexicon, though the differences are local, occurring only here and there. There is no *lingua franca* capable of expressing, in its entirety, the content of them all or even of any pair.

With much reluctance I have increasingly come to feel that this process of specialization, with its consequent limitation on communication and community, is inescapable, a consequence of first principles. Specialization and the narrowing of the range of expertise now look to me like the necessary price of increasingly powerful cognitive tools. What's involved is the same sort of development of special tools for special functions that's apparent also in technological practice. And, if that is the case, then a couple of additional parallels between biological evolution and the evolution of knowledge come to seem especially consequential. First, revolutions, which produce new divisions between fields in scientific development, are much like episodes of speciation in biological evolution. The biological parallel to revolutionary change is not mutation, as I thought for many years, but speciation. And the problems presented by speciation (e.g., the difficulty in identifying an episode of speciation until some time after it has occurred, and the impossibility, even then, of dating the time of its occurrence) are very similar to those presented by revolutionary change and by the emergence and individuation of new scientific specialties.

The second parallel between biological and scientific development, to which I return again in the concluding section, concerns the unit which undergoes speciation (not to be confused with a unit of selection). In the biological case, it is a reproductively isolated population, a unit whose members collectively embody the gene pool which ensures both the population's self-perpetuation and its continuing isolation. In the scientific case, the unit is a community of intercommunicating specialists, a unit whose members share a lexicon that provides the basis for both the conduct and the evaluation of their research and which simultaneously, by barring full communication with those outside the group, maintains their isolation from practitioners of other specialties.

To anyone who values the unity of knowledge, this aspect of specialization – lexical or taxonomic divergence, with consequent limitations on communication – is a condition to be deplored. But such unity may be in principle an unattainable goal, and its energetic pursuit might well place the growth of knowledge at risk. Lexical diversity and the principled limit it imposes on communication may be the isolating mechanism required for the development of knowledge. Very likely it is the specialization consequent on lexical diversity that permits the sciences, viewed collectively, to solve the puzzles posed by a wider range of natural phenomena than a lexically-homogenous science could achieve.

Though I greet the thought with mixed feelings, I am increasingly persuaded that the limited range of possible partners for fruitful intercourse is the essential precondition for what is known as progress in both biological development and the development of knowledge. When I suggested earlier that incommensurability, properly understood, could reveal the source of the cognitive bite and authority of the sciences, its role as an isolating mechanism was prerequisite to the topic I had principally in mind, the one to which I now turn.

This reference to "intercourse", for which I shall henceforth substitute the term "discourse", bring me back to problems concerning truth, and thus to the locus of the newly restored bite. I said earlier that we must learn to get along without anything at all like a correspondence theory of truth. But something like a redundancy theory of truth is badly needed to replace it, something that will introduce minimal laws of logic (in particular, the law of non-contradiction) and make adhering to them a precondition for the rationality of evaluations (Horwich, 1990). On this view, as I wish to employ it, the essential function of the concept of truth is to require choice between acceptance and rejection of a statement or a theory in the face of evidence shared by all. Let me try briefly to sketch what I have in mind.

Ian Hacking, in an attempt (1982) to denature the apparent relativism associated with incommensurability, spoke of the way in which new "styles" introduce into science new candidates for true/false. Since that time, I've been gradually realizing (the reformulation is still in process) that some of my own central points are far better made without speaking of statements as themselves being true or as being false. Instead, the evaluation of a putatively scientific statement should be conceived as comprising two seldom-separated parts. First, determine the status of the statement: is it a candidate for true/false? To that question, as you'll shortly see, the answer is lexicon-dependent. And second, supposing a positive answer to the first, is the statement rationally assertable? To that question, given a lexicon, the answer is properly found by something like the normal rules of evidence.

In this reformulation, to declare a statement a candidate for true/ false is to accept it as a counter in a language game whose rules forbid asserting both a statement and its contrary. A person who breaks that rule declares him or herself outside the game. If one nevertheless tries to continue play, then discourse breaks down; the integrity of the language community is threatened. Similar, though more problematic, rules apply, not simply to contrary statements, but more generally to logically incompatible ones. There are, of course, language games without the rule of non-contradiction and its relatives: poetry and mystical discourse, for example. And there are also, even within the declarative-statement game, recognized ways of bracketing the rule, permitting and even exploiting the use of contradiction. Metaphor and other tropes are the most obvious examples; more central for present purposes are the historian's restatements of past beliefs. (Though the originals were candidates for true/false, the historian's later restatements – made by a bilingual speaking the language of one culture to the members of another – are not.) But in the sciences and in many more ordinary community activities, such bracketing devices are parasitic on normal discourse. And these activities – the ones that presuppose normal adherence to the rules of the true/false game – are an essential ingredient of the glue that binds communities together. In one form or another, the rules of the true/false game are thus universals for all human communities. But the result of applying those rules varies from one speech community to the next. In discussion between members of communities with differently structured lexicons, assertability and evidence play the same role for both only in areas (there are always a great many) where the two lexicons are congruent.

Where the lexicons of the parties to discourse differ, a given string of words will sometimes make different statements for each. A statement may be a candidate for truth/falsity with one lexicon without having that status in the others. And even when it does, the two statements will not be the same: though identically phrased, strong evidence for one need not be evidence for the other. Communication breakdowns are then inevitable, and it is to avoid them that the bilingual is forced to remember at all times which lexicon is in play, which community the discourse is occurring within.

These breakdowns in communication do, of course, occur: they're a significant characteristic of the episodes *Structure* referred to as "crises". I take them to be the crucial symptoms of the speciation-like process through which new disciplines emerge, each with it own lexicon, and each with its own area of knowledge. It is by these divisions,

I've been suggesting, that knowledge grows. And it's the need to maintain discourse, to keep the game of declarative statements going, that forces these divisions and the fragmentation of knowledge that results.

I close with some brief and tentative remarks about what emerges from this position as the relationship between the lexicon – the shared taxonomy of a speech community – and the world the members of that community jointly inhabit. Clearly it cannot be the one Putnam (1978, pp. 123–38) has called metaphysical realism. Insofar as the structure of the world can be experienced and the experience communicated, it is constrained by the structure of the lexicon of the community which inhabits it. Doubtless some aspects of that lexical structure are biologically determined, the products of a shared phylogeny. But, at least among advanced creatures (and not just those linguistically endowed), significant aspects are determined also by education, by the process of socialization, that is, which initiates neophytes into the community of their parents and peers. Creatures with the same biological endowment may experience the world through lexicons that are here and there very differently structured, and in those areas they will be unable to communicate all of their experiences across the lexical divide. Though individuals may belong to several interrelated communities (thus, be multilinguals), they experience aspects of the world differently as they move from one to the next.

Remarks like these suggest that the world is somehow mind-dependent, perhaps an invention or construction of the creatures which inhabit it, and in recent years such suggestions have been widely pursued. But the metaphors of invention, construction, and mind-dependence are in two respects grossly misleading. First, the world is not invented or constructed. The creatures to whom this responsibility is imputed, in fact, find the world already in place, its rudiments at their birth and its increasingly full actuality during their educational socialization, a socialization in which examples of the way the world is play an essential part. That world, furthermore, has been experientially given, in part to the new inhabitants directly, and in part indirectly, by inheritance, embodying the experience of their forebears. As such, it is entirely solid: not in the least respectful of an observer's wishes and desires; quite capable of providing decisive evidence against invented hypotheses which fail to match its behavior. Creatures born into it must take it as they find it. They can, of course, interact with it, altering both it and themselves in the process, and the populated world thus altered is the one that will be found in place by the generation which follows. The point closely parallels the one made earlier about the nature of evaluation seen from a developmental perspective: there, what required evaluation was not belief but change in some aspects of belief, the rest held fixed in the process; here, what people can effect or invent is not the world but changes in some aspects of it, the balance remaining as before. In both cases, too, the changes that can be made are not introduced at will. Most proposals for change are rejected on the evidence; the nature of those that remain can rarely be foreseen; and the consequences of accepting one or another of them often prove to be undesired.

Can a world that alters with time and from one community to the next correspond to what is generally referred to as "the real world"? I do not see how its right to that title can be denied. It provides the environment, the stage, for all individual and social life. On such life it places rigid constraints; continued existence depends on adaptation to them; and in the modern world scientific activity has become a primary tool for adaptation. What more can reasonably be asked of a real world?

In the penultimate sentence, above, the word "adaptation" is clearly problematic. Can the members of a group properly be said to adapt to an environment which they are constantly adjusting to fit their needs? Is it the creatures who adapt to the world or does the world adapt to the creatures? Doesn't this whole way of talking imply a mutual plasticity incompatible with the rigidity of the constraints that make the world real and that made it appropriate to describe the creatures as adapted to it? These difficulties are genuine, but they necessarily inhere in any and all descriptions of undirected evolutionary processes. The identical problem is, for example, currently the subject of much discussion in evolutionary biology. On the one hand the evolutionary process gives rise to creatures more and more closely adapted to a narrower and narrower biological niche. On the other, the niche to which they are adapted is recognizable only in retrospect, with its population in place: it has no existence independent of the community which is adapted to it (Lewontin 1978). What actually evolves, therefore, is creatures and niches together: what creates the tensions inherent in talk of adaptation is the need, if discussion and analysis are to be possible, to draw a line between the creatures within the niche, on the one hand, and their 'external' environment, on the other.

Niches may not seem to be worlds, but the difference is one of viewpoint. Niches are where *other* creatures live. We see them from outside and thus in physical interaction with their inhabitants. But the inhabitants of a niche see it from inside and their interactions with it are, to them, intentionally mediated through something like a mental representation. Biologically, that is, a niche is the world of the group which inhabits it, thus constituting it a niche. Conceptually, the world is *our* representation of *our* niche, the residence of the particular human community with whose members we are currently interacting.

The world-constitutive role assigned here to intentionality and mental representations recurs to a theme characteristic of my viewpoint throughout its long development: compare my earlier recourse to gestalt switches, seeing as understanding, and so on. This is the aspect of my work that, more than any other, has suggested that I took the world to be mind-dependent. But the metaphor of a mind-dependent world – like its cousin, the constructed or invented world – proves to be deeply misleading. It is groups and group-practices that constitute worlds (and are constituted by them). And the practice-in-the-world of some of those groups *is* science. The primary unit through which the sciences develop is thus, as previously stressed, the group, and groups do not have minds. Under the unfortunate title, "Are species individuals?", contemporary biological theory offers a significant parallel (Hull, 1976, provides an especially useful introduction to the literature). In one sense the procreating organisms which perpetuate a species are the units whose practice permits evolution to occur. But to understand the outcome of that process one must see the evolutionary unit (not to be confused with a unit of selection) as the gene pool shared by those organisms, the organisms which carry the gene pool serving only as the parts which, through bi-sexual reproduction, exchange genes within the population. Cognitive evolution depends, similarly, upon the exchange, through discourse, of statements within a community. Though the units which exchange those statements are individual scientists, understanding the advance of knowledge, the outcome of their practice, depends upon seeing them as atoms constitutive of a larger whole, the community of practitioners of some scientific specialty.

The primacy of the community over its members is reflected also in the theory of the lexicon, the unit which embodies the shared conceptual or taxonomic structure that holds the community together and simultaneously isolates it from other groups. Conceive the lexicon as a module within the head of an individual group member. It can then be shown (though not here) that what characterizes members of the group is possession not of identical lexicons, but of mutually congruent ones, of lexicons with the same structure. The lexical structure which characterizes a group is more abstract than, different in kind from, the individual lexicons or mental modules which embody it. And it is only that structure, not its various individual embodiments, that members of the community must share. The mechanics of taxonomizing are in this respect like its function: neither can be fully understood except as grounded within the community it serves.

By now it may be clear that the position I'm developing is a sort of post-Darwinian Kantianism. Like the Kantian categories, the lexicon supplies preconditions of possible experience. But lexical categories, unlike their Kantian forebears, can and do change, both with time and with the passage from one community to another. None of those changes, of course, is ever vast. Whether the communities in question are displaced in time or in conceptual space, their lexical structures must overlap in major ways or there could be no bridgeheads permitting a member of one to acquire the lexicon of the other. Nor, in the absence of major overlap, would it be possible for the members of a single community to evaluate proposed new theories when their acceptance required lexical change. Small changes, however, can have large-scale effects. The Copernican Revolution provides especially well-known illustrations.

Underlying all these processes of differentiation and change, there must, of course, be something permanent, fixed, and stable. But, like Kant's *Ding an sich*, it is ineffable, undescribable, undiscussible. Located outside of space and time, this Kantian source of stability is the whole from which have been fabricated both creatures and their niches, both the "internal" and the "external" worlds. Experience and description are possible only with the described and describer separated, and the lexical structure which marks that separation can do so in different ways, each resulting in a different, though never wholly different, form of life. Some ways are better suited to some purposes, some to others. But none is to be accepted as true or rejected as false; none gives privileged access to a real, as against an invented, world. The ways of being-in-the-world which a lexicon provides are not candidates for true/false.

REFERENCES

Biagioli, M. (1990), "The Anthropology of Incommensurability," *Studies in History and Philosophy of Science*, 21: 183–209.

Hacking, I. (1982), "Language, Truth and Reason," in *Rationality and Relativism*, M. Hollis and S. Lukes (eds.). Cambridge: MIT Press, pp. 49–66.

Horwich, P. (1990), *Truth*. Oxford: Blackwell.

Hull, D.I. (1976), "Are Species Really Individual?", *Systematic Zoology,* 25 : 174–191.

Kuhn, T.S. (1983a), "Commensurability, Comparability, Communicability," *PSA 1982, Volume Two.* East Lansing: Philosophy of Science Association, pp. 669–688.

—— (1983b), "Rationality and Theory Choice," *Journal of Philosophy,* 80: 563–570.

—— (1987), "What are Scientific Revolutions?" in *The Probabilistic Revolution, Volume 1: Ideas in History,* L. Krüger, L.J. Daston, and M. Heidelberger (eds.). Cambridge: MIT Press, pp.7–22.

—— (1990), "Dubbing and Redubbing: the Vulnerability of Rigid Designation," in *Scientific Theories,* Minnesota Studies in the Philosophy of Science, XIV, C.W. Savage (ed.). Minneapolis: University of Minnesota Press,: pp. 298–318,

Lyons, J. (1977), *Semantics, Volume I.* Cambridge: Cambridge University Press.

Lewontin, R.C. (1978), "Adaptation," *Scientific American* 239:212–30.

Putnam, H. (1978), *Meaning and the Moral Sciences.* London: Routledge & Kegan Paul.

Part IV
The Boundaries of Science

10 Four Models for the Dynamics of Science*
Michel Callon

"We must explain why science – our surest example of sound knowledge – progresses as it does, and we must first find out how in fact it
does progress" (Kuhn, 1970, p. 20). Many answers have been proposed
to these two questions. In choosing to organize this chapter in terms
of different models of scientific development, I have deliberately sought
to emphasize the collective character of work in science studies. My
aim is to avoid the repetitive and controversial step of taking a few selected books by a number of great authors – the science studies canon
– as the point of departure. To be sure, my way of presenting the arguments has its drawbacks. For instance, the debates that have driven the
field as it has grown do not come into focus. However, the theoretical
structure of arguments and choices is made clearer, as is the fact that
analysts are always struggling with a series of different dimensions. It
is thus impossible to give a definition of, for example, the nature of
scientific activity, without at the same time suggesting a certain interpretation of the overall dynamics of development and establishing the
identity of the actors involved. Even the most philosophical works imply a conception of the social organization of science, and reciprocally
the purest sociological analyses assume views of the nature of scientific
knowledge.

Also my approach draws attention to the overall coherence of what
may appear to be different approaches to STS. It turns out that, once a
decision has been made about the character of scientific findings, certain consequences for the description of the institutions and dynamics
of science necessarily follow. Though it is true that authors often escape
from the logic of a single model by combining several together, the use
of the models reveals the way in which authors from different schools
and disciplines sometimes in reality share a common framework of assumptions.

* Chapter 2 in S. Jasanoff, G.E. Markle, J.C. Petersen and T. Pinch (eds.), *Handbook of
Science and Technology Studies* (Thousand Oaks, CA: Sage Publications, 1995), pp. 29–63.

I have distinguished four models, each of which emphasizes a central issue. The first is that of science as rational knowledge where the object is to highlight what distinguishes science from other forms of knowledge. The second is that of science as a competitive enterprise where the main concern is the organizational forms that science takes. The third is the sociocultural model and particularly the practices and tacit skills that it brings into play. The fourth model, that of extended translation, attempts to show how the robustness of scientific statements is produced and simultaneously how the statements' space of circulation is created.

Each model is characterized by its answers to six questions that lay out the social and cognitive dimensions of scientific development. Though the list of questions might be considered fragmented, from a practical point of view the schema appears to work. The questions are these: (a) What does scientific production consist of? (b) Who are the actors and what competence do they have? (c) How does one define the underlying dynamic of scientific development? (d) How is agreement obtained? (e) What forms of social organization (internal or external) are assumed? (f) How are the overall dynamics of science described?

MODEL 1: SCIENCE AS RATIONAL KNOWLEDGE

This model seeks to clarify what distinguishes science from other human activities. It focuses on scientific discourse and explores the links it establishes with the reality of which it speaks.

Nature of Scientific Production

The outcome of research activity consists of statements and networks of statements. The classification of these statements and the characterization of their relations is a central issue.

The most common classification is one that opposes observational statements (or empirical ones) and theoretical statements. This distinction accounts for the dual dimension of science: experiments and data collection, and also conjectures and generalizations.

If one takes the following statements:

(a) Any electron placed in an electric field is subject to a force proportional to its charge.
(b) In the circuit C situated in this laboratory, the intensity of the current is 50 amperes.

(c) The needle on the ammeter placed in the circuit C points to the figure 100.

These three statements are independent and their vocabulary is different in each case. In statement (a) the entities referred to are not directly observable by human beings by means of the five senses alone – nobody has ever seen an electron and even less so an electric field – which is why these notions are said to be abstract. In statement (b), the repertoire is similarly abstract although certain entities – for instance, "the circuit C" or "this laboratory" – are directly observable. It is with statement (c) that we enter into the realm of the senses. The figure 100 may be seen, as may the ammeter itself. The fact that the needle points to the figure 100 may be agreed upon after visual inspection only.

How does one go from statement (a) to statement (c)? The notion of translation may be used to describe these moves (strictly speaking, it should be called "limited translation" so as to maintain the distinction between this and the extended notion of translation in Model 4). The translations that permit statements (a), (b), and (c) to be related are far from obvious. Several means for creating these translations have been suggested, and all take the form of a sort of abstract calculation: for example, correspondence rules, coordinated definitions, dictionaries, or elaboration of an interpretative system.[1] In all cases it is generally recognized that it is not possible to move from one kind of statement to another by means of logic alone (Grünbaum and Salmon, 1988). Whatever the particular strategy, it leads to the creation of a third family of statements that associate terms from both observational and theoretical statements and consequently act as translation operators.[2]

With the proliferation of intermediate statements, the distinction between theoretical and observational statements is by no means clear. A first position, which may be termed *reductionist*, is to minimize the distance between both types of statements. It covers two extreme forms: (a) Theoretical statements are derived from observational statements (positivism and logical empiricism); such a doctrine can be mobilized either to provide a criterion of validity (the so-called inductionist theory) or to establish demarcation criteria between statements that have meaning and those that do not;[3] (b) observational statements are shaped by theoretical considerations without which they have no meaning:[4] This is the so-called theory-ladenness of observations.

A second position is a refusal to establish hierarchical links between theoretical and observational statements. Thus, though there are indeed connections, in the first instance it is assumed that the different categories of statements are relatively independent from one another. Under these circumstances, it is possible to test empirical predictions derived from theoretical statements or to decide whether one theory explains a set of observations better than another.

In this model, knowledge production is basically reduced to the production of statements between which translation relationships can be established. Translation is confined to its linguistic meaning – translation is not an exit from the universe of statements. This explains the natural drift toward philosophical and ontological questions. How can one avoid discussing what is "represented" by the statements and the essences they bring into play?[5]

Actors

The relevant actors are essentially the researchers but reduced to the role of statement utterers. Technicians with their skills, disseminators of knowledge, manufacturers of instruments, teachers, and pieces of experimental apparatus are totally absent; society is rarified, reduced to its simplest expression. A consequence of this work of purification is the attribution of wide-ranging competence to the (rare) actors involved. The more narrowly the group is circumscribed, the more the task entrusted to its survivors is complex.

The competencies assumed by the researchers are sensory and cognitive. The scientist must be capable of articulating statements that integrate her observations. She is thus dependent on her five senses, and particularly on sight (observation is always mentioned). The scientist must also be capable of imagining statements that are not directly linked to observation and of introducing translations between them. Her ability to produce metaphors and analogies is emphasized by authors like Holton (1973; Hesse, 1974). Others insist on aesthetic sensitivity; certain theories or reasoning seduce by their simplicity, their elegance, or their beauty.

To these cognitive competencies is added what we might call the rational dimension. The notion of rational activity rests on the capacity to make justifiable decisions. The rules for such justification have been described in many different, and indeed contradictory, ways. They may have to do with the promise of a given theory, its generality, its robustness, the extent to which it fits experimental data, its ability to resist rigorous testing, or its simplicity. However, such judgments don't take place in a vacuum but depend on existing theories and statements. In talking of this accumulation of objective knowledge – knowledge distributed in books, articles, libraries, or the memories of computers – Popper (1972) talks of a "third world." One could say, to paraphrase Boltanski and Thévenot (1991), that the particular competence of the scientist lies in the ability to justify why one statement is to be preferred rather than another in a specific set of circumstances.

In this model, the scientist is a monstrous being – one that incorporates a range of diverse competencies, normally thought to be distributed between different members of society.

Underlying Dynamic

Why does science advance? Or to formulate it in terms of the model, why do scientists tirelessly add new statements to existing ones? And why do they transform, amend, and invalidate them?

First, the critical and ongoing work on statements assumes that the scientist is endowed with a solid moral commitment. This is not so much to keep her from the temptation of fraud, for the debate between specialists is sufficient to eliminate fraud – honesty results from mutual scrutiny. Rather, it is to encourage her to produce ever more statements, which she must be prepared to test and possibly to abandon. The scientist is caught in a double bind; on the one hand, she must devise and produce an increasing number of statements but, on the other hand, she must submit these to the ruthless constraint of selection.

Second, elements of an answer can be found in the institution of science; this is where the complementarity of this model with Model 2, of science as a competitive enterprise, is most obvious. The reward system of science is essential, for even the scientist with the most acute moral sense will not strive to produce new statements without encouragement. Scientific institutions act to channel the force driving scientists – whether a passion for truth, the desire to participate in the collective enterprise of knowledge, the wish to control nature, or the relentless search to resolve problems or contradictions. The scientist is but an operator by whom statements are brought into existence and confronted with one another. The model brings into play a sort of Darwinism extended to statements.

Agreement

In this model, agreement covers statements of all kinds – for instance, theoretical or observational – and also the constructions and arrangements they engender.

Agreement is first to be explained by the fact that actors share similar competencies. They are able to agree on strict observational statements like these: The thread breaks or the curve peaks (which of course implies existing agreement on what a thread or a peak is). They are also capable of inference (a logical contradiction is obvious to all), able to establish the extent to which a statement is general[6] and/or make decisions based on what at the time is considered to be good reason.

Beyond shared competencies, what plays an essential role in the construction of agreement is the existence of a field of discussion where statements can be confronted. We thus find another way of defining rational activity that dates back to ancient Greece at least. Vernant (1990) suggests that science is but the continuation of political debate in a different arena – its transposition from the social to the cosmos. Reason comes into being where arguments occur. And because it is centered on statements, Model 1 treats very seriously the existence of the space of debate in which statements are expressed and their robustness tested. Discussion between scientists takes place in colloquia, in journals, or more informally between colleagues around the laboratory bench before publication or presentation. It also takes the more subtle form of interior dialogue when a scientist debates with herself to anticipate objections and simulate probable debate. But this self-imposed rigor is similar to that exerted by her colleagues – or that which she exerts on them in turn. Private and public space, or formal relations and informal ones, are not opposed to one another. If indeed there is a boundary, it is the one between the errors that are kept to oneself or one's immediate colleagues, and so may be repaired without damage, and those that can only be admitted to by losing part of one's credibility. This is more a question of reputation than a difference between modes of argumentation and discussion.[7]

But, to take the vocabulary used earlier, how are the translations that turn observational statements into theoretical (and vice versa) made robust enough to survive in the debate? There are many answers to these questions. Some have to do with tests and their interpretation. To be convincing and to overcome the subjectivity of a statement means subjecting its validity to experimental checks and to the criticism of colleagues who can verify that the tests are meaningful and correctly interpreted. Others suggest that theoretical statements need to be presumptively true about the world. Conformity with experience, which is always problematic, is less important than, for instance, the ability to make verifiable predictions. Yet others adopt a pragmatic view: the robustness of a translation is measured by applying criteria or rules that have been developed over the course of centuries. Examples of such standards include the extent to which a statement is general, the economy of the entities that it mobilizes, its ability to resist demanding tests, and its fertility in leading to unexpected applications. Finally, others consider criteria as pure conventions that convince only those who are already committed to them.

Whatever the solution adopted, the existence of explicit and shared standards is admitted, whether they are hypothetical or categorical or are reasonable conventions.

Social Organization

This model imposes severe constraints on the social organization of scientific work. The paradox is that the more one insists on the cognitive and discursive dimension of knowledge, the more one increases the demands on social organization. Those who utter statements can only undertake their work – that of discussing, testing, experimenting, selecting, falsifying, and so on – if they are protected by society as a whole and by particular institutions.

Without the public space of (free) discussion, science degenerates into beliefs stamped with subjectivity. Science is synonymous with democracy or, to use Popper's (1945) expression, open society. In an open society, institutions are the revisable creations of human activity; the critical mind knows no limits – gods, Caesars, or tribunes. Questioning is permanently renewed and no state of rest is satisfactory. The individual is privileged because she both introduces and judges novelty. There is an analogous concern in the writing of Habermas (1987) with pressing science into a space for public discussion and communication.

But it is not enough that science should be immersed in an open society; powerful institutions that guarantee the smooth functioning of critical debate are also necessary. What will be said in the presentation of Model 2 applies here without restriction; with respect to social organization, Model 2 can be considered a natural complement to Model 1.

Dynamics of the Whole

The development of science is expressed in the proliferation of statements that are the result of a dialogue between man and nature. A silent man, faced by an equally silent nature, could neither accumulate statements nor produce revisable knowledge. So a scientist does not simply read the great book of Nature; she transcribes it, translating it into statements inscribed in linguistically shaped argument. Putting the universe into words is the essential task of scientific knowledge. Science is thus developed in the form of a dual dialogue, first between scientists and Nature (observational statements and theoretical statements) and second between scientists themselves. These two dialogues are interdependent; they take the form of a triangle in which one of the protagonists (Nature) is content to reply in a cryptic way to the questions it is posed. As in any confrontation, contradictions and incomprehension constantly develop. What exactly is nature's message? How should experiments be conceived and the results interpreted? What theoretical statements can be put forward? What do they explain? These gaps and divergences restart the machine of knowledge.

Such a vision necessarily implies the notion if not of progress then at least of progression. There are statements, forever more statements that hold Nature as close as possible and ply it with ever more awkward and precise questions. So investigation has no end, yet it continues! Nature's reality is asserted and the statements produced are seen as an increasing theoretical approximation or a better experimental description. Or, alternatively, one may not express any opinion about this reality and simply concentrate on the endless production of more and more robust or reliable statements.[8] Whether the statements that are produced tend toward the truth, relate together ever-increasing numbers of empirical observations, or increase our ability to control and manipulate the world, the tragic beauty of Model 1 is that it is scientists and scientists alone who have to choose which statements to preserve and which to discard.

MODEL 2: COMPETITION

There are numerous variations of this model, but they all share the two fundamental tenets: (a) Science produces theoretical statements whose validity depends on the implementation of appropriate methods. (b) The evaluation of knowledge – an evaluation that leads to its certification – is the result of a process of competition or, more generally, of a struggle usually described with categories borrowed from economics or sociobiology.

Nature of Scientific Production

In this model nothing is said about the content of scientific work. It is simply assumed that the research scientist develops knowledge that is submitted to the judgment of colleagues. Knowledge is generally transmitted in the form of publications that are disseminated without any particular restriction.

These publications are in principle intelligible to specialists in the field. One can make use of the notion of information to speak of their contents. This knowledge or information is characterized by its novelty, its originality, or perhaps its degree of generality. An evaluation of its utility, as perceived by others – scientists or nonscientists – is also possible. This model does not exclude the existence of tacit skills, but this is alluded to without being turned into a specific component.

Actors

For this model, the actors in the production of scientific knowledge are the research scientists themselves. A distinction is made between the world of scientists (the specialists) and that of the layperson. Technicians are reduced to an instrumental role on the same level as the experimental apparatus. Science is first and foremost an intellectual adventure, and its practical and technical dimensions are eclipsed.

Research scientists are social beings whose individual competencies are not defined or analyzed as such. Their membership in a discipline or specialty determines their aims and ambitions, together with their theoretical and experimental choices. Therefore the rationality of scientific activity results from the interaction between scientists, and in particular from their competition, not from any particular inherent predisposition that distinguishes them from other human beings.

The motivations behind scientists' actions are not theirs alone. Various suggestions have been made by different authors. Merton (1973) insists on the role of norms that define permissible behavior and on reward systems that institutionalize the production of knowledge. Bourdieu (1975) sees scientists as agents guided by their habitus who develop strategies for positioning in a field structured by the interlacing of these different strategies. Hagstrom (1966) conceives of scientists as striving to maintain the confidence of their colleagues.

Scientists therefore have a dual role. This resembles a Darwinian struggle in which they are both judges and litigants. Every researcher judges his or her colleagues (Is the knowledge new and robust? Is the information useful?) but is similarly judged by them in the same way.

Underlying Dynamic

What are the mechanisms responsible for this organized and collective search for knowledge? Why are scientists led to produce more and more knowledge?

The answers provided by this model draw their inspiration from different versions of economic theory. One may first, like Hagstrom, conceive of an exchange economy. The scientist who is evaluated positively by her colleagues receives recognition, and this, in turn, bolsters confidence in her. This is a gift economy.

The model may be that of neoclassical economics. Here the scientist is comparable with an entrepreneur. The product she offers her colleagues is knowledge, which the latter evaluate as a function of its utility and quality. This evaluation is measured in the form of symbolic awards (see the section "Social Organization," below). Each scientist is supposed to maximize her personal profit, that is, the recognition granted her. A climate of competition is thus created, which, as in the neoclassical market, channels individual passions and selfish interests into a collective, rational, and moral enterprise (Ben-David, 1991; Cole, 1973; Hull, 1988; Merton, 1973).

Or the model may be that of a capitalist economy as described by Marxists. Scientists are not so much interested in recognition per se as in the possibility of obtaining ever more of it – the object is one of accumulation or circulation. Here one encounters Bourdieu's (see 1975) works and Latour's first analyses (Latour and Woolgar, 1979). Researchers have no choice: If they want to survive among their colleagues, they have to accumulate credit or credibility, which constitutes their capital. Without capital they cannot obtain support for new programs. On the other hand, the more capital they have, the more they are able to carry out research, the results of which would increase their initial endowment. Scientists are thus caught up in a logic of success.

One of the features of the economic metaphor is that the psychological motivations or aims of scientists are not important. Competition coordinates individual behaviour. So this is science without a knowing subject – a prospect dear to philosophers as different as Popper (1972) and Althusser (1974). The fact that some authors present a Darwinian metaphor, whereas others are committed to one drawn from economics, makes no fundamental difference to the interpretation proposed here.

Agreement

In this model the production of agreement implies what could be called free discussion between scientists. Science is caught in a double movement of openness and closure. Openness guarantees that all points of view can make themselves heard, and closure signifies that reaching agreement is the objective assigned to these discussions.

The openness of debate must not go beyond the scientific community. In this model researchers monopolize discussion and any interference from outside is a potential source of disorder. Not all exchange with the sociopolitical environment is excluded. External requests may be formulated, and preoccupations and convictions may be brought to the attention of the scientific community. The model tolerates, for example, the idea that industry or political decision makers ask questions and orient programs (Merton, 1938/1970). It also admits that metaphysical concerns and philosophical convictions motivate researchers. Thus Freudenthal (1986) has linked Newtonian physics to Hobbes's political philosophy. Yet such influences do not (or should not) go right to the heart of scientific activity. They contribute to formulating problems or putting them in a hierarchy. For this reason they play an important role in creating the preconditions for scientific agreement. This agreement, to be complete, must also relate to the technical matters that are immune to outside influence. Agreement cannot be determined therefore by positions of power or arguments from authority. There is an irreducible internal core that is the responsibility of the scientific community. The scientist can only be convinced by statements that in the last analysis draw on method for their robustness. What counts as an acceptable method could vary over time but is considered taken for granted during any given period.

Social Organization

Organization is one of the central variables of the model. Indeed, the viability of the scientific enterprise rests upon an organization that strictly separates the inside from the outside.

Internal Organization
The incentive system plays a vital role, driving scientists to produce knowledge. It is based on a double trigger mechanism. "Discoveries" (or, more broadly, "contributions") are identified and attributed to certain scientists, who are rewarded according to the quality of these contributions.

In this model, what counts as a discovery is the outcome of a social process. In the incessant flow of scientific production, how does one isolate identifiable units of knowledge that are more or less independent of one another? And how does one then decide on the origin of each of these elementary contributions? There is no universal answer. The delineation of contributions as well as their imputation (Gaston, 1973; Merton, 1973) often gives rise to controversies and to revaluation (Brannigan, 1981; Woolgar, 1976).

The identification of discoveries and their "authors" could not be stabilized without material devices and rules that codify the formulation of knowledge and its transmission. Thus the scientific article in its present form makes it possible to delimit a piece of information precisely, to organize its dissemination, to identify the authors who produced it, to date their contributions, and to mention what has been borrowed from other authors by means of quotations and citations (Price, 1967).[9]

The functioning of the reward system depends upon the identification and attribution of contributions. Their importance, their quality, and their originality are all evaluated simultaneously. And the possible forms of reward are varied, so that they may be adapted to the supposed importance of the contribution: promotion, prizes that vary from the most modest to the most prestigious, election to an academy, and eponymy (giving the name of the scientist to a result that has been attributed to him, as in the case of Ohm's law). The researchers – the actors in scientific development – are thus encouraged to contribute to the advancement of knowledge.

The model suggests that these rewards are symbolic in character; evaluation is not directly related with possible economic gain. Science is a public good.[10] Statements like "the structure of DNA is a double helix" are nonrival and nonappropriable. That Mr. Jones mobilizes the statement for his own activities does not diminish its usefulness to Mr. Brown and does not prevent the latter from doing the same thing. So with these public goods market mechanisms don't work efficiently. And this is why conventional economic incentives (valuing goods on the market) are replaced by another reward system – one that urges researchers to produce knowledge that they make public. Thus publication enables contributions (discoveries) to be identified and imputed, but it also ensures that these public goods are disseminated – which explains why publication is considered to be a cornerstone of science.

Another essential issue is the maintenance of free access to discussion. The social organization should encourage scientists to produce knowledge but must also favor open debate: seminars, colloquia, or the right to reply in journals. This ensures that any scientist wishing to participate is able to do so. Free access is a basic principle that is inscribed in the norms of science and its institutional forms.

The model stresses the role of individuals (the researchers). Yet scientific activity is more and more a matter of teamwork. So how is the emergence of joint research sites like laboratories explained? And what is their role? In this model the question is as problematic as the existence of firms was to economics (Coase, 1937). Even if some authors examine the organizational structures, performance, and strategies of laboratories (Whitley, 1984), in this model the laboratory is an anomaly. Its existence can simply be seen as a consequence of technical constraints: the management of large-scale equipment or experimental work that depends on the division of labor to secure economies of scale.

Relations with the Environment

Model 2 explores the relationship between science and its environment but does so by establishing a clear boundary between inside and outside. When this boundary is crossed, the norms, rules of the game, incentives, and types of resources break down. The notion of a scientific institution, with its own goals, values, and norms (Merton, 1973), together with the notion of a scientific field (Bourdieu, 1975), mark the existence of territory. Numerous historical analyses have shown how this social space governed by its own laws has become autonomous and how the role of the professional scientist has gradually emerged and been consolidated (Ben-David, 1971).

The existence of autonomy does not exclude exchange and influence with the outside world. For instance, Bourdieu conceives of two markets: a restricted market, limited to specialists, where scientific theories are debated, and a general market that transmits the products, thus stabilized, to the external actors interested in them – firms, state agencies, and the educational system (Bourdieu, 1971). Between the two markets there are mechanisms of control. The value of a product (a theory) on the external market depends partly on the value it is given by the internal market (and vice versa). Again David, Mowery, and Steinmueller (1992), adopting the economist's point of view, consider science to produce nonappropriable information that is reused by economic agents. The latter, in turn, produce the specific (and appropriable) information that they need, in a more predictable and less costly manner. And yet again Rip (1988) proposes a generalization of the cycles of credibility by introducing the "fundability" of research projects, linking the logic of scientific development to that of the politico–economic actors.

The duality of organizational forms is crucial in this model. The border between the internal and external is essential to science and protects its core yet must be sufficiently permeable to transmit the influences that nourish science and ensure its social utility.[11] The organizations that link science to its environment (industrial research centers and state agencies for encouraging research) play a crucial role in managing these exchanges in a proper way (Barnes, 1971; Cotgrove and Box, 1970; Kornhauser, 1962; Marcson, 1960).

Dynamics of the Whole

This model depends on a regular process of growth against which it is possible to explain historical "accidents" and decreasing returns. This growth is explained by the fact that scientists work in those areas of research where the anticipated symbolic profits are likely to be highest because the problems being tackled are considered important, and where there are still many areas of ignorance. Accordingly, everything that fosters mobility also tends to favor overall growth, whereas anything that impedes it tends to reduce the productivity of science (Ben-David, 1971, 1991; Mulkay, 1972). If free debate is hindered, if the incentive system malfunctions, and if positions of monopoly come into being, then that productivity may decrease. Here again one finds arguments that are fairly close to those used in the analysis of economic growth.

Moreover, if society does not guarantee the boundaries, and if it does not support the internal organization of science and its rules, then research as a whole will break down, as when the Nazis advocated a racialized and nationalized science or when the Soviet Communist party rejected Mendelian genetics. So the model goes beyond mere considerations of production. Science produces knowledge, but the institution that supports it has an essential function, that of enabling rational knowledge to develop. When the dynamics of science are hindered, reason is affected.

MODEL 3: SCIENCE AS SOCIOCULTURAL PRACTICE

This model says that science does not really differ from other activities and the certainties it leads to do not enjoy any particular privilege –

an argument based on the fact that science is much more than the simple translation of statements. The third model suggests that science must be considered to be a practice whose cultural and social components are as important as the constraints that arise from the order of discourse.

Nature of Scientific Production

Model 1 was content to limit its investigation to statements and assumed that these were transparent, their meanings lying simply in the system of statements. Yet, as the pragmatists of language have taught us, a statement has no meaning without a context. Model 3 adopts this position and emphasizes the importance of nonpropositional elements (tacit skills) in the production of knowledge.

The contribution of authors such as Kuhn (1962) and Wittgenstein (1953) is essential. The notions of rules and how they may be followed, language games, forms of life, and learning by examples underline the importance of tacit knowledge – a notion developed by Polanyi (1958) to account for the transmission of noncodified information. Certain knowledge – for example, knowledge linked to the functioning of instruments or the interpretation of data supplied by these instruments – cannot be expressed in the form of explicit statements. In this view science is an adventure that depends on local know-how, on specific tricks of the trade, and on rules that cannot easily be transposed. Formal statements can only travel and be understood if their instrumental environment and the knowledge incorporated in human beings is the same. This theme was developed brilliantly by authors such as Fleck (1935) and then Ravetz (1971): "In every one of its aspects, scientific inquiry is a craft activity depending on a body of knowledge which is informal and partly tacit" (p. 103). Collins enriched this argument considerably in several studies. For instance, in his study of the construction of the TEA laser, he showed that the diffusion of knowledge could not be reduced to the mere transmission of information: "The major point is that the transmission of skills is not done through the medium of written words" (Collins, 1974). Collins thus distinguished the algorithmic and the enculturation models. In the former, science consists of the production of codified transparent information; in the latter, tacit skills and learning are important – a scientific statement is always opaque, its meaning reducible neither to what it states nor to what is said by the system of statements to which it belongs. The distinction between algorithmic and enculturational models becomes essential when the question of replication of experiments is considered. The reproduction of an experiment always implies close interaction between scientists and experimental arrangements; an entire culture is transmitted with this know-how, these ways of seeing and interpreting, these observational statements.[12] As Collins (1974) says: "Only those scientists who spent some time in the laboratory where the success has been achieved prove capable of successfully building their own version of the laser." The affirmation of the enculturation model has a general implication: that practices incorporated in human beings (those who manipulate and interpret) are interwined with experimental apparatus, protocols, and observational or theoretical statements. To extract statements from this whole and to transform them into a privileged object of scientific production is to take them out of their context and strip them of their meaning.

Actors

The actors involved in the dynamics of developing scientific knowledge are not limited to experimentalists and theoreticians. In a highly suggestive article, Collins and Pinch (1979) introduce a distinction between what they call the constitutive and the contingent forums. They show how groups outside the scientific community may be mobilized in the production of knowledge. The list of these groups depends on the particular situations under study: the manufacturers and distributors, the media, state agencies, firms with their engineers, or even external pressure groups (philosophers, ethical committees, and so on) – any or all of these may participate. The border between insiders and outsiders fluctuates and is negotiable. But what is analytically important is to explore the mechanisms by which constraints, demands, and interests outside the circle of researchers influence scientific knowledge. In an exemplary work devoted to the Great Devonian Controversy in geology, Rudwick (1985) follows the different actors who were directly or indirectly interested in the debate during the 1830s about the existence of a geological stratum (the Devonian). He gives real depth to all these characters, reconstituting the network of relations and locating them in the institutional frameworks of the period. Wise's (Wise and Smith, 1988) work on Lord Kelvin, Schaffer's (1991) on astronomers, MacKenzie's (1981) on the emergence of statistics, and Pestre's (1990) on Neel are other examples of such analyses.

Attention is also paid to those who work in laboratories. In Models 1 and 2, technicians are present everywhere, but in the form of transparent shadows. They carry out experiments, collect samples, and determine measurements; yet their work has no influence on the content of knowledge and they have the same status as instruments. The sociocultural model repairs this omission. Just as it emphasizes experimental work, it also brings into play those who carry out the experiments and prepare the samples. Shapin (1989), in a highly instructive article, has greatly contributed to this rehabilitation. Knorr (Knorr Cetina, in press) stresses the particular role of Ph.D. students in laboratory life.

To be sure, the researchers are not forgotten. Their competences are diversified and include the capacity not only to formulate and interpret coded statements and algorithms but also to elaborate and control tacit skills or the rules of the art. The researchers (technicians must be included in this category) manipulate, decipher, inspect, tinker with, interpret, and reason (Knorr, 1981; Latour and Woolgar, 1979; Lynch, 1985). They are further more capable of learning and memorizing. The notion of learning, although central to this model, has largely been left unexamined. Several different approaches to learning exist in literature. Bayesian analyses insist on the probabilistic character of knowledge and on the role of experiments in strengthening or transforming subjective probabilities (Hesse, 1974); others refer to Piagetian theories or those of artificial intelligence (Mey, 1982) or to Gestalt psychology (Kuhn, 1962). This offers a wide field for research. Whatever the theoretical stance adopted, the underlying hypotheses are clear; the learning capacity of actors endows them with both historical depth (they guarantee a certain continuity of knowledge) and a (permanent) faculty for invention, that is, for redefining routines and rules for coordinating action, which enables one to understand why science is not limited to repetition.

The stress laid on tacit skills and learning mechanisms leads to the social group. Interaction is only developed within the framework of a shared culture and scientific activity is no exception. This hypothesis has its source in the notion of a paradigm proposed by Kuhn, who refers on the one hand to the group and on the other to the scientific competence and production of each of its members. For Collins it is the core set that is the fundamental actor responsible for the production and transmission of knowledge. It groups together researchers who share the same problems and culture. Collins also refers to Granovetter (1973) to suggest that a researcher's impact is greatest if she enters into unusual or atypical social relationships (see Mulkay, 1972). Schaffer (1991) adopts an analogous point of view: "The coordination between these two networks was crucial, because it showed that observatory managers and experimental astronomers might collaboratively extend their control beyond the boundaries of celestial mechanics" (p. 6). Scientific groups are structured like social networks – they can become denser, close in on themselves, fragment, or merge (Crane, 1972; Mullins, 1972). The dynamics of these networks depend on the strategies of relationship building followed by their members, and each transformation of the social network implies a cultural transformation.

In extending the field of analysis by analyzing all the social groups that intervene in the process of creating knowledge (the "constituency of interest"), the defenders of Model 3 give the description a distinctively sociological flavor, without sinking into reductionism. For the first time, sociology treats the contents of science with the same degree of depth and the same concern for detail as any other human activity.

Underlying Dynamic

To account for the dynamics of scientific activity, there is no need to invent new sociological explanations. Barnes (1977) provides the clearest and most systematic presentation of this point of view. Inspired by the Marxist tradition, of which we may also find traces in Habermas's work, he writes: "Knowledge grows under the impulse of two great interests – an overt interest in prediction, manipulation and control, an overt interest in rationalization and persuasion" (Barnes, 1977, p. 38). Thus, in the phrenological controversy studied by Shapin (1979), we find a mixture of sociopolitical and cognitive interests. The endeavor to clarify the possible existence of the frontal sinuses is as much to score points in the class struggle in Edinburgh as to learn anything about the brain. These two families of interests are to be found in all societies; if certain ones like our own have developed science, it is for contingent historical reasons. Interests linked to prediction and control have been intensified and then inscribed in specific institutions.

More generally in Model 3, the explanation of the underlying scientific dynamics depends on the particular sociological models used. We have just evoked Barnes's macrosociology but there are more microsociological possibilities. In Pickering's recent texts, we find an explanation that makes no distinction between a scientist and any other goal-oriented social actor: "Doing science is real work" (Pickering, 1990). Science is a practice and is analyzed like all practices; a researcher has resources, tries to reach her goals, and seeks to create coherence between the disparate and sometimes intractable elements that make up her environment (instruments, theoretical, and experimental models), some of which resist all reorganization. Knorr (1992), relying upon Merleau Ponty's philosophy, gives an illuminating description of what she calls the "epistemic cultures" of high-energy physics and molecular biology. She stresses the disunity of scientific practices that depend on "their orientation toward and treatment of signs, on their relations to themselves, on the forms of alignments they institute between subjects and natural objects, on their general approach to capturing and engaging truth effects in inquiry" (p. 3). And there are other possibilities, including ethnomethodology (Lynch, 1985), symbolic interactionism (Clarke and Gerson, 1990; Fujimura, 1992; Star, 1989), or cultural anthropology (Hess, 1992; Traweek, 1988). All these studies rest on the same assumption: that science is a human activity, one that is specific but that does not merit a change of analytical instruments. Possible explanations for the development of science are as numerous as sociological theories!

Agreement

Agreement between scientists must be explained in the same terms as consensus between social actors anywhere. The principles of Bloor's (1976) "Strong Programme" are the methodological translation of this hypothesis. Because nothing distinguishes science from other human activities, and because scientists are like other social actors, agreement, disagreement, success, and failure need not be explained in different terms.

This argument can be illustrated by the exemplary work of Collins on gravitational waves. As Golinski (1990) puts it, for Collins,

> experiment is potentially open ended. At no point, in his view, does nature force a particular interpretation upon experimenters...The evidence is always too much to fit within interpretive scheme and too little to determine the choice between any number of possible alternative schemes...Controversy can be continued as long as a critic can find the resources to sustain these objections...Sufficient differences between two versions of an experiment could always be found by a critic who wished to deny that a proper replication had been achieved. (p. 494)

Collins (1985) calls this type of dispute the "experimenter's regress." What remains to be explained is why protagonists with different interests, know – how, and practices end up considering that the debate is closed.

The answers given by the sociocultural model tend to fall into several classes. First, there are fairly traditional macrosociological explanations. Because agreement never rests on indisputable evidence, its construction depends solely on the state of social forces and particularly those outside the scientific community, or the group of researchers involved in the debate. The Edinburgh School (Barnes, Bloor, Shapin, and MacKenzie) carried out many case studies in which the influence of political, economic, or cultural interests created a balance of power favorable to a particular outcome. This approach can sometimes appear determinist and mechanistic. Thus it is occasionally claimed that identifiable external groups or social classes add their own weight to that of the scientists with whom they agree. Alternatively, scientists may be left to choose their allies themselves – in this case such alliances do not pervert science because nature is sufficiently ambiguous and experiments are sufficiently complex to support the different opinions and judgments. As there are never uncontroversially good reasons to choose one theory rather than another, there is room for sociological explanation without endangering the autonomy of scientific work (Barnes and Shapin, 1979; Wallis, 1979).

The alternative approach uses notions such as confidence. For instance, in his work on solar neutrinos, Pinch (1986) elegantly shows the importance of the creation of a climate of confidence throughout the design and conduct of experiments. By associating the representatives of several disciplines, by taking into account objections as they are raised, the project becomes a collective enterprise based on relations of reciprocity (exchange of information and so on); the agreement on the results is the fruit of this growing confidence. The nature and extent of relations formed during the conception and realization of experiments and during the elaboration of theories largely determine the likelihood of agreement as opposed to a continual experimental regress. The relationship between this type of analysis and developments in game theory (Axelrod, 1984; Kreps and Wilson, 1982) has been little explored.

Agreement may be facilitated by operations on the instruments themselves. Collins's research has shown that the difficulty with replication is largely ascribed to differences between pieces of experimental apparatus. As Collins suggests, standardization and the calibration of instruments reduce the likelihood of divergence and favor agreement. If this calibration is not achieved, one returns to the situation so well described by Schaffer (1989) with respect to Newton's experiments on the refraction of light in prisms:

Newton's "law" did not compel experimenters such as Rizetti: "it could be a pretty situation," the Italian exclaimed, "that in places where experiment is in favor of the law, the prisms for doing it work well, yet in places where it is not in favor, the prisms for doing it work badly." For such critics, Newton's prisms never became transparent devices of experimental philosophy (p. 100).

This transparency of instruments, created by scientists, which became important in the second half of the nineteenth century, "let nature speak for itself" (Daston and Galison, in press), led to the black-boxing of experimental methods and to their standardization (Latour, 1987). Of course, this agreement in turn depends on collaboration and compromise, which must again be explored. Once it is achieved, however, it is inscribed in the calibrated instruments and provides a solid basis for new agreements. This may prompt a distinction between passive instruments (Fleck, 1935), which are not reconsidered, and active instruments, which evolve and become controversial. Passive instruments form the common ground on which arguments and counterarguments can be deployed. They furnish a common measure. G. Bachelard's (1934) notion of "phénoménotechnique" greatly contributed, in his time, to emphasizing the importance of agreement sealed by instruments. Because the instruments come to embody the theories they are used to support, disagreement is made more difficult (Latour and Woolgar, 1979); the refutation of a statement implies the refutation of the instruments and their calibration. The theories are "hardwired," to use Galison's (1987) nice expression.

Finally, the sociocultural model allows the use of all available means. The possible mechanism for "closure" and the possible studies of these mechanisms are endless.

Social Organization

The sociocultural model is, paradoxically, only moderately interested in questions of organization and institutional forms. This observation applies as much to the internal organization of scientific activity as to its relations with the sociopolitical environment.

The notion of rules is probably one of the best suited to account for a social organization capable of managing scientific practice in its entirety. Rules are both implicit and explicit; they are not outside action but are interpreted, elaborated, and transformed within action. Again, they are both social and technical, ensuring a minimum of coherence and making anticipation and discussion possible. They are compatible with the proliferation of social groups and the diversity of their identities. Rules, which are more or less local and specific, form the fixed point around which relationships of power and influence can be developed. Sociological and economic work on the appearance of rules or conventions could be usefully mobilized to enrich the sociocultural model (Bloor, 1992; Favereau, in press; Lynch, 1992; Vries, 1992).

Emphasizing the role of learning, the model consequently stresses the importance of skills transmission and training. This produces relationships of dependence between masters and disciples, and also within laboratories between different actors with different types of skill. Shapin (1989) gives a good illustration of this type of analysis by highlighting the crucial position held by technicians in early laboratories. This sociology reintegrates more traditional considerations of power and domination into the world of science (Schaffer, 1988).

Finally, Model 3 considers boundaries between science and its environments as constructed by actors themselves in various hybrid settings. Jasanoff's (1990) study of regulatory science, Abir-Am's (1982) investigation of Rockefeller Foundation policy in molecular biology, Dubinskas's (1988) work on high-technology organizations, Wynne's (1992) analysis of the entanglement of science and policymaking in environmental issues are a few examples of this growing field of analysis.

Dynamics of the Whole

The sociocultural model challenges the idea of continuity in the development of scientific knowledge.

If science does not progress in a linear way, it is because it is involved in social relations that have their own logic. Barnes's notion of interest is very useful from this point of view. Scientific knowledge can be seen always as a response to one kind of interest, that of prediction and control, but its contents are organized and structured according to different and changing social configurations. Knowledge is marked by the conditions of its production; Kuhn's approach is exemplary in its insistence on the incommensurability of skills and paradigms. The historicity of science is expressed in the problems it asks itself and can be seen as a function of global history.

A more subtle analysis is also possible. Collins, for example, notes that the diffusion of knowledge cannot take place without transposition and adaptation to local circumstances. No replication has ever resembled the experiment that inspired it, even when the instruments have been perfectly calibrated and procedures highly standardized. Transfer involves loss and creation, elimination and addition. This view leads to the original interpretation of Kuhn's work by Masterman, which links paradigms and "Wittgensteinian" family resemblances. The argument is that any new instantiation of the paradigm creates a discrepancy from the original exemplar. The distance from the original grows from one instantiation to the next, and the paradigm ends up betraying itself (Masterman, 1970). The dynamics of science are born of these successive discrepancies – discrepancies that are nothing more and nothing less than the research process itself. It is because diffusion inherently involves transformation and transposition that science is forever developing. This leads to a conception in which the dynamics of science create "a genuinely historical process; facts and phenomena, concepts and theories as well as the instruments and institutions of science are bound to the wheel of what happened" (Pickering, 1990).

MODEL 4: EXTENDED TRANSLATION

We have seen in Model 1 that the notion of translation can be used to explain the establishment of links between different statements. Model 4 develops this definition beyond the domain of codified knowledge. *Translation* refers here to all the operations that link technical devices, statements, and human beings. The notion of translation leads to that of *translation networks*, which refers to both a process (that of translations that are joined together) and a result (the temporary achievement of stabilized relations). This model seeks to explain the proliferation of scientific statements and their broadening sphere of circulation. Finally, it calls for a deep reformulation of social theory.

The Nature of Scientific Production

Manufacturing Statements
Like Model 1, the extended translation model assumes that the prime objective of scientific activity is to produce statements. But, like Model 3, it stresses the process of production and the role of nonpropositional elements in this process. Take the following two statements: (a) "The structure of DNA is a double helix." (b) "The facade of the pension where Father Goriot lived had been covered with a layer of poor quality pink paint which bad weather had caused to crack." The difference between

statements (a) and (b) does not lie in the statements themselves but in the extent to which the reader is able to work his or her way up the chain of elements that support the statements. Statement (a) refers to other statements, other objects, and other time spaces, which it sums up and condenses, and to which it gives access. The second statement refers to nothing other than texts and the inescapable fiction of the world of the novel. The notion of a translation chain describes the series of displacements and equivalences necessary to produce a particular type of statement.

Insofar as science is concerned, translation chains combine heterogeneous elements of which the most important are statements, technical devices, and the tacit skills that can rightly be called embodied skills. To understand how relationships can be built between these different elements, one must first introduce Latour's notion of inscription, which refers to all written marks (Latour, 1987; Latour and Woolgar, 1979). Inscriptions include graphic displays, laboratory notebooks, tables of data, brief reports, lengthier and more public articles and books. The notion of inscription points to the importance of writing and to its diversity. Thus the division between instruments (i.e., experiments) and statements (i.e., observations), implied by the preceding models, is replaced by a range of inscriptions, from the crudest marks to the most explicit and carefully crafted statements. From marks to diagram, from table to graph, from graph to statement, and from statement to statement – each is a translation.

Translation chain:

$\rightarrow$ instrument $\rightarrow$ marks $\rightarrow$ diagrams $\rightarrow$ tables $\rightarrow$ curves $\rightarrow$ observational statement 1 $\rightarrow$ theoretico–observational statement 2 $\rightarrow$ theoretical statement 3 $\rightarrow$ and so on

Writing devices are important in all scientific fields and beyond. For example, Foucault (1975) analyzes the hospital as a device that places the individual in a "network of writing."[13] As entities are translated, resistance encountered, and answers gathered, the devices progressively take on form and materiality. Although the task of writing is general, experience shows that a charmed quark, a suffering body, a replicated gene, a humiliated social group, or a geological stratum and its fossils cannot be written in the same way.[14]

Science is a vast enterprise of writing, but to move from an inscription to a statement, and from one statement to another, requires embodied skills and/or technical devices. Without them the manufacture of knowledge (Knorr, 1981) would be unproductive. Thus it is the constant interaction between inscriptions, technical devices, and embodied skills that leads to the development of statements. These interactions may be observed in the composition of experiments (Hacking, 1983), in the interpretations of inscriptions (Amann and Knorr Cetina, 1988a, 1988b; Lynch, Livingstone, and Garfinkel, 1983; Pinch, 1985), in conversations between scientists, or between scientists and technicians, and the writing and the rewriting of articles or reports (Myers, 1990). All these interactions are translations, and they all contribute to the production of statements – a process that Law (1986b) calls heterogeneous engineering. Ethnographic research has described many of them, and graphic methods such as those developed by Fujimura (n.d.) or Gooding (1992) make these easier to depict.

Taking Statements Out of Laboratories

Scientific activity is not simply a matter of manufacturing statements; often (if not always) it seeks to take statements out of the laboratory. But this challenges the conventional distinction between the content of knowledge and the context of production. The notion of translation makes it possible to understand how context and content are simultaneously reconfigured.

Translation leads to the identification and shaping of allies and to seeking their support. It means establishing an equivalence between, say, the biochemical study of an obscure polymer and its absorption by certain body organs and many other agents in society, for example, the groups and institutions that support the struggle against cancer, the field of biochemistry interested in such a polymer, or the pharmaceutical industry and the medical profession (Law, 1986a). A team of biochemists can define other actors and suggest the following translation: We want what you want, so ally yourselves with us by endorsing our research and you will have a greater chance of obtaining what you want (Callon, 1980). Such translations are always tentative and in certain cases postulate completely new actors, which are then brought into existence. The translations might be inscribed in texts stating explicitly the contribution of the projected work, in material substances, or in skills and instruments. These translations might require huge investments. They link closely the definition of very technical problems with the constitution of a space of circulation for the knowledge that is produced.

Translation Networks

The notion of translation network refers to a compound reality in which inscriptions (and, in particular, statements), technical devices, and human actors (including researchers, technicians, industrialists, firms, charitable organizations, and politicians) are brought together and interact with each other. The networks vary in length and complexity. Some only rarely leave the laboratories or their communities of specialists and act primarily via instruments and statements. Others stabilize some of these entities and mobilize them to multiply connections with nonspecialists. Wise (1988), for instance, describes how machines act as material and durable mediators between engineering, industry, and the esoteric concerns of particular domains of research. Yet other networks are active on both fronts and enter into a dynamic of expansion, where each translation within the laboratory leads the network outside to be lengthened. In all cases it can be said of scientific activity that it establishes translation networks.

Inversion of Translations

When a network is established, scientists talk not only on behalf of electrons or DNA, which they translate in their laboratories, but also for the countless external actors they have interested and that have become the context for their actions. Their ability to act as legitimate spokespersons is due to the series of representations that have been set up. This led Star to propose the notion of *re-representation* (Star and Griesemer, 1989). For translation is also representation. In the system produced by Galileo to translate gravitational forces, there is a succession of representations: The clepsydra represents time; the angle of incline re-represents the difference in drop; the table re-re-represents the sphere's course; the curve $(re)4$-presents the table; and the mathematical formula $(re)5$-presents the curve. As with elections, one can talk of representation to the nth degree. But actors attracted to scientific work are also re-presented. Biochemists seek to re-present chemotherapy and the fight against cancer. The argument is that the scientist's particular strength is that she is able to accumulate both types of representation: to re-present herself as spokesperson of both nature and society.

This analysis sheds new light on the standard problem of reference. Thus the statement "the structure of DNA is a double helix" is the last link in a chain that, from translation to translation, refers to other inscriptions, embodied skills, and technical devices. Statements do not talk of an outside reality; they are simply one location point in a long and teeming network. There is no one "reference" but an entanglement of "microreferences": The statement refers to a table that refers to a trace; the trace refers to a technical device, and its interpretation refers to embodied skills. So it is only when attention is focused on the final statement that the translation chain is split, that one can talk of out-thereness. Then one has inversion: The pulsar is said to be the cause of the statement while it is present at each point of the chain of translation but in various forms including statements (Latour and Woolgar, 1979; Woolgar, 1988). Similarly, context cannot be dissociated from scientific content unless we put the translations that define it between brackets. So the notion of translation is preferable to that of reference, even though the etymology is close. This is because, when it is said that a statement translates DNA, or that biochemists translate chemothera-pists' projects, no hypotheses about reality or correspondence are made. One is instead reminded that reference is nothing more than an effect of a translation chain, and that its robustness depends entirely upon the latter.

Actors

The extended translation model substitutes the notion of an actor with that of an actant (a notion borrowed from semiotics; Latour, 1987, 1988). *Actant* refers to any entity endowed with the ability to act. This at-tribution may be produced by a statement (the statement "somostatine inhibits the release of the growth hormone" attributes the property of inhibition of the actant growth hormone to the actant somostatine), by a technical artifact (a chromatographer gives gases the ability to diffuse in a column having elements that are themselves defined as obstacles to this progression; it also implies a researcher inspecting the signs of diffusion as well as other technical artifacts required in its functioning), or by a human being who creates statements and constructs artifacts.

The notion of an actant is particularly important in the study of scientific activity. This is because the latter permanently modifies the list of entities making up the natural and social world. Out of labora-tories come quarks, enzymes, and proteins, all new actants that did not exist before being brought into play by statements, tables, machines, or embodied skills. But within laboratories, social groups interested in scientific production are also being formed – groups that make up the famous social context. Before Einstein wrote to Roosevelt, politicians could not want the atom bomb; afterward, they wanted it very much. The actant "Roosevelt-who-wants-the-atom-bomb to combat the powers menacing the free world" is no less a laboratory creation than "somostatine-which-inhibits-the-growth-hormone." This, then, is the attraction of the notion of actant. It is sufficiently supple to account for the proliferation of entities that all contribute to scientific production:

electrons and chromatographers, the president of the United States and Einstein, physicians with their assistants, the cancer research campaign, electron microscopes and their manufacturers – all are actants.

The list of actants and their definitions are liable to change, and these changes often give rise to debate. If it is claimed in another laboratory that somostatine also exists in the pancreas and does not inhibit the growth hormone but inhibits the production of insulin, then somostatine's definition changes (Latour, 1987). The very identity of the actant somostatine is transformed, even if its name stays the same. But this is also the case for Roosevelt if he is convinced of the impracticability of the Manhattan project. Actants may more or less successfully resist definition imposed on them and act differently. Then their identity depends on the state of the network and the translations under way, that is, on the history in which they are participating. Society and nature fluctuate like the networks that order them (Callon, 1986, 1989; Latour, 1987, 1991a) – existence precedes essence. The latter has variable geometry, changing as time passes. And this is why the model rejects broad divisions, both between nature and society and between human and non-human. It does not challenge the existence of differences. On the contrary, it multiplies them by allowing the observer to register them all and follow them as they change. The analysis of science is a wonderful laboratory. It is a place where one may study social links in the making.

Underlying Dynamic

The extended translation model gives a broad definition of action. An actant may be a pharmaceutical firm that aims at developing anticancer drugs, a political party that supports cruise missiles, a technician working on a mass spectrometer, a researcher interpreting data charts, or an electron that does not interact with a flow of protons. All these actants are brought into play, mobilized in statements, instruments, or embodied skills. Each new translation may modify, transform, contradict, or alternatively strengthen former translations. Each, that is, may modify or stabilize the actants' universe. To translate is to describe, to organize a whole world filled with entities (actants) whose identities and interactions are thereby defined. In this model the notion of action disappears in favor of that of translation. What, then, is the explanation of scientific change?

To translate a device into an inscription, an inscription into a statement, or a statement into embodied skills, is to create a discrepancy, a betrayal. In short, equivalence is the exception. It is only obtained with difficulty and at great expense. Divergence between translations and the proliferation of entities is the rule, not the exception. The chromatograph traces a curve, the technician draws up charts, the scientist goes from one statement to another, her competence is reinscribed in an experimental device that produces new marks, and so on. Every new translation produces a discrepancy in relation to previous translations, which it then threatens.

So why are there so many proliferating translations? One doesn't have to imagine that actors are steeped in power, trying to impose their equivalences at all costs (though this is not impossible). The notion of action is distributed to all the actants. It is enough to imagine that even the most modest actant, the humblest electron microscope, the most docile technician, and the least imaginative researcher, all produce slightly differing translations. The proliferation of discrepancies lies in these small betrayals. The universe of translation is polytheist. History is an accumulation of such betrayals, and, as the sciences are nothing more than a set of extended translations, their dynamics are no different. This is another way of saying that uncertainty lies at the heart of scientific production. But it is also a way of saying that nature is neither more nor less active or malleable than society.

Agreement

The extended translation model does not talk of assent and dissent. Rather, it speaks more generally of alignment or dispersion of translation networks.

To speak of consensus as in Models 1 and 2, or of closure of debate as in Model 3, is to privilege the discursive dimension of science. By contrast, the translation model, even if it emphasizes the production of statements, assigns great importance to the hidden side of debates – to all that is not discussed but the presence of which allows dialogue to be established.[15] All controversies, even the most fierce and relentless, depend upon a tacit agreement about what is important and what is not. Collins himself shows this in his study of gravitational waves. The discussion between Weber and his colleagues, the exchange of arguments and counterarguments, would have been impossible without deeper agreement about the meaning of Einsteinian theories, the capabilities of computers, the character of mathematical tools, or the nature of torsional moment. For there to be disagreement on the interpretation of a recording, a whole invisible infrastructure of embodied skills, of known and recognized technical artifacts, is needed. Its existence makes discussion possible.

In Model 4 the meaning of a statement – the possibility it has of being taken up or discussed – depends on the chain of translation in which it is located. The explanation of a statement's force – its ability to convince – is no different than the explanation of its meaning. Again, it depends on translation chains and the references these create. Force, then, is a function of the robustness of chains and more particularly of the morphology of the networks they constitute. An isolated statement has no more force than it has meaning. It follows, then, that networks with differentiated elements, which have translated one another, are most forceful. And so are those with many intertwinings. This is because any attempt to question the network is rapidly confronted with a dense network of translations that all support one another. The translation network and the heterogeneity of its components (technical devices, statements, inscriptions, embodied skills, social groups outside laboratories) explain the robustness of arguments.

Such an interpretation is to be found in the work of Pickering or Hacking, though they are concerned mainly with laboratory translations. Pickering distinguishes three categories of elements: models of phenomena, experimental procedures, and interpretative models. The chains of translation are stabilized (another method for defining robustness) when these three subsets are given coherence, that is, when "the interpretation model affecte(d) a smooth translation between the material procedure and one of the two contending phenomenal models" (Pickering, 1990). It is their assembly and the translations that make them converge – Pickering calls this the mangle of practice (Pickering, 1995) – that lead to robustness and stability. Hacking is concerned with the way in which "the laboratory sciences tend to produce a sort of self-vindicating structure that keeps them stable" (Hacking, 1992). This leads him to explore the interactions between a series of heterogeneous elements that strengthen one another – elements that he regroups into three large families of ideas, things, and marks[16] (see also Ackerman, 1985). Such groups and the iterative process giving significance and force to statements by making them coherent are another, novel, way of defining learning.

The robustness of networks depends on the alignment and interlacing of translations created in laboratories. But it extends far beyond these factors. Fujimura (1992), for example, highlights the multiplicity of links that contribute to creating long and robust networks. As elegantly summarized by Pickering (1992):

Her examples include the cells that circulate between the operating room and medical and basic researchers, the recombinant-DNA techniques that flow between the different laboratories that constitute the various fields of technical practice, the computerized data bases that transport findings from one social world to the next, ... and the oncogene theory that serves to organize conceptual, social and material relations between all the social worlds involved. (p. 13)

This leads the translation model to propose a local definition of the universal. According to this definition, statements, experimental devices, and incorporated know-how go no further than the translation networks they compose and in which they circulate. So the universality of science lies in the extension and extent of these networks. Model 4 thus accounts for the character of science stressed by Model 1: universality, capitalization, closure of dissent.

Social Organization

In the translation model, organization is seen from two different perspectives – either from the standpoint of the overall dynamics of networks or in terms of their internal management.

The creation and development of networks depend on a set of conditions that either facilitate or hinder the deployment of translations. Sometimes translations and devices in which they are inscribed may trigger opposition that they do not have the strength to overcome. Can any recording device be used to make an embryo write? Can a human being suffer so that the limits of his resistance may be studied? Is research on bacteriological warfare acceptable? Obviously, seeking answers to these questions is sometimes thought to be illegitimate. And the limits, in principle always revisable, are embodied in the protests, rules, or technical devices that together restrict the field of tolerated translations.

Other obstacles to the proliferation of translations lie in the more or less explicit arrangements that define the circulation of statements, instruments, and embodied skills or that distribute property rights (Cambrosio, Keating, and MacKenzie, 1990). Thus rules of confidentiality may hinder the ramification of networks, while exclusive rights to certain results limit the possibility of connection (as, for example, in the case of patents that could protect the identification of human genes). Finally, the mechanisms for designating the legitimate spokesmen (the actants authorized to speak on behalf of the networks) also influence the character of possible translations. This applies, for example, to the evaluation procedures of researchers, to the composition of commissions responsible for defining research programs, and to the conditions for exercising expertise.

Who is authorized to make whom talk? Who may ally herself with whom? Who speaks on whose behalf? The answers to these three questions define the space for the development of translation networks.

The model is also concerned with the internal management of networks and organizational forms in which they are embodied. The extension of networks and the diversity of their translations mean that the organization of interaction between their heterogeneous elements is an important strategic matter. New analytical tools are needed to study the distribution and links between instruments, statements, and embodied skills and, more generally, all the mobilized actants. Both the contents and the modes of circulation of what is produced depend on the dynamics of these interactions. Some recent studies (which are still small in number: Cambrosio and Keating, 1992; Cambrosio, Keating, and MacKenzie, 1990; Knorr Cetina, in press; Law, 1993; Vinck, Kahane, Larédo, and Meyer, 1993) highlight the variety of configurations and emphasize the increasing importance of networks of laboratories that are linked with firms, state agencies, or hospitals. The study of their organization and, notably, of their multiple forms of coordination (market, hierarchy, trust, technique, and so on) are of particular importance for the extended translation model.

Dynamics of the Whole

The notion of translation network suggests that it is not only the distinction between nature and society that is outdated, but that the conventional opposition between macro- and microanalysis (between global change and local action) is inappropriate.

In the past the opposition between society and nature was used to distinguish a world of passive entities from a world of human actors capable of imagination, invention, and expression. Translation networks establish a continuum between these two extremes – extremes that in practice are never reached. If one still wants to talk of nature and society, it is better to say that translation networks weave a socionature, an in-between that is inhabited by actants whose competence and identities vary along with the translations transforming them. Both passive beings and genuine actors are found there, but the dividing line is not laid down. The history of science is mixed up with the history of these socionatures, which are as varied and come in as many forms as the networks that shelter them.

Size and structural effects are properties of networks. Three concepts make it possible to describe the tension between local action and global change: irreversibility, lengthening, and variety (Callon, 1991, 1992; Callon, Law, and Rip, 1986; Latour, 1991b; Law, 1991).

A network becomes *irreversible* to the extent that its translations are consolidated, making further translations foreseeable and inevitable. Under such circumstances, embodied skills, experimental devices, and systems of statements become increasingly interdependent and complementary. The collective learning that takes place makes accumulation possible. A development ends up by following a perfectly determined sociotechnical path that progressively reduces the room for manoeuver of the actants involved. Other developments and other configurations are always possible in which the reversibility of networks are maintained and the translations remain open.

A translation network is *lengthened* to the extent that it enrolls an increasing number of diverse actants. These may come inside or outside the laboratories for what is important is the number of entities that are associated. The lengthening of a network is generally accompanied by "black-boxing" in which entire chains of translation are folded up and embodied in sentences, technical devices, substances, or skills. Indeed, this process of black-boxing lies at the heart of scientific dynamics (Latour, 1987). In this way, preceding extended networks are punctualized in a new actant; they are maintained, but in an easily manipulable and durable form. Furthermore, they contribute to the production of ever more statements, themselves doomed to pursue their existence silently in the bodies or machines that ensure the enterprise's continuity.

A translation network creates its own coherence. Where there are many diverse and disconnected networks, there are many translations. Conversely, when networks are strongly interconnected to form a system, the level of *diversity* is low. This level is obviously a product of history. But there are two elements that are particularly important in maintaining some degree of diversity. First, certain actors (e.g., state authorities) encourage the proliferation of translation networks. Second, the existence of boundary objects (Star and Griesemer, 1989) or mediators (Wise, 1988) may enable translation networks to coexist peacefully and may mean that one does not necessarily eliminate the others. Such boundary objects or mediators serve to link disjointed translation networks, which thus join together without necessarily fusing them into one. They are sufficiently ambiguous (polysemous in the case of notions and statements, multifunctional in the case of technical devices, complex in the case of embodied skills[17]) to serve as points of departure for divergent translation chains, to which they serve as gateways. Sometimes the weak links formed by boundary objects may strengthen, in which case fusion follows; connections multiply and the same statements, competencies, and technical devices circulate freely between the different points of the new network.

The model of extended translation does not oppose local and global nor does it negate agency and passive behaviors. Rather, it describes the dynamics of networks of different lengths, degrees of irreversibility, diversity, and interconnectedness. This double challenge to the opposition between micro and macro and the distinction between nature and society is to be found in the debate on the environment. Is global change linked to perfecting the design of the catalytic exhaust pipe? Does the feasibility of society depend on the creation of bacteria that have been programmed to destroy themselves? These questions are new because they blur the distinction between science and politics, and the one between human and nonhuman. Here it is clear that translation networks become both the protagonists and the subjects of debate. So it is that the eternal problem of political philosophy is posed: Who has the right to speak on behalf of whom? But the terms in which the question are answered are new. Unlike Model 3, it is not an application of politics to science, but science is now the source of new ideas and concepts of political philosophy.

CONCLUSION

The models presented here have allowed us to regroup scattered works into four coherent units. The argument suggests that, to understand the dynamics of science and its growth, we need to explore both its contents and its organization. The way each model grants priority to different questions, and the way in which it tackles other issues, depends on the way in which it treats these questions. We have explored this coherence, which is sometimes not visible when authors are considered individually.

It would be unfair not to mention that these models fail to capture some of the most promising new developments in social studies of science and technology. For example, work on discourse analysis (Ashmore, Myers and Potter, 1995) attempts to grasp the irritating problem of reflexivity and to imagine new literary forms (Fox Keller, 1995) and last but not least politically crucial work on gender (Wajcman, 1995; Fox, 1995) cannot be expected to be integrated with this presentation. That these works do not fit in with the four models is probably a consequence of focusing mostly on such general issues as the cultural and political place of scientific knowledge in modern societies rather than on the specific dynamics of science.

The schema adopted dramatizes difference rather than convergence. Yet convergences do exist to some extent. For instance, Models 4 and 1 share the notion of translation. Models 3 and 4 stress the role of instruments and incorporated skills in the dynamics of science; they also recognize the importance of entrenched networks of notions and/ or technical standards for taming controversies. And to be sure, there are other points in common. But more important, in my view, each model has enriched the preceding one. The question of the soundness of such a progressiveness remains open. The reader should understand that giving a rhetorically plausible answer is difficult for the author! Each model's strength obviously hinges upon the number of allies it is able to enlist for its backing and defense. Model 4 is devised to satisfy simultaneously those who are obsessed with the need to explain a statement's robustness, those who see science as a competition between knowledge claims, and finally those who consider science to be a heterogeneous sociocultural practice. How successful such an attempt will be depends on the reader and not on the author.

Be that as it may, future research could be undertaken in two directions:

1. Each model is strong in certain areas and weak in others. For example, in Model 2, the borrowings from economics are limited to the most general theories and, admittedly, the oldest. The concepts of industrial economy are not used. Notions such as barriers to entry, differentiated return on investments, imperfect competition, diversification or differentiation strategies would certainly enrich the analysis. More generally, the historical investigation of the emergence and the evolution of so-called scientific institutions deserves to be carefully scrutinized. Those who are committed to Model 1 might wish to explore how some new criteria for assessing the robustness of statements emerge and are accepted. The supporters of Model 3 might deepen their research on the establishment of agreement and develop a more articulated cultural history of scientific practices, paying attention to the boundary construction between science and its environment (Gieryn,

Chapter 11 in this volume). And those of Model 4 have at present little to say about the organizational forms accompanying or hindering translation networks.

2. The models presented in this chapter have much to say about the relationships between statements, technical artifacts, and embodied skills as well as the substitution or complementarity to which these give rise. But there is little work on the links between either the translation networks of science on one hand and technology (Bijker and Pinch, 1987) and economics on the other. Such investigations might show how networks develop in which statements, technical devices, money, embodied skills, confidence, and commands all circulate. If this is done, then a link will be built with neighboring disciplines, in particular with the economics of technical change whose recent results show a remarkable convergence with those of the sociology of science and technology. Such, at any rate, is an exciting possibility.

NOTES

Author's Note: I received a lot of stimulating comments while this chapter was in preparation. In addition to two anonymous referees, I thank François Jacq, Bruno Latour, Dominique Pestre, Trevor Pinch, Vivian Walsh, Yuval Yonay, and all my colleagues from the Centre de sociologie de l'innovation. Without John Law's support and help, this chapter probably would never have been completed. This version was completed while I was a member at the Institute for Advanced Study in Princeton, which I thank for its support and hospitality.

1. For a clear presentation, see Jacob (1981).
2. Two opposite arguments on the conditions of possibility of such translations are proposed by D. Davidson (1984) and Quine (1969).
3. As in Carnap's radical theory, which rejects any meaning for statements that cannot be directly related to observation: This permits one to say that the correctly constructed statement "Caesar is a prime number" has no meaning. This point of view is also defended by Wittgenstein in the Tractatus (Wittgenstein, 1921).
4. This position was vigorously defended by Bachelard (1934) and taken up again by Hanson (1965). In general it corresponds to philosophical realism in which it is argued that theoretical progress increases access to natural reality. Curiously, however, philosophical relativists such as Quine insist on the theoretical (and hence arbitrary) character of all observation statements.
5. Here one finds all the debates between realism, pragmatism, positivism, and relativism. Realists insist that statements increasingly approximate real-

ity (Putnam, 1978); positivists argue that the accumulation of observation statements extends and increases the precision of our knowledge (Carnap, 1955). For pragmatists, science is treated as a reliable tool that makes it possible to act on and control nature (Laudan, 1990). And relativists insist that statements teach us nothing about reality "out there" (Feyerabend, 1975; Quine, 1953).

6. For example, the statement "gravitational force is an inverse function of a power of distance" is less demanding than the statement "gravitational force is an inverse function of the square of distance."

7. The sociology of Goffman can be usefully applied to explain how actors try to avoid losing face (and try to keep their opponents from losing it too). See Wynne (1979) on this point.

8. For an appealing presentation, see Laudan (1990).

9. The article is only one of the ways of identifying and imputing discoveries; the laboratory notebook where experiments and their dates are written down is also important as is the strict separation between technicians and researchers. These elements are held together by norms, rules, or organizations. There are no journals without publishing houses, just as there are no articles without referees; there is no agreement about the date of experiments without commissions of experts who establish a chronology. The separation between technicians and researchers, essential if discoveries are not to be imputed to teams, is maintained by a system of diplomas, recruitment procedures, a strict hierarchy of occupational roles, and so on. A precise history of this system remains mostly to be written.

10. Goods are said to be public when their use by one person does not exclude their use by another person. According to economists, this property is an intrinsic characteristic of certain goods.

11. A detailed presentation and genealogy of this "eclectic" position is given by Shapin (1992a).

12. Numerous empirical studies have supported this hypothesis. Cambrosio (1988), in an article neatly titled "Going Monoclonal," is outstanding, for one sees everything that should be learned on the job, all the details that matter but that never appear in the texts or spoken word though they are essential for producing monoclonal antibodies by the world's best scientist. Cambrosio even notes amusingly that the tacit part of the practices is so important that certain superstitions are developed to account for the success or failure of an experiment that cannot be explained by explicit knowledge.

13. "The examination which places individuals in a field of surveillance also situates them in a network of writing; it involves them in a whole series of documents which capture and bind them. The examination procedures were immediately accompanied by a system of thorough registration and documentary accumulation" (Foucault, 1975, p. 191).

14. For a detailed analysis of the variety of the devices of translations in physics and molecular biology, see Knorr Cetina (in press).

15. Paradoxically, Model 3, which emphasizes the role of tacit knowledge, focuses the analysis of closure on explicit controversies.

16. Each of these three groups includes five different items. We find in them statements about anything from general theories to the modeling of instruments; the material elements mobilized including the targets (pro-

cessed natural substances, laboratory animals, samples, and so on) subjected to checks but also tools and other generators of data; inscriptions
(or "marks") produced by the generators of data and on which operations
are carried out (evaluation, reduction, analysis, and interpretation). It is
from the converging of these items, which is always difficult and constitutes the thread of science, that the robustness of knowledge and its stability results.

17. A statement may be a boundary object, as, for instance, in the case of the
Lorentz equation that established a link between Newtonian and Einsteinian mechanics. Again, instruments are often powerful mediators as well
as human beings (Downey, in press).

REFERENCES

Abir-Am, P. (1982). "The Discourse of Physical Power and Biological Knowledge in the 1930s: A Reappraisal of the Rockefeller Foundation's 'Policy' in
Molecular Biology", *Social Studies of Science*, 12, pp. 341–382.

Ackerman, R. (1985). *Data, Instruments and Theory: A Dialectical Approach to
Understanding Science*, Princeton: Princeton University Press.

Althusser, L. (1974). *La philosophie spontanée des savants*, Paris: Maspero.

Amman, K. and Knorr Cetina, K. (1988a). "The fixation of (visual) evidence"
(Special issue: Representation in Scientific Practice, M. Lynch and S. Woolgar, eds.), *Human Studies*, 133–169.

—— (1988b). "Thinking through talk: An ethnographic study of a molecular
biology laboratory", in Lowell Hargens, R.A. Jones and Andrew Pickering
(eds.), *Knowledge and Society: Studies in the Sociology of Science Past and Present*, Greenwich, CT: JAI.

Ashmore, M. Myers, G., and Potter, J. (1995). "Discourse, Rhetoric, Reflexivity:
Seven days in the library", in S. Jasanoff, G.E. Markle, J.C. Petersen and T.
Pinch (eds.), *Handbook of Science and Technology Studies*, Thousand Oaks:
Sage Publications, pp. 321–342.

Axelrod, R. (1984). *The Evolution of Cooperation*, New York: Basic Books.

Bachelard, G. (1934). *Le Nouvel Esprit Scientifique*, Paris: PUF

Barnes, B. (1971). "Making out in industrial research", *Science Studies*, 1, pp.
157–175.

—— (1977). *Interests and the Growth of Knowledge*, London: Routledge & Kegan
Paul.

Barnes, B. and Shapin, S. (eds.) (1979). *Natural Order: Historical Studies in Scientific Culture*, London: Sage.

Ben-David, J. (1971). *The Scientist's Role in Society: A Comparative Study*, Englewood Cliffs, NJ: Prentice-Hall.

—— (1991). *Scientific Growth: Essays on the Social Organization and Ethos of
Science*, G. Freudenthal (ed.), Berkeley: University of California Press.

Bijker, W.E. and Pinch, T.J. (1987). "The social construction of facts and artefacts: Or how the sociology of science and the sociology of technology might
benefit each other", in Wiebe E. Bijker, Thomas P. Hughes and Trevor J.
Pinch (eds.), *The Social Construction of Technological Systems: New Directions*

in the Sociology and History of Technology, Cambridge, MA: MIT Press, pp. 17–50.

Bloor, D. (1976). *Knowledge and Social Imagery,* London: Routledge & Kegan Paul.

—— (1992). "Left and right Wittgensteinians", in A. Pickering (ed.), *Science as Practice and Culture,* Chicago: University of Chicago Press, pp. 266–282.

Boltanski, L. and Thévenot, L. (1991). *De la justification: Les économies de la grandeur.* Paris: Gallimard.

Bourdieu, P. (1971). "Le marché des biens symboliques", *L'Année Sociologique,* 22, pp. 49–126.

—— (1975). "La spécificité du champ, scientifique et les conditions sociales du progrès de la raison", *Sociologie et Sociétés,* 7(1).

Brannigan, A. (1981). *The Social Basis of Scientific Discoveries,* Cambridge: Cambridge University Press.

Callon, M. (1980). "Struggles and negotiations to decide what is problematic and what is not: The socio-logics of translation" in Karin Knorr, Roger Krohn and Richard Whitley (eds.), *The Social Process of Scientific Investigation,* Dordrecht: Reidel, pp. 197–219.

—— (1986). "Some elements of a sociology of translation: Domestication of the scallops and the fishermen of St Brieux Bay", in John Law (ed.), *Power, Action and Belief: A New Sociology of Knowledge? (Sociological Review Monograph),* London: Routledge & Kegan Paul, pp. 196–229.

—— (1991). "Techno-economic networks and irreversibility", in J. Law (ed.), *A Sociology of Monsters: Essays on Power, Technology and Domination (Sociological Review Monograph),* London: Routledge & Kegan Paul, pp. 132–164.

—— (1992). "Variety and irreversibility in networks of technique conception and adoption," in D. Foray and C. Freeman (eds.), *Technology and the Wealth of Nations,* London: Frances Printer.

Callon, M. (ed.) (1989). *Le Science et ses Réseaux: Genèse et circulation des faits scientifiques* (Anthropologie des sciences et des techniques), Paris: La Découverte.

Callon, M., Law, J. and Rip, A. (eds.) (1986), *Mapping the Dynamics of Science and Technology: Sociology of Science in the Real World,* London: Macmillan.

Cambrosio, A. (1988). "Going Monoclonal: Art, science, and magic in the day-to-day use of hybridoma technology", *Social Problems,* 35, pp. 244–260.

Cambrosio, A. and Keating, P. (1992). "A matter of FACS. Constituting novel entities in immunology", *Medical Anthropology Quarterly,* 6, pp. 362–384.

Cambrosio, A., Keating, P. and MacKenzie, M. (1990). "Scientific practice in the courtroom: The construction of sociotechnical identities in a biotechnology patent dispute", *Social Problems,* 37, pp. 301–319.

Carnap, R. (1955). 'Testability and meaning,' in H. Feigl and M. Brodbeck (eds.), *Readings in the Philosophy of Science,* New York: Appleton, Century, Crofts, pp. 47–92.

Clarke, A. and Gerson, E. (1990). "Symbolic interactionism in social studies of science", in H. Becker and M. McCall (eds.), *Symbolic Interaction and Cultural Studies,* Chicago: University of Chicago Press, pp. 179–214.

Coase, R. (1937). "The nature of the firm", *Economica,* 4, pp. 386–405.

Cole, J. (1973). *Social Stratification in Science*, S. Cole (ed.), Chicago: University of Chicago Press.

Collins, H.M. (1974). "The TEA set: Tacit knowledge and scientific networks", *Science Studies*, 4, pp. 165–186.

—— (1985). *Changing Order: Replication and Induction in Scientific Practice*. London: Sage.

Collins, H.M. and Pinch, T. (1979). "The construction of the paranormal, nothing unscientific is happening", in R. Wallis (ed.), *On the Margins of Science: The Social Construction of Rejected Knowledge (Sociological Review Monograph)*, Keele: University of Keele.

Cotgrove, S. and Box, S. (1970). *Science Industry and Society*, London: George Allen & Unwin.

Crane, D. (1972). *Invisible Colleges*, Chicago: University of Chicago Press.

Daston, L. and Galison, P. '*The Image of Objectivity*', *Representations*, 40 (Fall), pp. 80–128.

David, P.A., Mowery, D.C. and Steinmueller, W.E. (1992). "Analysing the economic payoffs from basic research," *Economics of Innovation and New Technology*, 2, 73–90.

Davidson, D. (1984). *Truth and Interpretation*, Oxford: Oxford University Press.

Downey, Gary L. (in press). "Training Engineers as Boundary Subjects", *Science as Culture*.

Dubinskas, F. (ed.) (1988). *Making Time: Ethnographic Studies of High-technology Organization*, Philadelphia: Temple University Press.

Favereau, O. (in press). "Règles, organisation at apprentissage collectif", in A. Orléan (ed.), *Analyse Économique des Conventions*, Paris: Presses Universitaires de France.

Feyerabend, P. (1975). *Against Method*, London: New Left Books.

Fleck, L. ([1935]1979). *Genesis and Development of a Scientific Fact*, Chicago: University of Chicago Press.

Foucault, M. (1975). *Surveiller et Punir*, Paris: Gallimard.

Fox, M.F. (1995). 'Women and scientific careers', in S. Jasanoff, G.E. Markle, J.C. Petersen and T. Pinch (eds.), *Handbook of Science and Technology Studies*, Thousand Oaks: Sage Publications, pp. 205–223.

Freudenthal, G. (1986). *Atom and Individual in the Age of Newton*, Dordrecht: Reidel.

Fujimura, H. (n.d.), "A Tool for Dynamic Analysis of Situated Scientific Problem Construction", manuscript submitted for publication.

Fujimura, J.H. (1992). "Crafting science: Standardized packages, boundary objects and 'translation'", in A. Pickering (ed.), *Science as Practice and Culture*, Chicago: University of Chicago Press, pp. 168–211.

Galison, P. (1987). *How Experiments End*, Chicago: University of Chicago Press.

Gaston, J. (1973). *Originality and Competition in Science*, Chicago: University of Chicago Press.

Golinski, JV. (1990). "The theory of practice and the practice of theory: Sociological approaches in the history of science", *ISIS*, 81, pp. 492–505.

Gooding, D. (1992). "Putting agency back into experiment", in A. Pickering (ed.), *Science as Practice and Culture*, Chicago: University of Chicago Press.

Grünbaum, A. and Salmon. W. (eds.) (1988). *The Limitations of Deductivism*. Berkeley: University of California Press.

Granovetter, M.S. (1973). "The strength of weak ties", *American Journal of Sociology*, 78, pp. 1360–1380.

Habermas, J. (1987). *Théorie de l'agir communicationnel, 2. Pour un critique de la raison fonctionaliste*, Paris: Fayard.

Hacking, I. (1983). *Representing and Intervening: Introductory Topics in the Philosophy of Natural Science*, Cambridge: Cambridge University Press.

—— (1992). "The self-vindication of the laboratory sciences", in A. Pickering (ed.), *Science as Practice and Culture*, Chicago: University of Chicago Press, pp. 29–64.

Hagstrom, W.O. (1966). *The Scientific Community*, New York: Basic Books.

Hanson, N.R. (1965). *Patterns of Discovery*, Cambridge: Cambridge University Press.

Hess, D.J. (1992). "Introduction: The new ethnography and the anthropology of science and technology", in D.J. Hess and L. Layne (eds.), *Knowledge and Society: The Anthropology of Science and Technology* (vol. 9), Greenwich, CT: JAI.

Hesse, M. (1974). *The Structure of Scientific Inference*, London: Macmillan.

Holton, G. (1973). *Thematic Origins of Scientific Thought: Kepler to Einstein*, Cambridge, MA: Harvard University Press.

Hull, D. (1988). *Science as a Process: An Evolutionary Account of the Social and Conceptual Development of Science*, Chicago: University of Chicago Press.

Jacob, P. (1981). *De Vienne à Cambridge*, Paris: Gallimard.

Jasanoff, S. (1990). *The Fifth Branch: Science Advisors as Policy Makers*, Cambridge, MA: Harvard University Press.

Keller, E.F. (1995). "The Origin, History, and Politics of the Subject called 'Gender and Science': A first person account", in S. Jasanoff, G.E. Markle, J.C. Petersen and T. Pinch (eds.), *Handbook of Science and Technology Studies*, Thousand Oaks: Sage Publications, pp. 80–94.

Knorr Cetina, K. (1981). *The Manufacture of Knowledge: An Essay on the Constructivist and Contextual Nature of Science*, Oxford: Pergamon (rev. edn., 1984, *Die Fabrikation von Erkenntnis*, Frankfurt: Suhrkamp).

—— (1992). *Liminal and Referent Epistemologies in Contemporary Science: An Ethnography of the Empirical in Two Sciences*, paper presented at the Thursday Seminar, Princeton Institute for Advanced Study, Princeton University.

—— (in press). *Epistemic Cultures: How Scientists Make Sense*, Bloomington: Indiana University Press.

Kornhauser, W. (1962). *Scientists in Industry: Conflict and Accommodation*, Berkeley: University of California Press.

Kreps, D. and Wilson, R. (1982). "Reputation and imperfect formation", *Journal of Economic Theory*, 27, pp. 253–279.

Kuhn, T.S. (1962). *The Structure of Scientific Revolutions*, Chicago: University of Chicago Press.

—— (1970). "Logic of discovery or psychology of research?", I. Lakatos and A. Musgrave (eds.), *Criticism and the Growth of Knowledge*, Cambridge: Cambridge University Press, pp. 1–23.

Laudan, L. (1990). *Science and Relativism: Some Key Controversies in the Philosophy of Science*, Chicago: The University of Chicago Press.

Latour, B. (1987). *Science in Action: How to Follow Scientists and Engineers through Society*, Cambridge, MA: Harvard University Press.

—— (1988). *The Pasteurization of France*, trans. A. Sheridan and J. Law, Cambridge, MA: Harvard University Press.

Latour, B. (1991a). *Nous n'avons jamais été modernes: Essai d'anthropologie symétrique*, Paris: La Découverte.

—— (1991b). "Technology in society made durable", in J. Law (ed.) *A Sociology of Monsters: Essays on Power, Technology and Domination* (*Sociological Review Monograph*, London: Routledge & Kegan Paul, pp. 103–130.

Latour, B. and Woolgar, S. (1979). *Laboratory Life: The Social Construction of Scientific Facts*, Princeton: Princeton University Press.

Law, J. (1986a). "Laboratories and texts", in M. Callon, J. Law and A. Rip (eds.), *Mapping the Dynamics of Science and Technology*, London: Macmillan.

—— (1986b). "On the methods of long-distance control vessels navigation and the Portuguese route to India", in J. Law (ed.), *Power, Action and Belief: A New Sociology of Knowledge?* (*Sociological Review Monograph* 38). Keele: University of Keele, pp. 234–263.

—— (1991). "Power, discretion and strategy", in J. Law (ed.), *A Sociology of Monsters: Essays on Power, Technology and Domination* (*Sociological Review Monograph*), London: Routledge & Kegan Paul, pp. 165–191.

—— (1993). *Modernity, Myth and Materialism*, London: Blackwell.

Lynch, M. (1985). *Art and Artifact in Laboratory Science: A Study of Shop Work and Shop Talk in a Research Laboratory*, London: Routledge & Kegan Paul.

—— (1992). "Extending Wittgenstein: The pivotal move from epistemology to sociology of science", in A. Pickering (ed.), *Science as Practice and Culture*, Chicago: University of Chicago Press, pp. 215–265.

Lynch, M., Livingstone, E. and Garfinkle, H. (1983). 'Temporal order in laboratory work', in K. Knorr Cetina and M. Mulkay (eds.), *Science Observed: Perspectives on the Social Study of Science*, London: Sage.

MacKenzie, D. (1981). *Statistics in Britain: 1865–1930*, Edinburgh: Edinburgh University Press.

Marcsun, S. (1960). *The Scientist in American Industry*, New York: Harper.

Masterman, M. (1970). "The nature of a paradigm", in Imre Lakatos and Alan Musgrave (eds.), *Criticism and the Growth of Knowledge*, Cambridge: Cambridge University Press, pp. 59–90.

Merton, R.K. (1938/1970). *Science, Technology and Society in Seventeenth-Century England*, New York: Harper & Row (originally published in *Osiris*, 1938).

—— (1973). *The Sociology of Science: Theoretical and Empirical Investigations*, N.W. Sorter (ed.), Chicago: University of Chicago Press.

Mey, M. De (1982). *The Cognitive Paradigm*. Dordrecht: Reidel.

Mulkay, M. (1972). *The Social Process of Innovation*; London: Macmillan.

Mullins, N. (1972). "The development of a scientific speciality: The Phage Group and the origins of molecular biology", *Minerva*, 10(1), pp. 51–82.

Myers, G. (1990). *Writing Biology: Text and the Social Construction of Scientific Knowledge*, Madison: University of Wisconsin Press.

Pestre, D. (1990). *Louis Neel: Le magnétisme et Grenoble* (Vol. Cahier d'Histoire du CNRS), Paris: CNRS.

Pickering, A. (1990). "Knowledge, practice and mere construction", *Social Studies of Science*, 20, pp. 682–729.

—— (1995). *The Mangle of Practice*, Chicago: The University of Chicago Press.

Pickering, A. (ed.) (1992). *Science as Practice and Culture*, Chicago: University of Chicago Press.

Pinch, T. (1985). "Towards an analysis of scientific observation: The externality and evidential significance of observation reports in physics", *Social Studies of Science*, 15, pp. 167–187.

—— (1986). *Confronting Nature: The Sociology of Neutrino Detection*. Dordrecht: D. Reidel Publishing Co.

Polanyi, M. (1958). *Personal Knowledge*, London: Routledge & Kegan Paul/Chicago: University of Chicago Press.

Popper, K.R. (1945). *The Open Society and Its Enemies*, London: Routledge & Kegan Paul.

—— (1972). *Objective Knowledge: An Evolutionary Approach*, Oxford: Clarendon.

Price, D.J. de Solla (1967). "Networks of Scientific Papers", *Science*, 149, pp. 510–515.

Quine, W.V.O. (1953). "Two dogmas of empiricism", in W.V. Quine, *From a Logical Point of View*, Cambridge, MA: Harvard University Press, pp. 20–46.

—— (1969). *Ontological Relativity and Other Essays*, New York: Columbia University Press.

Ravetz, J.R. (1971). *Scientific Knowledge and its Problems*; Oxford: Oxford University Press.

Rip, A. (1988). "Contextual transformation in contemporary science", in A. Jamison (ed.) *Keeping Science Straight: A Critical look at the Assessment of Science and Technology*, Gothenburg: University of Gothenburg, Department of Theory of Science.

Rudwick, M.J.S. (1985). *The Great Devonian Controversy: The Shaping of Scientific Knowledge Among Gentlemanly Specialists*, Chicago: University of Chicago Press.

Schaffer, S. (1988). "Astronomers mark time: Discipline and the personal equation", *Science in Context*, 2, pp. 115–145.

—— (1989). "Glass works, Newton's prisms and the uses of experiment", in D. Gooding, T. Pinch and S. Schaffer (eds.), *The Uses of Experiments: Studies in the Natural Sciences*, Cambridge: Cambridge University Press, pp. 67–104.

—— (1991). *Where Experiments End: Table-top Trial in Victorian Astronomy*, unpublished manuscript, Cambridge.

Shapin, St. (1979). "The politics of observation: Cerebral anatomy and social interests in the Edinburgh phrenology disputes", in R. Wallis (ed.), *On the Margins of Science: The Social Construction of Rejected Knowledge* (*Sociological Review Monograph*, 27), London: Routledge & Kegan Paul, pp. 139–178.

—— (1989). "The invisible technician", *American Scientist*, 77, pp. 553–563.

—— (1992) "Discipline and Bounding: The history and sociology of science as seen throught the externalism–internalism debate", *History of Science*, 30, pp. 333–369.

Star, S.L. (1989). *Regions of Mind: Brain Research and the Quest for Scientific Certainty*, Stanford, CA: Stanford University Press.

Star, S.L. and Greisemer, J. (1989). "Institutional ecology, 'translations' and boundary Objects: Amateurs and professionals in Berkeley's Museum of Vertebrate Zoology, 1907–1939", *Social Studies of Science*, 19, pp. 387–420.

Traweek, S. (1988). *Beamtimes and Lifetimes: The World of High Energy Physicists*, Cambridge, MA: Harvard University Press.

Vernant, J.-P. (1990). "La formation de la pensée positive dans la Grèce archaïque", in J. Vernant and P. Vidal-Naquet (eds.), *La Grèce ancienne*, Paris: Seuil, pp. 196–228.

Vinck, D., Kahane, B. Laredo, P. and Meyer, J. (1993). "A network approach to studying programmes: Mobilizing and coordinating public responses to HIV/AIDS", *Technology Analysis and Strategic Management*, 5(1), pp. 39–54.

Vries, G. de (1992). *Wittgenstein and the Sociology of Scientific Knowledge: Consequences to a Farewell Epistemology*, mimeo.

Wajcman, J. (1995). "Feminist Theories of Technology", in S. Jasanoff, G.E. Markle, J.C. Petersen and T. Pinch (eds.), *Handbook of Science and Technology Studies*, Thousand Oaks: Sage Publication, pp. 189–204.

Wallis, R. (ed.) (1979). *On the Margins of Science: The Social Construction of Rejected Knowledge (Sociological Review Monograph, 27)*, Keele: University of Keele.

Whitley, R. (1984). *The Intellectual and Social Organization of the Sciences*, Oxford: Oxford University Press.

Wise, N. (1988). "Mediating Machines", *Science in Context*, 2, pp. 77–113.

Wise, N., and Smith, C. (1988). *Energy and Empire*, Cambridge: Cambridge University Press.

Wittgenstein, L. (1921). *Tractatus Logico-Philosophicus*, London: Routledge & Kegan Paul.

—— (1953). *Philosophical Investigations*, Oxford: Blackwell.

Woolgar, S. (1976). "Writing an intellectual history of scientific development: The use of discovery accounts", *Social Studies of Science*, 6, pp. 395–422.

—— (1988). *Science: The Very Idea*, London: Tavistock.

Wynne, B. (1979). "Between Orthodoxy and Oblivion: The normalisation of deviance in science", in R. Wallis (ed.), *On the Margins of Science: The Social Construction of Rejected Knowledge (Sociological Review Monograph)* Keele, UK: University of Keele.

—— (1992). "Uncertainty and environmental learning: Reconceiving science and policy in the preventive paradigm", *Global Environmental Change*, 2, pp. 137–154.

11 Boundaries of Science*
Thomas F. Gieryn

The working title of this *Handbook* presumed three neatly bounded territories: science, technology, and society. This chapter makes those territories and especially their borders into objects for sociological interpretation and seeks to recover their messiness, contentiousness, and practical significance in everyday life. Its focus is on the "boundary problem" in science and technology studies: Where does science leave off, and society – or technology – begin? Where is the border between science and non-science? Which claims or practices are scientific? Who is a scientist? What *is* science?

The chapter begins with two perspectives on the boundary problem, essentialism and constructivism. Essentialists argue for the possibility and analytic desirability of identifying unique, necessary, and invariant qualities that set science apart from other cultural practices and products, and that explain its singular achievements (valid and reliable claims about the external world). Constructivists argue that no demarcation principles work universally and that the separation of science from other knowledge-producing activities is instead a contextually contingent and interests-driven pragmatic accomplishment drawing selectively on inconsistent and ambiguous attributes. Research in the sociology of science has raised doubts about the ability of any proposed "demarcation criteria" to distinguish science from non-science. Attention has thus shifted from criticisms of essentialism to examinations of when, how, and to what ends the boundaries of science are drawn and defended in natural settings often distant from laboratories and professional journals – a process known as "boundary-work."[1] Essentialists *do* boundary-work; constructivists *watch* it get done by people in society – as scientists, would-be scientists, science critics, journalists, bureaucrats, lawyers, and other interested parties accomplish the demarcation of science from non-science.

The second part discusses four theoretical precursors and extensions of the elementary constructivist idea that the boundaries of science are

* Chapter 18 in S. Jasanoff, G.E. Markle, J.C. Petersen and T. Pinch (eds.), *Handbook of Science and Technology Studies* (Thousand Oaks, CA: Sage Publications, 1995), pp. 393–407; 424–443.

social conventions. Each draws upon scholarly literatures and theoretical traditions initially developed outside STS: sociological studies of professions, symbolic interactionism and its recent interest in social worlds, anthropologically driven studies of cultural history focused on the significance of classification practices, and feminist theories of knowledge.

The chapter concludes with empirical research illustrating constructivist studies of the boundary problem. I examine four episodes of boundary-work, different in the *goals* sought by those who claimed their place *inside* science. The dispute between Hobbes and Boyle (Shapin and Schaffer, 1985) illustrates pursuit of a monopoly over cultural authority through exclusion of those offering discrepant and competitive maps of the place of science in the intellectual landscape. D'Alembert and Diderot's *Encyclopédie* (Darnton, 1984) is an example of how insider-scientists construct other knowledge-producing systems as foils to legitimate the expansion of science. Third, the affair surrounding Cyril Burt's alleged misconduct (Gieryn and Figert, 1986) points to a familiar kind of boundary-work: expulsion of deviant scientists, as a means of social control and of preserving professional reputation and public confidence. Finally, Jasanoff's (1990) study of science advisers involved in government policy making presents the contested border between science and politics as an occasion for scientists to protect their autonomy and authority from usurpation or control by outsiders – government bureaucrats, elected officials, and lobbyists.

ESSENTIALISM AND ITS CONSTRUCTIVIST CRITIQUE

The analytic problem of demarcating science from non-science has attracted the attention of three major figures in science studies – Karl Popper in philosophy, Robert Merton in sociology, Thomas Kuhn in history – and each opted for an essentialist solution. No doubt a desire for definitional tidiness led them to the demarcation problem. Each was an interpreter of *science*, and how could one know what to look at – and what to look away from – without a workable definition of the phenomenon at hand? Demarcation was perhaps vital as well for efforts to explain what these three took as a singular achievement of science: an improving validity and reliability in its models of the world. Criteria of demarcation became, in effect, explanations for the superiority of science (among knowledge-producing practices) in producing truthful claims about the external world. There was, as well, an ideological dimension in these classic demarcation tries. Merton's wish to dismiss Nazi science as perverse, or Popper's wish to dismiss psychoanalysis and Marxism as pseudoscience, created the need for a standard to identify the real McCoy.

Falsifiability

Popper gave the demarcation problem its name, and his philosophical solution – falsifiability – remains the most familiar one. He arrived at demarcation while dealing with *logical* problems discerned in philosophers' prior attempts to "justify" theories or generalizations. For Popper, demarcation is an epistemological matter, to be resolved by finding something in the *Methodology* of science – in its "epistemic invariants" (Laudan, 1983, p. 28) – that accounts for the superiority of science in providing reliable and valid knowledge about the world.

When Popper started out in the 1920s, a once-popular philosophy held that science seeks justification for its theories through the accumulation of confirmatory empirical evidence. Scientists try to verify theories with corroborating facts, and, if they succeed, science progresses toward truth. Popper recognized that this verificationist strategy ran aground on logical problems of induction. The accumulation of corroborating evidence could tell scientists what *was*. But reliable predictions of the future (something that general theories in science are expected to offer; Popper, 1972, pp. 349 ff.) are uncertain: The next observation could always in principle yield a refutation. Popper's watershed response not only offered an escape from the inductivist illogic of verification but a criterion for demarcating scientific from nonscientific statements.

In place of verification, Popper prefers falsification; in place of certain truth, he offers conjecture. Science advances toward truth (though never arriving at certainty) by a combination of bold conjecture and severe criticism. Scientists work from problems to theoretical generalizations and basic statements that are in principle *falsifiable*; potentially, some empirical observation could logically contradict or refute them (Popper, 1959, pp. 40 ff). The bolder such conjectures, the more rapidly science moves ahead: Theories that are more easily falsified (i.e., a larger number and wider variety of observations are potentially able to refute the claim) contribute more to the advancement of objective knowledge than more timid claims. Bold conjecture alone does not make science: Scientists must also subject such conjectures to severe critical scrutiny as they try their hardest to refute the theory. No theory is immune from such criticism. Science thus is not a confirmation game (looking for evidence to corroborate a generalization) but a refutation game (looking for evidence to shoot it down). The result is not certain truth but ever bolder conjectures that have (for the time being) survived critical refutation, and thus assume better (but still fallible) approximations to reality and enlarged credibility.[2] Scientific knowledge grows – progresses – more through the elimination of error (actual falsification) than through the cumulative repetition of corroborative evidence.

Popper made falsifiability do double duty. First, in contrast to verification and its problems with induction, conjecture-and-refutation put science on a steadier logical footing and appeased some nervous philosophers. Second, falsifiability provided Popper with a wedge to drive between science and non-science. Any of three conditions would suffice to dump a claim, a practice, a belief, or their adherents into non-science: if the claim were not potentially falsifiable (i.e., no conceivable empirical observation would logically contradict the assertion), if there was no sincere and severe effort to refute the claim, or if a claim is not rejected when refuting empirical evidence is presented. Nonscience includes metaphysics, ideology, and pseudoscience, and Popper insists that these activities may have meaning or practical significance, but they simply are not *scientific* (which is why Popper can put such worthy but non-empirical pursuits as mathematics and logic outside science; Popper, 1963, pp. 253 ff.). This same demarcation criterion enabled Popper (1972) to locate the historical turn from prescience to science in ancient Greece of the fifth or sixth century BC, where a "critical attitude" (falsification and the rest) replaced a "dogmatic hanging on of the doctrine in which the whole interest lies in the preservation of the authentic tradition" (p. 347).

Popper (1963) traces his interest in demarcation to the intellectual swirl of 1919 Vienna, involving Marx's theory of history, Freudian psychoanalysis, and Adlerian "individual psychology" (pp. 34 ff.). He asked himself: What is *wrong* with these theories, and what made them so different than those in physics? The theories were not "wrong" because they lacked verification; indeed, it was precisely the abundance of confirming observations that made Popper suspicious: "Once your eyes were thus opened you saw confirming instances everywhere" (Popper, 1972, p. 35). Popper's eventual conclusion that Marxism and psychoanalysis lacked falsifiability – and thus were outside empirical science – was reached via astrology. Astrologers did indeed accumulate much confirming evidence, but they put their theories and prophecies in such a vague language that any evidence could be made to fit. That soothsaying is then pinned on Marxists, who reinterpret not just the theory but the evidence as well in efforts to salvage their general theory of history: "Marxism has established itself as a dogmatism which is elastic enough... to evade any further attack" (Popper, 1963, p. 334). Psychoanalysis shares this inadequacy; both Freud and Adler cast their original theories in a form that could not be falsified by any observation of human behavior. Popper sought a methodology for science that would at once rescue its theories from the logical problems of induction and permit the demarcation of empirical science from impostors and from other non-empirical but still meaningful belief systems such as metaphysics. Falsifiability, conjecture, and refutation fit the bill.

Constructivist criticism of Popper's try at essentialist demarcation centers on the *reproducibility* of falsifying empirical evidence. Falsifiability is a logical condition but falsification is a practical accomplishment of observation and experiment, and on this Popper and his constructivist critics agree. They disagree on whether falsification is a straightforward unambiguous lineup of evidence against theory, or whether the process is shrouded in interpretative ambiguities that get resolved only through complex social negotiations (that go beyond logic and methodology). Popper is no naive empiricist, and the idea of "theory-laden facts" is as much a part of his legacy as falsification. Science cannot be built from immediate and unstructured sense perceptions; rather, observations are guided by problems-at-hand and by theories that act as nets for sifting through the infinite details of reality. And though observations must be sufficiently independent of theories if they are to perform their occasionally necessary role of falsification, the question remains: *When* does observation justify the abandonment of a theory? Popper worried about the possibility that worthy theories would be too quickly dumped in the face of any old falsification report: "We shall take [a theory] as falsified only if we discover a *reproducible* effect which refutes [it]. In other words, we only accept the falsification if a low-level empirical hypothesis which describes such an effect is proposed and *corroborated*" (Popper, 1959, p. 86, italics added). Nothing hard about that, for Popper: "Any empirical scientific statement can be presented (by describing experimental arrangements, etc.) in such a way that anyone who has learned the relevant technique can test it" (p. 99).

No one has done more than Harry Collins (1985) to expose the problematic nature of what Popper takes as open and shut – namely, *when* is an empirical claim successfully replicated, reproduced, or corroborated? Collins argues that there is no unambiguous and impersonal algorithm for reproducing an experimental procedure. Scientists routinely face the problem of deciding when a replication is competent and authentic.[3] "Experimenters regress" is a paradox for those like Popper who want to "use replication as a test of the truth [falsification] of scientific knowledge claims" (p. 2) because negotiation of the competence of a replication attempt is, at once, negotiation of the reality of phenomena at hand. In his study of gravity wave experiments, Collins reports that scientists' judgments about the competence of a replication experiment hinged on whether the results of that experiment were consistent with their theoretical assumptions going in. This research challenges an inherent part of Popper's demarcation criteria, for how can refutation or falsification occur if scientists sometimes exploit available rhetoric (human error, machine failure, extraneous circumstances, technical infelicity) to neutralize potentially falsifying observations by attributing them to incompetent or unauthentic replications? Thus Collins sets the stage for post-essentialist efforts to ascertain *how* some replications are deemed authentic and authoritative, a cultural and rhetorical game that is often tantamount to labeling them "scientific." Popper's demarcation criteria become a matter for scientists and others to negotiate.

Social Norms of Science

Merton's (1973, pp. 267–278) four social norms of science require only brief rehearsal here, for they have energized debate in sociology of science for a half century. His argument is as essentialist as Popper's, with the institutionalized ethos of science replacing falsifiability as a criterion for demarcating science from non-science. For Merton (1973b), "the institutional goal of science is the extension of certified knowledge," that is, "empirically confirmed and logically consistent statements of regularities (which are, in effect, predictions)" (p. 270). The theoretical problem is to identify a social and cultural structure for science that aids pursuit of this goal. Part of that structure is methodological, encompassing technical norms of empirical evidence and logical consistency. Another part is "moral" or social, consisting of four affectively toned norms "held to be binding on the man of science" (p. 269). These norms take the form of prescriptions and proscriptions, they are communicated and internalized during the socialization of scientists, and they are reinforced by sanctions levied against transgressors and by rewards heaped on successful conformists. Empirical evidence for their existence is found in moral indignations expressed by scientists in reaction to violations, along with their positive behavioral and rhetorical endorsement of the norms.

Scientists of course will violate this moral code on occasion, but here is what they are institutionally expected to live up to as an ideal. *Communism* asks scientists to share their findings, and the institution promises "returns" only on "property" that is given away. *Universalism* enjoins scientists to evaluate knowledge claims using "preestablished impersonal criteria" (say, prevailing theoretical or methodological assumptions), so that the allocation of rewards and resources should not be affected by the contributor's race, gender, nationality, social class, or other functionally irrelevant statuses. The norm of *disinterestedness* does not demand altruistic motivations of scientists, but channels their presumably diverse motivations away from merely self- interested behavior that would conflict with the institutional goal of science (extending certified knowledge). *Organized skepticism* proscribes dogmatic acceptance of claims and instead urges suspension of judgment until sufficient evidence and argument are available.

I have not yet found the words *demarcation* or *boundary* in Merton's classic paper (originally written in 1942), but the ideas behind them are expressed in its rationale. Merton (1973) creates an image of wolves at the door of science: "local contagions of anti-intellectualism" and "a frontal assault on the autonomy of science" (pp. 267, 268). Few readers in 1942 could think of anything but Aryan Science serving Naziism (see Hollinger, 1983). Such "incipient and actual attacks upon the integrity of science have led scientists to recognize their dependence on particular types of social structure" (Merton, 1973, p. 267). The social structure on which science *depends* is one where the four social norms are institutionalized. Without reading too much into the word *dependence*, this casts the norms as a kind of institutional sine qua non, as essential for extending certified knowledge as, say, falsifiability. In effect, the four social norms of science save the autonomy of science from external political or cultural interferences by arguing that such intrusions compromise the necessary moral conditions, which in turn make possible the extension of certified knowledge. The very words *Aryan science* contradict universalism, for example.

If the norms are read as demacration criteria, then knowledge-producing activities not ensconced in that institutionalized moral frame must be non-scientific. And so they are, when Merton uses the norms to distinguish scientific claims from mere ideology, such as racist assertions from Hitler's Germany. "The presumably scientific pronouncement of totalitarian spokesmen on race or economy or history are for the uninstructed laity of the same order as newspaper reports of an expanding universe or wave mechanics....The borrowed authority of science bestows prestige on the *unscientific* doctrine" (Merton, 1973, p. 277). What makes those claims unscientific? Not their substantive content so much as the anything-but-disinterested ambitions of their promoters: The "authority [of science] can be and is appropriated for *interested* purposes" (p. 277). When politics get inside the door of science, the autonomy and regulative force of the norms is breached, and the resulting claims are unscientific. The implication, of course, is that when the same claim is served up in the court of real science – with scientific ethos intact – then error will be exposed: "The criteria of validity of claims to scientific knowledge are not matters of national taste or culture. Sooner or later, competing claims to validity are settled by universalistic [and disinterested] criteria" (p. 271, n. 6).

Of the many criticisms of Merton's theory of the normative structure of science, one in particular is akin to Collins's criticism of Popper: It takes the supposed essential qualities of science – those that distinguish it from non-science – and makes them into matters for people in society to construct, interpret, negotiate, and deploy. Cicourel (1974, pp. 11–41) argues that rules or norms in general are not things for definition by sociological analysts but are available for definition by actors in everyday life. He suggests that a tacit layer of interpretative or basic norms guide actors as they try to decide which surface rules (i.e., "structural" norms) are relevant and appropriate for the situation at hand. In this scheme, Merton's social norms of science are surface rules that do not translate into behavior patterns in an immediate and direct way. Rather, the process is mediated by interpretative norms through which actors collectively decide what universalism would mean in a given setting or even whether universalism is pertinent at all for the evaluation of a scientific claim or action.

Mulkay extends Cicourel's general argument into a criticism of the norms of science in particular. Using Woolgar's study (1976) of the discovery of pulsars, Mulkay suggests that scientists' decisions to make findings public – as they are implored to do by the norm of communism – are caught up in momentary and situational contingencies such as "any result which might be of interest to the press must be kept as secret as possible" or "researchers must be particularly careful not to release information in such a way that the first achievement of a graduate student is jeopardized" (Mulkay, 1980, p. 121). These mitigating circumstances are more than "counternorms" (Mitroff, 1974) balancing out the injunctions of the institutionalized norms identified by Merton. Rather, interpretative work (grounded in identifiable tacit assumptions or interests) is required before the meaning of a surface norm such as communism is decided, and before reaching a judgment on whether it is pertinent as a guide for one's own behavior or as a standard for evaluating the behavior of others. Mulkay does not argue that scientific norms do not exist; something like communism may well be used rhetorically in certain situations as scientists make decisions and justify their practices to others. The norms could become stable elements of some scientists' boundary-work, useful at the time for separating their science from others' non-science.

Paradigmatic Consensus

It may be difficult to argue that Kuhn sits comfortably as an essentialist-demarcationist alongside Popper and Merton. The distance between Kuhn and Popper is substantial, as measured by Kuhn's (1977, p. 272) argument that critical refutation and incessant falsification attempts may simply not be present in "normal science," and by Popper's (1970, pp. 51–58) dismissal of Kuhn as a historical relativist and (worse) one who would allow the sociology and psychology of science to settle issues (like demarcation) that are more properly settled by logic and methodology. And some readers see Kuhn's emphasis on the moral force of *cognitive* norms in opposition to Merton's emphasis on *social* norms (see Pinch, 1982), although the principals involved see their work as complementary (Kuhn, 1977, p. xxi; Merton, 1977, p. 107). Moreover, Kuhn (1970) has expressed doubts about the possibility of demarcation criteria: "We must not, I think, seek a sharp or decisive demarcation criterion" (p. 6). Still, looking back on Kuhn from a perspective shaped by ten years of constructivist empirical studies of boundary-work, a case can be made that he set that line of inquiry in motion more as foil than as pioneer.

Kuhn (1962 and 1970) sets up the argument in *The Structure of Scientific Revolutions* with two puzzles, each announcing that the demarcation issue will run as leitmotif throughout the book. While hanging around social and behavioral scientists, Kuhn noticed that their arguments differed fundamentally from those he observed among physicists. Social scientists could not agree on first principles; they fought incessantly to distinguish significant from trivial problems for study, acceptable from unacceptable solutions, legitimate from illegitimate frameworks. Physicists at any particular time seemed to agree on first principles and reached consensus on answers to those domain questions. So, Kuhn wondered, what did the hard sciences have that the soft ones lacked? The second puzzle arose from obvious historical changes in the content of certified scientific claims – what was held as fact or a good explanation *then* is *now* preposterous, simplistic, crude, or just wrong. Can those old, now-rejected beliefs about nature (and associated investigative practices) be called scientific? Kuhn is no Whig: Rather than dumping rejected beliefs and practices into the non-scientific dustbin of myth, error, superstition, and ignorance, he asks instead how reasonable people *doing science* at earlier times could accept as valid beliefs and practices that look so obviously wrong to us today (Kuhn, 1962 and 1970, pp. 2–3, 1977, p. xii). Kuhn needs criteria for science independent of the content of provisionally valid beliefs and legitimate practices, because something identifiable as science persists through often revolutionary changes in its content.

Both puzzles are solved with Kuhn's "discovery" of paradigms, and the degree and kind of consensus among practicing scientists that paradigms engender. In place of a progressive, linear, and cumulative movement toward the present and best comprehension of nature, science for Kuhn moves through fits and starts along distinctive historically specific trajectories. Periods of normal science are punctuated by revolutions and these, with time, are closed off by a new normality – which compares with earlier normal science not as a progressive improvement toward more accurate or encompassing models of nature but as an incommensurable way of thinking about nature and how it should be understood. Kuhn's model of the history of science hinges on the multivalent concept of paradigm (Masterman, 1970), which, in a sense consists of background assumptions about the way the natural world works – coupled with methodological and theoretical exemplars or models that translate those deep assumptions into working rules to guide the selection of problems and acceptable procedures for their solution. Research in normal science is puzzle solving where the perimeter frame, the cut-out pieces, and the spaces to be filled in are specified by a paradigm. On occasion, some anomalous pieces cannot be made to fit, and when this happens at a time when another puzzle paradigm becomes available, science undergoes a temporary period of revolutionary alternation between frames of meaning. Eventually, a new paradigm attracts most practitioners and wins, not by convincing scientists with its superior logic or empirical evidence but through a non-rational gestalt-switch conversion grounded more in the psychology of perception and the sociology of commitment than in methodology.

The periodically successful achievement of paradigmatic consensus within a research community separates mature science from immature science, social science, Baconian science, art, technological craft work, astrology, and other realms of non-science. Referring to investigators of electricity (Cavendish, Coulomb, Volta) in the late eighteenth century, Kuhn (1962 and 1970) writes: "They had...achieved a paradigm that proved able to guide the whole group's research....it is hard to find another criterion that so clearly proclaims a field, a science" (p. 22). And of Newtonian physics, he writes: "Paradigms prove to be constitutive of science" because they "provide scientists not only with a map but also with some of the directions essential for map-making" (Kuhn, 1962 and 1970, p. 109). Paradigmatic consensus is both unique to science and essential for its successes at filling in the puzzles (though not sufficient; Kuhn, 1977, pp. 60–65, 231, n.3). We arrive at a third essentialist demarcation principle: "Work under the paradigm can be conducted in no other way, and to desert the paradigm is to cease practicing the science it defines" (Kuhn, 1962 and 1970, p. 34, cf. pp. 17, 76, 60). "That commitment and the apparent consensus it produces are prerequisites for normal science, i.e., for the genesis and continuation of a particular research tradition" (p. 11).

In *The Structure*, Kuhn spends considerable time discussing knowledges and practices that are not part of mature and normal science, to show that the presence or absence of paradigmatic consensus can be used to distinguish real science from something else. He constructs the intriguing category of "sort of" science (Kuhn, 1962 and 1970, p. 11), which, at the level of individual practice, mimics mature science but in its collective accomplishments fails to achieve the coherent progress that characterizes fully developed science (filling in the puzzle). Speaking of pre-Newtonian optics, Kuhn argues that "though the field's practitioners were scientists, the net result of their activity was something less than science," because they could "take no common body of belief for granted, [and] each writer...felt forced to build his field anew from its foundations" (p. 13; cf. pp. 101, 163). The lack of consensus is generally apparent in the "interschool debates" of "immature sciences," and in the "prehistory of science" (p. 21). Moreover, when a paradigm does come along to offer a more rigid definition of a field, those who choose not to conform "often simply stay in the departments of philosophy from which so many of the special sciences have been spawned" (p. 19). And philosophy is not science because "there are always competing schools, each of which constantly questions the very foundations of the others." Once a paradigm is in place, for a researcher to abandon its worldview without hopping to an alterative puzzle is tantamount to leaving science: "To reject one paradigm without simultaneously substituting another is to reject science itself" (p. 79). Kuhn's demarcation of science from social science suggests *how* (in part) paradigmatic consensus makes real science mature. The paradigm provides an agenda of interesting research problems for scientists to attack and insulates their work from possibly competing agendas coming into science from the wider society. With no paradigm in place, social scientists lack this insulation, which, by extension, feeds their incessant and ideological arguments over first principles and prevents their coherent progress (Kuhn, 1962 and 1970, pp. 21, 37). Finally, Kuhn agrees with Popper that astrologers were non-scientific, but not because they failed to put their claims in a falsifiable form; rather, "they had no puzzles to solve and therefore no science to practice" (Kuhn, 1977, p. 276).

Having worked hard to set up paradigmatic consensus as an essentialist demarcation principle, it is puzzling then to see Kuhn dismiss the practical problem of *defining* science. He asks rhetorically: "Can very much depend upon a *definition* of science? Can a definition tell a man whether he is a scientist or not? If so, why do not natural scientists or artists worry about the definition of the term?" (Kuhn, 1962 and 1970, p. 160). Kuhn implies that scientists rarely worry about definitions of themselves. But as the upcoming review of constructivist studies of boundary-work will make abundantly clear, scientists do on occasion worry about the definition of science, because they use those definitions to tell others that they are not scientific, in episodes where much depends on how a definition is played out. Kuhn is an essentialist not only because he offers paradigmatic consensus as a demarcation principle but because he dismisses as unimportant, merely "semantic," those questions that animate constructivist studies of boundary-work.

What Kuhn chose not to consider is that the degree of consensus in science itself might be a matter of interpretation, negotiation, and settlement – by scientists and sometimes other involved parties. When does a research community reach the level of paradigmatic consensus that moves it automatically from immature to mature science? And who answers that question, participants or their analysts? Gilbert and Mulkay's study of the reception of chemiosmotic theory in biochemistry treats consensus as part of scientists' interpretative work as they construct and give meaning to the history of their field. Consensus cannot be treated "as a typical collective phenomenon, that is, as a potentially measurable aggregate attribute of social groupings which exists separately from the interpretative activities of individual participants"

(Gilbert and Mulkay, 1984, p. 139). Instead, consensus is a contextually contingent product of scientists' variable interpretative procedures, which means that, for Kuhn to conclude analytically that consensus *exists* in a research community at a designated time, he must ignore potentially wide discrepancies in scientists' own sense of the degree and kind of consensus they supposedly share.

In particular, Gilbert and Mulkay suggest that scientists must themselves solve three interpretative problems as they consider what consensus means and whether their field has it. First, they must decide the limits of membership of their research community, for inclusion or exclusion of certain individuals could easily affect their conclusions about the extent of consensus. Second, they must reach judgment on the changing beliefs of other scientists in regard to chemiosmosis: Who accepts it, and when did their conversion to the new framework occur? Third, scientists must decide the cognitive content of the chemiosmotic theory: If there is consensus, just what does the community agree on? From interviews, Gilbert and Mulkay find considerable variation among scientists (and even within a scientist in different situations) in how these interpretative problems are worked through. The boundaries of the research community were discrepantly drawn, beliefs were attributed to other scientists in varying ways, and chemiosmotic theory came to mean quite different things. Gilbert and Mulkay's (1984) findings raise doubts about whether Kuhn's paradigmatic consensus can essentially distinguish science from non-science if "a given field at a particular point in time cannot be said to exhibit a specifiable degree of consensus" (p. 140). The stage is set once again for empirical investigations of how consensus, among many other possible attributions, is used by people in society to construct a border between science and something else.

From Demarcation to Boundary-Work

Constructivist studies of scientific knowledge and practice[4] raise doubts about the ability of criteria proposed by Popper, Merton, and Kuhn to demarcate science from non-science – but in their wake comes a paradox. If there is nothing inherently, universally, and necessarily distinctive about the methodology, institution, history, or even consequences[5] of science, then why and how is science today routinely assigned a measure of "cognitive authority" rarely enjoyed by other cultural practices offering different accounts of reality? Paul Starr (1982) defines *cultural authority* as "the probability that particular definitions of reality... will prevail as valid and true" (p. 13). Few would doubt, in modern Western societies, that science has considerable cognitive authority: "Science is next to being *the* source of cognitive authority: anyone who would be widely believed and trusted as an interpreter of nature needs a license from the scientific community" (Barnes and Edge, 1982, p. 2). *On what grounds* is this authority warranted, if not for some epistemological or social quality essential to science and not found outside it? The challenge is to explain the cognitive authority of modern science without attributing to it essential qualities found by sociologists to be anything but essential.

The boundary problem in science studies is, in effect, an attempt to get around this paradox. The object of sociological study is no longer practices of scientists at the lab bench or their accounts of nature in professional journals; no explanation for the cultural authority of science could be found there without succumbing to the essentialism of Popper, Merton, or Kuhn. Instead, attention shifts to representations of scientific practice and knowledge in situations where answers to the question "What is science?" move from tacit assumption to explicit articulation. The task of demarcating science and non- science is reassigned from analysts to people in society, and sociological study focuses on episodes of "boundary-work": "the attribution of selected characteristics to the institution of science (i.e., to its practitioners, methods, stock of knowledge, values and work organization) for purposes of constructing a social boundary that distinguishes some intellectual activity as non-science" (Gieryn, 1983, p. 782).

Boundary-work occurs as people contend for, legitimate, or challenge the cognitive authority of science – and the credibility, prestige, power, and material resources that attend such a privileged position. Pragmatic demarcations of science from non-science are driven by a social interest in claiming, expanding, protecting, monopolizing, usurping, denying, or restricting the cognitive authority of science. But what *is* "science"? Nothing but a *space*, one that acquires its authority precisely from and through episodic negotiations of its flexible and contextually contingent borders and territories. Science is a kind of spatial "marker" for cognitive authority, empty *until* its insides get filled and its borders drawn amidst context-bound negotiations over who and what is "scientific."

Put another way, the authority of science is reproduced as agonistic parties fill in the initially empty space with variously selected and attributed characteristics, creating a cultural map that, if accepted as legitimate, advances their interests. In these cartographic contests over *distributions* of scientific authority among diverse people, practices, and knowledge claims, the link between authority and the space marked "science" is made ever more secure. Whatever ends up as inside science or out is a local and episodic accomplishment, a consequence of rhetorical games of inclusion and exclusion in which agonistic parties do their best to justify their cultural map for audiences whose support, power, or influence they seek to enroll. Crucially, the "essential features" of science are provisional and contextual *results* of successful boundary-work, not determinants of who wins. Why are some maps more persuasive than others? There are no *general* determinants of success at cultural cartography, but clearly it helps if your depiction of the edges of science makes the interests of powers-that-be congruent with your own plans.

"Unique" features of science, qualities that distinguish it from other knowledge-producing activities, are to be found not *in* scientific practices and texts but in their representations. Boundary-work stands in the same relationship to what goes on in laboratories and professional journals as a topographic map to the landscape it depicts; both *select* for inclusion on a cultural or geographic map those features of reality most useful for achieving pragmatic ends (legitimating authority to knowledge claims or hiking through wilderness). Geographic cartographers often make new maps without constant reinspection of reality outdoors by "copying" and selectively editing extant maps to suit changing needs or wants, and the same is so for cultural cartographers. Those contesting the borders of science select from and creatively reconstruct past episodes of boundary-work, often using old maps to legitimate the validity of their own. Importantly, neither actual scientific practice and discourse in labs or journals nor earlier maps showing the place of science in the culturescape *determine* how the boundaries of science will get drawn next time the matter comes up for explicit debate. Rather, opposing maps are better understood in terms of immediate (and dynamic) interests and goals of their cartographers and the uses to which they are put (i.e., convincing people of one's cognitive authority or denying it to somebody else). In this sense, then, the space for science is empty because, at the outset of boundary-work, nothing of its borders and territories is given or fixed by past practices and reconstructions in a deterministic way.

But that idea could easily be exaggerated into a silly conclusion that every episode of boundary-work occurs de novo, and that there are no patterns at all from one episode to the next. Scientific practices and antecedent representations of it form a *repertoire* of characteristics available for selective attribution on later occasions. That repertoire is presumably not limitless; it might be extremely challenging these days to persuade others that eye of newt and toe of frog make witches purveyors of "good laboratory practice." Interpretative flexibility in the boundaries of science need not imply infinitely pliable;[6] some maps of science are easier than others to defend as bona fide representations, in part because some cartographers are more easily able to point to specific concrete practices or to earlier mappings as rhetorical "evidence." Indeed, some maps achieve a provisional and contingent obduracy that may preempt boundary-work. Borders and territories of cultural spaces sometimes remain implicit, matters of personal belief or of such apparent tacit intersubjective agreement that people working together need not explicate "what everybody knows" about the meaning of science.

Such stability in the borders of science could itself be overestimated, as if the issue were settled centuries ago once and for all. *Boundary-work abounds* simply because people have many reasons to open up the black box of an "established" cartographic representation of science – to seize another's cognitive authority, restrict it, protect it, expand it, or enforce it. A survey of historical instances of boundary-work would turn up a science with no consistent shape, no necessarily enduring features. One would find massive diversity in the characteristics attributed to science and used to demarcate it from something else (e.g., theoretical, empirical, certain, uncertain, useful, useless, finite, endless, quantitative, qualitative, precise, imprecise, inductive, deductive) just as one would find in "something else" a massive diversity of nonsciences (pseudoscience, amateur, bad science, fraud, Marxism, popularizations, politics, technology, management, religion, philosophy, art, mechanics, craft, social science, and so on). It is precisely the emptiness of science – a space waiting for edging and filling – that best accounts for its historically ascendent cultural authority.

The turn from essentialist studies of demarcation to constructivist studies of boundary-work is well under way[7] as the next section will report...

EDITOR'S NOTE

In a section omitted here, "Theoretical Nourishment," Gieryn describes theoretical studies concerning the sociology of professions, "social worlds," history of cultural classifications, and the influence of feminist studies to the boundary question. Although of interest to the professional sociologist, these four theoretical frameworks were developed outside the sociology of science proper, and thus are not directly germane to our project.

EMPIRICAL STUDIES OF BOUNDARY-WORK: FOUR SPECIMENS

Rather than survey empirical studies of the boundary problem in STS, I have chosen exemplary works that display four types of boundary-work: monopolization, expansion, expulsion, and protection. Together, they suggest how precipitous Barnes (1982) was to declare – after theoretically setting up the boundary problem in 1974 – that "from a sociological point of view there is little more to be said...about the boundary of science in general. The boundary is a convention" (p. 93). Much has been learned about science since, from continuing scrutiny of its margins.

Monopolization: A Cartographic Contest for Cultural Authority

As Shapin and Schaffer tell it (1985), the debate between Robert Boyle and Thomas Hobbes in the 1660s was "about" an air pump in the same way that *Moby Dick* is "about" a whale. At stake was the delineation of authentic and authoritative knowledge: How was it made? Who could make it? What was it for? The dispute was more than one between Boyle's experimentalism and Hobbes's rationalism, for at issue as well was the constitution of the social order itself in Restoration England. The debate is a classic specimen of a kind of boundary-work involving science, where contending parties carve up the intellectual landscape in discrepant ways, each attaching authority and authenticity to claims and practices of the space in which they also locate themselves, while denying it to those placed outside. What makes *Leviathan and the Air-Pump* such a useful guide to the controversy is its sustained attention to places and spaces, borders and territories, insiders and outsiders – as the authors note, "the cartographic metaphor is a good one" (Shapin and Schaffer, 1985, p. 333).

Hobbes versus Boyle was a battle of the maps – cultural maps on which authoritative and authentic knowledge could be assigned or denied to knowledge producers by *where* they were located (and how). Crucially, there was no single map, as if Boyle and Hobbes glared at each other across a common frontier that moved to and fro depending upon who was "winning." Each sought to occupy and control a space for authentic and authoritative knowledge, but it does not follow that Boyle's "outside" (excluded and delegitimated kinds of knowledge making) mirrored Hobbes's "inside" (and vice versa). The two maps depicted *alternative* cultural universes, with important places and landmarks given different labels and with distinctive grounds for locating a border here or there. The maps guided presumed users to where they could find authentic and credible knowledge and told them why they could not find it outside that space.

Boyle's map put authenticity and authority on the side of his experimental physiology, carefully bounded from such unworthies as metaphysics, politics, and religion (Shapin and Schaffer, 1985, p. 153). "Science"(the inside space) sought matters of fact – fallible, provisional, corrigible – whose authenticity was decided by nature, as the community of competent interveners and observers collectively witnessed its goings-on. Assent to an experimental fact was, at once, a collective accomplishment of those committed to a certain discursive style *and* the result of procedures through which collective judgments were objectified into questions that "nature decides." Experimental discourse was limited to matters of fact; carefully contained dissent over theoretical explanations of the observed was tolerated but such interpretations and hypotheses were treated as undecidable. Facts gained authenticity through their collectively being seen, via a multiplicity of witnesses extended through three means: opening up the house of experiment – the nascent laboratory – for public view, so that like-minded experimentalists could see for themselves; replicating the experiment by building air pumps throughout Europe; and allowing for "virtual witnessing" of elaborate textual and faithfully detailed graphic representations of the experimental apparatuses and procedures. This community of experimenters saw themselves both as humble craftsmen, modest in the range of what they said they knew, and as priests poring over the book of nature. Both the technical and the priestly were arguments legitimating the cultural authority of Boyle and his fellow experimenters.

Boyle's "outside" was a raucous mix of illegitimate forms of knowledge making, held together only by their tendency to exacerbate dissent rather than contain it. Knowledge grounded in private and personal experience – what came from the alchemist's closet or from the passionate dogmatism of secretists and religious enthusiasts – was excluded. Because proper experimental discourse was nescient about causes (say, of the spring of air), all conjecture and speculation were located out there as well. If the matter was not decidable on experimental grounds, it was put outside – and such was the case for politics and human affairs in general, and in particular those philosophies proffering certain truth (described on this map as "tyrannical dogmatism"). Hobbes was a behemoth landmark in that region of imperious, egotistical dogmatists – the systematists, the authorities who would not submit their claims to the discriminating trial of nature-in-experiment.

Hobbes's "inside" is not labeled tyranny and dogma, of course, but neither does his "outside" contain anything that would command assent to a claim (as Boyle thought collectively witnessed experiment could do). On the philosophy or "science" side of Hobbes's border was certain knowledge, rational deductions capable of securing irrevocable, universal, and obligatory assent. Geometry was the model for knowledge making that carried authenticity and authority: One must accept that any line drawn through the center point of a circle divides it equally. Philosophy sought not facts but explanations in causal form, where cause is traced back to the motions of contiguous bodies. The compelling force of such logical deductions does not reside in their objectification in nature but precisely in their conventional, artifactual qualities: People made knowledge certain by deducing inescapable explanations as they proceeded rationally from agreed-upon definitions. Once properly deduced, there was little room for continued dissent. These principles for the manufacture of certain knowledge were as applicable to politics as to mechanics – Hobbes had no boundary separating studies of nature from studies of human affairs. As Leviathan commands assent in matters political, so philosophy – causal, rational, deductive, certain – commanded assent in matters epistemological.

Hobbes's "outside" includes some territories also "outside" on Boyle's map. Neither made room for religion on the inside, but for different reasons. Theologians did not participate in the experimental form of life. Their knowledge was certain – not fallabilities grounded in nature but timeless truths grounded in proper readings of sacred texts – and that sounded sufficiently dogmatic to warrant exclusion by Boyle. Hobbes excluded theologians – along with all other exclusive professional groups including Boyle's circle of experimenters – because they *impeded* the pursuit of certain truth and the order (social and intellectual) that only a philosophy commanding universal assent could secure. The church, the professions, those who practice 'physics' and 'natural history' (where Hobbes put Boyle, with derision) were sources of divided authority; each offered competing grounds for assent and so together could never reach the settlement commanded by authentic philosophy. What Boyle did was simply not philosophy for Hobbes; its refusal to decide causes made it incomplete, the tentativeness of its claims (always provisional and corrigible) attested to its variability and weakness, the inductive move from experimental observation to causal explanation was obviously fallacious, and its reliance upon experiment anchored so-called knowledge on the shifting passions of personal experiences and professional interests (guarding one's exclusive turf). What anyone saw was inherently and endlessly contentious. Worse, the obvious technical ingenuity of experimenters did not count as philosophical wisdom, and so Boyle went on Hobbes's outside along with gardeners, apothecaries, workmen, quacks, machine minders, and other banausic pursuits possessing little moral authority.

Several generic features of boundary-work are illustrated by this episode. The maps are those of Boyle and of Hobbes, as read by Shapin and Schaffer (1985, p. 342). They are not interpreted in terms of how well or poorly they correspond to putatively universal or essential qualities of science or to the "science" we might map today. The Hobbes–Boyle debate is a contest for what kind of knowledge making would be accepted – at that moment – as authentic and authoritative. Shapin and Schaffer (1985) seek to recover cultural circumstances especially difficult for the modern person to grasp: It seems to us as if authoritative knowledge about nature – science – has always been tightly coupled with experiment, yet there was nothing inevitable or self-evident about that link in the seventeenth century (p. 13). By going back to when the bounds of authentic natural knowledge were less formed, Shapin and Schaffer discover particular historical conditions in which that connection between experiment and authoritative knowledge first took hold.

Importantly, the use of experiment to divide legitimate from illegitimate knowledge is an accomplishment of *representations* of Boyle's practices; it does not flow directly from those practices in some unmediated way. It was less important what Boyle "actually" did with his pump than how those activities and that machine were represented, described, located on a map drawn precisely to legitimate them. The authors state their purpose to explore "the historical circumstances ...in which experimentally produced matters of fact were *made into* the foundations of what counted as proper knowledge" (Shapin and Schaffer, 1985, p. 3, italics added; cf. pp. 52, 91). The actual practices of "science" – whether Boyle's experiments or Hobbes's philosophy – lay open for multiple interpretations and offer a bank of qualities from which cultural cartographers make selective withdrawals as they construct maps that give meaning (in this case, authority, credibility, authenticity) to those practices.[8] Actual scientific practices underdetermine the maps of it; the remainder of variance must be explained by circumstances of the boundary-work itself: What was at stake? Who needed to be convinced? What arguments did the adversary present?

The Boyle–Hobbes spat also shows cultural maps in *action*, as something more than coffeehouse chatter. Boundaries were not only drawn but policed (Shapin and Schaffer, 1985, p. 135) – a task Boyle made easier by translating his inside space into a physical place in the built environment (the nascent lab). The game for Boyle was not only to legitimate experiment but to deny authenticity and authority to the kind of knowledge Hobbes made. Success would be likely if Boyle could move everyone – rivals, audiences, bystanders – onto his playing field, with borders and territories that he drew and labeled (pp. 173–174). He did that in a crafty move, in effect arguing that only those who were *in* the experimenter's community – who went into Boyle's laboratory or built one for himself and performed competent experiments – could challenge claimed facts. But, catch-22, the price of admission to the lab (and to the Royal Society, as Hobbes found out) was a commitment to Boyle's program. "External critics" (p. 222) were instantly delegitimated! But Boyle's policing tactic opened up a new attack for Hobbes: Wasn't his exclusion from experimental space and the laboratory place evidence enough for the private nature of Boyle's sect? Hobbes challenged the claim that experimental facts achieved their authenticity from public witnessing by pointing to Boyle's insistence that, because it was his air pump, everyone had to play by his rules or go home (he was the 'master' of the lab; p. 39). The significance of this for studies of boundary-work is clear; maps alone do not win authority for those inside a space, but they provide rules for making real-world decisions – denying admittance or membership, for example – that selectively allocate cultural authority in terms more palpable than cartographic.

Shapin and Schaffer eschew the question: "Who won and why?" Boyle comes out a leg up on Hobbes: Although boundary-workers in the four centuries since the first air pump have used Hobbes's right reason to distinguish their science from others' unreason, Boyle's experiment has perhaps more often been employed as the sine qua non of science. If Boyle's experimental space has achieved legitimacy, it is not because its contours – against those of Hobbes – more closely corresponded to "real" science; nor should we assume that Boyle's cartographic arguments would necessarily work on any other occasion when the boundaries of science are contended (there are no universal determinants of success). Boyle "won" because his space was better able to hold the diverse interests of powers-that-be in Restoration England – indeed, the laboratory became an 'idealized reflection of the restoration settlement' (Shapin and Schaffer, 1985, p. 341). There, competent practitioners could publicly gather to manufacture facts in cool technical and professional calm, and agree to their validity in nature without lapsing into endlessly contentious debate over hypothetical causes. Experimental space was what Restoration England could become: a "peaceable society between the extremes of tyranny and radical individualism" (hadn't England had enough of both during the Commonwealth?) (p. 341). Hobbes was portrayed as the dogmatist whose certain knowledge would undo the fragile Restoration polity. Boyle rode to victory on the coattails of the Restoration as a political achievement, even as he and his experimental space contributed to that settlement. But his "victory" – as in any boundary-work – is a "local success," for there is no "unbroken continuum between Boyle's interventions [and his cultural cartography] and twentieth-century science" (p. 341). In no way did the borders and territories of science get settled for all time by Boyle and Hobbes.

Expansion: Enlightenment Encroachments

A second type of boundary-work occurs when insiders seek to push out the frontiers of their cultural authority into spaces already claimed by others. Such was the task of the *philosophes* in the eighteenth century, who sought to extend their mixture of rationalism and empiricism into a domain of questions and problems owned by religion and embodied in the institution and dogma of the church. A manifesto for the Enlightenment project was D'Alembert's *Preliminary Discourse*, written as a warm-up to Diderot's *Encyclopédie* and wonderfully interpreted by the historian of French culture Robert Darnton (1984). That text and its accompanying map are boundary-work nonpareil, spatial representations of a cultural landscape that provide grounds for a belief that "philosophy" – D'Alembert's inside authoritative space – could swallow up whatever counts as genuine knowledge, leaving only poetry and memory outside.

D'Alembert called his *Preliminary Discourse* a *mappemonde*, a map of the world of knowledge "to show the principal countries, their position and their mutual dependence, the road that leads directly from one to the other" (in Darnton, 1984, p. 195). The map has three continents, one much larger than the other two and with many more nested countries and provinces. In this "Detailed System of Human Knowledge," all understanding is divided into three spaces depending upon its source: memory and imagination (the two little continents) and reason (the big one). *Imagination* is the origin of poetry, both civil and sacred, and includes narrative, dramatic, and parabolic (allegorical) forms. *Memory* is the source of history, divided into civil (memoirs, antiquities, literary history), sacred (history of prophecies), ecclesiastical (conspicuously empty of further subdivision), and natural (uniformities and deviations of nature – "monstrous vegetables" – along with arts, trades, and manufactures). *Reason* (in the center, of course) is the fount of philosophy, which holds everything else, and divides into the science of being (metaphysics), the science of God (theology, religion, and, "whence through abuse, superstition"), the science of man (logic and ethics), and the science of nature (mathematics and physics – each subdivided into many tiny regions that have since grown to be 'sciences' on their own).

Only reason yields knowledge; imagination and memory must produce something else. D'Alembert put up a boundary between the unknowable (poetry and history) and the known (what science tells us). He shrinks the territory given to poetry and history: Ecclesiastical history has nothing identifiable within it, in sharp contrast to the highly differentiated – and visually more capacious – space for mathematics, divided first into pure and applied and then broken down into optics, ballistics, hydraulics, and dozens more. Most significant, however, are the kinds of understanding moved under the umbrella of philosophy; natural and revealed theology were now knowable in the same sense as botany and zoology, as were questions of good and evil. Matters of ethics and morals, questions about the spiritual, were located within the compass of reason, which put them under the control and the authority of those men who epitomized the rational spirit – D'Alembert, Diderot, and their kindred *philosophes*. The church had no claim on such issues as these and was in effect forced to choose between speaking unknowables or speaking the rational and empiricist tongue of the Enlightenment. "Pigeon-holing is therefore an exercise in power" says Darnton (1984, p. 192), and this mapping of knowledge identifies winners and losers: Philosophy is now in charge of problems once assumed to be ecclesiastical, as rationality and empiricism are extended to matters once known only through tradition or revelation.

It is one thing to claim cartographically an expansion of territory under the cultural authority of science; it is quite another to legitimate those bulging frontiers so that people in society accept them. D'Alembert shows himself the master rhetorician, moving through three steps to show *why* reason, philosophy, and science are justified in staking their new claim – and each step is a response to potential difficulties raised by the one before it. He begins by rewriting human history in a classically progressive idealist way – life is better now than before thanks to philosophers' accomplishments. Bacon, Descartes, and Locke, along with scientific allies Newton, Galileo, and Harvey, are not just Great Men but Heroes, carrying along the "progressive march of reason" right up to its culmination in the *Encyclopédie* itself. The *philosophes* and their precursors were "cast in the grand role" and became the "moving force in history" (Darnton, 1984, pp. 199, 206–209). The tune is familiar enough and played still as scientists engage in border wars – "better living through chemistry." But this legitimation of the expanded authority of science through appeals to its salutary effects on the human condition carries a risk: Isn't the progressive, idealist reconstruction of history nothing but a self-interested polemic designed to justify what is really just a sectarian grab for power by the *philosophes* from the churchmen? Because "no map could *fix* topography of knowledge" (Darnton, 1984, p. 195), D'Alembert's *mappemonde* was vulnerable to delegitimizing arguments that it was a not-so-transparent effort to render authoritative precisely that which the *philosophes* uniquely offered (rationality tempered with empiricism), and thus intended to put them on top of an expanded reign. How could D'Alembert lessen the appearance of self-interest?

Scientists then as now are skilled at objectifying the referent and making it real, and that is what D'Alembert needed to do with his map of knowledge. The second step in his legitimation of the expanded frontiers of philosophy is another familiar feature of scientists' boundary-work: drawing independent authority for one's own map by linking it to cartographic efforts of earlier generations of boundary-workers (objectification by attributing authorship *elsewhere*). If no map can fix the topography of knowledge, some achieve greater stability and obduracy from their repeated redrawings – at later times, in other places. D'Alembert's map is not merely of his own making; it is sufficiently close to Bacon's outline in *The Advancement of Learning* for some critics to accuse him of plagiarism. No matter: It is less the content of Bacon's map that D'Alembert exploited than his authority as a cartographer distant in time and place who represented the world of knowledge in similar ways.

Or *not* so similar – and that is as well an endemic feature of boundary-workers' reproduction of maps from the past. D'Alembert drew selectively from Bacon's map as he drafted his own and, in certain places, undermined it entirely. Where Bacon had made a large space for ecclesiastical history, D'Alembert's was tiny; where Bacon was at pains to separate divine learning from human learning, D'Alembert "submitted God to reason"; where Bacon showed natural history (i.e., science) as then deficient and embryonic, D'Alembert shows it in full flower. The discrepancies between Bacon's and D'Alembert's maps raised yet another problem for the latter's efforts to legitimate his *mappemonde*. His selective, even creative, redrawing of Bacon's outline is evidence itself of the arbitrariness and frailty of all exercises in cultural cartography. Darnton (1984) writes that "what one philosopher had joined another could undo" (p. 195), and D'Alembert needed once again to find a way to prevent the undoing of his own map.

His third step is legitimation through a kind of self-referential authority, yet another familiar move in boundary-work. *Who* has the authority to draw (or undo) maps of the world of understanding? D'Alembert in effect argues that the authority of his philosophy extends not just to morals, ethics, revelation, theology, logic, oratory, rhetoric, jurisprudence, mechanics, astronomy, geology, medicine, and falconry – but to cultural cartography as well. "Setting up categories and policing them is therefore a serious business," and "the boundary keepers turned out to be the philosophes" themselves (Darnton, 1984, pp. 193, 209). Drawing boundaries around the knowable is itself a knowable task, best accomplished through the same reason and evidence that marks all philosophical inquiry. D'Alembert's rearrangement of our "mental furniture" (Darnton, 1984, p. 193) worked reasonably well, for in retrospect one sees in them the seeds of the academic disciplines – some scientific, others left out – that came along in the next century. Enlightenment maps of the knowable – with contingent modifications, of course – then took on an even more obdurate and stable form as curricula that spread everywhere when modern universities were built.

Expulsion: Banishing Burt

A common kind of boundary-work involves insiders' efforts to expel not-real members from their midst. The labels attached by insider scientists to those booted out vary: deviant, pseudoscientist, amateur, fake. Those excluded typically give off the appearance of being "real" scientists, and may believe themselves to be so. But insiders define them as poseurs illegitimately exploiting the authority that belongs only to bona fide occupants of the cultural space for science. Such processes of social control no doubt foster a homogeneity of belief and practice within science by threatening insiders with banishment for perceived departures from the norm (Zuckerman, 1977). Sanctioning deviants is also an opportunity for corrective public relations campaigns, restoring among powerful constituencies elsewhere in society a belief that science on its own is capable of weeding out impostors (so hands off) and restoring confidence that science is really only what genuine insiders say it is (nothing dirty going on).

All this comes together in the posthumous trial of Sir Cyril Burt, where continuing negotiations over the propriety of his scientific conduct ramify into self-conscious anxieties among psychologists about whether their discipline itself belongs inside the space for science. Gieryn and Figert (1986) examine, as a specimen of boundary-work, *psychologists' responses* to Oliver Gillie's visible 1976 accusation in *The Times* (London) that Burt had engaged in fraud. A short list of Burt's supposed misconduct includes the following: He concocted statistics that had been presented as results of experiments evidently never carried out; he invented fictional authors as allies to bolster support for his claims; he made inappropriate use of research conducted by students; he exaggerated his stature in the history of psychology by claiming priority to another's discovery. Gillie's accusations forced psychologists to reconstruct the boundaries of science: On what grounds should Burt remain inside or be expelled? At the time Gieryn and Figert concluded their analysis,[9] Burt was widely recognized by insiders to have been guilty of fraud and was banished from real science. His first biographer L.S. Hearnshaw wrote the sentence:

> The verdict must be, therefore, that... beyond reasonable doubt, Burt was guilty of deception. He falsified the early history of factor analysis... ; he produced spurious data on MZ twins; and he fabricated figures on declining levels of scholastic achievement... Neither by temperament nor by training was he a scientist... His work often had the apperance of science, but not always the substance. (in Gieryn and Figert, 1986, p. 80)

So much for the man whose science was once so respected that he became the first from his discipline to be knighted.

Hearnshaw's summary judgment is of less interest to sociologists of boundary-work than the heated negotiations among psychologists that came before it. As is probably the case in most allegations of pseudoscience, amateur science, or deviant science, decisions about inside – outside forced debate over where that border shall be drawn – debate that becomes contentious because of diverse interests attached to the map that eventually wins out as (provisionally) accurate. Visions of Burt's misbehavior were refracted through the controversy between hereditarians and environmentalists over the determinants of intelligence. His data on twins lent support for genetic factors, and predictably those psychologists aligned with hereditarian theories argued that charges against Burt were trumped up. Several suggested that suspicions about Burt voiced by Leon Kamin (an environmentalist) even before the Gillie article were little more than a political attack on all research suggesting hereditary determinants of intelligence. But Kamin and others on the environmental side argued that the politics were all Burt's, that his views about intelligence were ideology masquerading fraudulently as science. In the mid-1970s, Burt's standing inside or outside genuine science depended upon who drew the map, a hereditarian or an environmentalist.

When it became clear to almost everybody – even hereditarians – that something fishy was going on, those who sought to defend Burt and his work redefined his conduct as on the margins of legitimately scientific practice – but not over the edge. There was much speculation about Burt's motivations: Was there intent to deceive (necessary, it seems, to sustain the charge of fraud) or was he simply a sloppy and negligent methodologist (reprehensible and even embarrassing perhaps, but hardly grounds for expulsion from science). One supporter even suggested that the oddities found in Burt's statistics would turn up with considerable frequency in examination of a random sample of similar articles in educational psychology. Though methodological sloppiness is hardly a flattering image for science, in the noisy context of the Burt controversy this lesser charge functioned for a time as a plea bargain (misdemeanors do not warrant death sentences).

When matters started to look even worse for Burt and guilt seemed the imminent conclusion, psychologists set aside their theoretical differences to salvage the cultural authority of psychology that would come from securing its place inside science. The rhetoric changed its focus from "What did Burt do and did he mean to do it?" to "What are the implications of the Burt affair for the science of psychology?" If Burt's guilt or innocence were simply a matter of the *politics* of intelligence, or if his alleged sloppiness is typical of psychological research, the professional risk for the discipline is clear: Is it really a science at all? Securing psychology's authoritative place within real science goes through three steps. First, responsibility for cultural cartography – where would Burt be located? – is claimed for psychologists themselves.

An outsider, like medical reporter Oliver Gillie, could not possibly know enough sophisticated psychology to judge Burt's conduct as proper or not, and he is condemned by psychologists for trying. Hearnshaw the insider is their preferred border guard, because a real science like psychology can patrol itself. Moreover, the authority of science itself is reproduced when its procedures alone are cast as the unique means to determine just what Burt was all about. One psychologist suggested that the whole affair be "thrashed out in the leading scientific journals" that "operate a proper refereeing system" (in Gieryn and Figert, 1986, p. 75). Such peer review would get to the scientific truth of the matter.

Second, psychologists located the cause for Burt's misbehavior in personal and idiosyncratic troubles (outside the borders) rather than in structural problems inside the science of psychology. Burt's life after World War II was a long string of crises, we are told, from failed marriages to (accidentally) destroyed research records, from debilitating Menière's disease to humiliation at being dismissed from the editor's post at a leading journal. The cognitive authority of scientific psychology is hardly touched by the misconduct of a "sick and tortured" man whose behavior was "not the act of a rational man" (in Gieryn and Figert, 1986, p. 79).

Finally, psychologists employed the rhetorical strategy that "the truth will out" (Gilbert and Mulkay, 1984) by distinguishing Burt's behavior and even his intentions (put outside science) from the possible validity of his claims and findings (inside). Hereditarians and environmentalists converged on the assumption that only further research on intelligence in twins would prove whether Burt's work has a lasting place in the science of psychology. Just as scientific claims are customarily detached from their authors (and linked instead to nature), Burt's data are detached from the seamy conditions of their creation: Science (as this episode of boundary-work concludes) is not in the making but in the made. The putatively self-correcting procedures of science guarantee that the truth will out; the rest is gossip (and not science). When people in society are persuaded of this belief, the cultural authority of science is not weakened by a potentially embarrassing case like Burt's, but strengthened.

Protection: Keeping Politics Near but Out

A final type of boundary-work by scientists involves the erection of walls to protect the resources and privileges of those inside. Successful boundary-work of this kind is measured by the *prevention* of the control of science by outside powers – or, put the other way, protection of the autonomous control *of* science *by* scientist–insiders. Threats are all around. When the Animal Liberation Front challenges the right to conduct certain experiments using laboratory animals, scientists work hard to distinguish their research from the use of animals in testing by the cosmetics industry, thereby appealing to higher values to justify their practices (new medicines or therapies save lives, but a new mascara will only make your eyes less itchy). In cases of technological failure such as the *Challenger* explosion, science is practically demarcated from "management decisions" or from "manufacturing" to shift blame for the disaster away from scientists and onto these other culprits (Gieryn and Figert, 1990). In these instances, boundary-work is employed in the struggle for *control* of science among scientists and outside powers, a struggle with high stakes for scientists; they stand to lose autonomy in setting research agendas or deciding among methodological strategies, and they risk loss of prestige, credibility, or even funding if blamed for unwanted technological developments.

Nowhere is this struggle for the control of science more apparent than in the endless negotiations of the boundary between science and politics, as recently examined in Jasanoff's (1990) study of science advisers in the American federal government. The cartographic challenge for scientists is to draw science near enough to politics (ideally, as adjacent cultural territories) without risking spillover of one space into the other or creating ambiguity about where the line between them should fall. That challenge is heightened (as in other kinds of boundary-work) by cartographic efforts of *others* seeking advantage from possibly different cultural maps – elected officials, government bureaucrats, journalists, interest groups, and many other folks who stand ready to draw science politics in ways congruent with their particular interests and programs.

For scientists, the mapping task is to get science close to politics, but not too close. Why? A key to the legitimation of scientists' cultural authority is the perceived pertinence of science for political decision making: As government officials turn to scientists for expert advice before promulgating regulations or statutes, they are simultaneously measuring and reproducing the authority of science over claims about reality (Mukerji, 1989). Too great a distance between science and politics threatens a critically important route for scientists' legitimation via their perceived political utility – and in particular their claim on government funding for their research. The relationship is symbiotic, of course; just as scientists draw legitimation from the use of science in government deliberations, so government officials (and others) are better able to legitimate their policy decisions by attaching to them the cultural authority of scientific expertise. The territories of science and politics converge not as a matter of some structural necessity or faceless rationality but because insiders to both have good reason to keep the other near at hand.

But not *too* close, and certainly not overlapping or interpenetrating. Only good fences keep politics and science good neighbors. Politicians, government bureaucrats, interest groups, and involved citizens keep science at arms length to preserve their discretion and thus their power. If policy can be fully determined by facts under control of scientists, what place is left for political choice – whether democratic, bureaucratic, or judicial? The dilemma for policy makers (in the broadest sense) is clear: Bring science near enough so that political choices are legitimated by their perceived grounding in authoritative and objective understandings of the facts as only science provides, but not so close that choices and futures become exclusively 'technical' and beyond the grasp and thus control of non-scientists. Scientists also need to keep the fence on their "politics" frontier well mended. After all, what makes scientific knowledge useful for politics is not just its content but its putative objectivity or neutrality.[10] Science can legitimate policy only if scientists are *not* treated as just another interest group and their technical input is *not* defined as just another opinion. Spillover in the other direction – from politics into science – is just as dangerous for scientists' autonomy: When politicians themselves make facts, the professional monopoly of scientists over this task is threatened. An even more likely threat is the capture of science by policy-making powers – a loss of scientists' control over their research agendas and, in the limiting case, over what is represented as "scientific" knowledge.

Jasanoff's book moves from these general observations to specifics of boundary-work in scientific advisory proceedings. She neatly summarizes the above tensions:

> When an area of intellectual activity is tagged with the label "science," people who are not scientists are *de facto* barred from having any say about its substance; correspondingly, to label something "not science" [e.g., mere politics] is to denude it of cognitive authority. (Jasanoff, 1990, p. 14)

Although those on both sides have reason to keep the two cultural territories close but not too close, Jasanoff finds over again that the "social construction" of the science/politics border is a crucial strategy through which distinctive interests of diverse players are advanced or thwarted. Scientists may or may not have a stake in any particular policy outcome, but they do have a professional interest in protecting their claim to authority over fact making, which thus shapes how they variously distinguish scientific matters from political (and how they respond to others' maps as well). Politicians, bureaucrats, and others involved in policy making may worry little about the cultural authority of scientists except insofar as it legitimates preferred policies and programs without preempting their discretionary choices about which ones to enact – and so they too draw maps showing variously located borders between science and politics.

Four examples will illustrate occasions for science/politics boundary-work – and the diverse interests this rhetorical game can serve – as scientists are brought near (if not fully within) the policy-making machinery. First, industry lobbyists or citizens' interests groups use boundary-work to discredit or delegitimate an unwelcome policy initiative by challenging the credibility of the science on which that policy is based – all the while preserving (and, obviously, reproducing) the cultural authority of science as an abstract space. The trick is to create a different cultural space for ersatz scientific practices and findings – variously called "regulatory science" or, bluntly, "bad science" – that do not measure up to standards established across the border in 'real' science (pure, academic, basic). Such a move leaves science untouched as the cultural space capable of producing expert, credible, and authoritative knowledge pertinent for policy making and, moreover, points out the road to policy initiatives more salutary for the interests of the critic: more *better* science.

In one case considered by Jasanoff, boundary-workers challenging the science behind a policy initiative turned out to be other scientists themselves. The Environmental Protection Agency (EPA) was asked to assess public health risks at several sites used by Hooker Chemicals for dumping toxic waste, including the infamous Love Canal neighborhood near Niagara Falls. The EPA commissioned a scientific study from a private consulting firm, whose results were leaked to the press and seemed to indicate unusually high rates of chromosomal abnormalities among neighborhood residents. The findings were immediately challenged by panels of scientific experts, who argued that proper research protocols were not followed (e.g., some critical controls were absent) and concluded that the consultant's report was not worthy of "science." Jasanoff (1990) writes that "in a politicized environment such as the U.S. regulatory process, the deconstruction of scientific facts into conflicting, socially constrained interpretations seems more likely to be the norm than the exception" (p. 37). It is not clear whether these deconstructions of facts from inside-real-science critics were shaped mainly by their interest-based opposition to the policy implications of initial findings, or (possibly in addition) by their concern as professional border guards for the EPA's unauthorized use of the authoritative adjective "science" to legitimate their policy initiative by grounding it in research that falls short of insiders' standards of good science. Importantly, the sociological issue is not some essentialist set of standards for good science (forever up for negotiation) but struggles between scientists and a government agency over who has the power to draw boundaries between good science and bad, and thus control the allocation of cultural authority attached to that space.

Jasanoff (1990) sets up the second example of boundary-work when she writes: "Participants in the regulatory process often try to gain control of key issues by changing their characterization from science to policy or from policy to science" (p. 14). Here, specific regulatory tasks or responsibilities are moved around the map as different players – agency bureaucrats, politicians, the courts, scientific advisors, scientists working as consultants to industry or interest groups, the outside scientific community – vie for authority to control a decision and steer it toward their own interests. When industry representatives depicted the promulgation of risk-assessment guidelines as a scientific matter, they were simultaneously defining the borders of science to include that task and announcing who alone could do the job.

> Industrial groups were convinced that these technocratic organizations [such as the National Academy of Sciences] would reach conclusions that were scientifically more conservative, hence more sympathetic to business interests, than those advocated by administrative agencies or the courts. (Jasanoff, 1990, p. 59)

On other occasions, responsibility, and thus control, is moved away from scientists back to agency administrators. For a time, the EPA administrators argued that the development of a "criteria document" for assessing ozone risks was a *separate* matter from "standard setting," and then used the argument to limit involvement of their Scientific Advisory Board to the former task while retaining discretionary control over the latter. Standard setting for ozone risks was mapped onto the politics side, so that EPA administrators could proceed without approval or review from scientists (Jasanoff, 1990, pp. 107–113). In still other instances, border disputes over the location of designated responsibilities result in a loss of control of the task by *all* of the players involved. That happened in a controversy over the risks of Alar, a growth regulator used on apples. As the EPA and Uniroyal argued over maps of good/bad science, *60 Minutes* and Meryl Streep (testifying before Congress) rendered their arguments largely irrelevant (p. 149).

The third example, in contrast to the others, suggests that porous and ambiguous fences *sometimes* make better neighbors than impermeably crisp ones. Looking at the regulatory activities of the Food and Drug Administration (FDA), Jasanoff suggests that blurring boundaries between science and politics can be effective for attaching the authority of science to policy initiatives while retaining political, administrative, or judicial control over their direction. "The line of demarcation between FDA's decision-making authority and that of its scientific advisers [was left] ill defined." For example, in considering possible approval for certain antiarrhythmic drugs (for treating angina pectoris), the FDA's scientific advisory network did not limit its work to a review of available scientific literature; it also "prepared the equivalent of a preliminary position statement": Isn't that "politics"? Although the agency exploited the cognitive authority of this scientific position statement as it justified its eventual policy, it did not cede ultimate control to its scientific advisers. The FDA employed "creative boundary-work to expand the extent of its authority and gracefully deviate from committee recommendations." Jasanoff (1990) concludes: "The ambiguity of the boundary between science and politics is strategically useful to FDA, permitting the agency to harness the authority of science in support of its own policy preferences" (p. 178).

But the fourth example is a reminder that hyperexis can occur in boundary-work as in most other aspects of social life; if blurring the boundary between science and politics is a good thing, there can – at times – be too much of it. Attention shifts now to scientists working for government agencies such as the EPA or the FDA, and their efforts to *distance* their science from the agency's politics. Boundary-work by these scientists is motivated less by an abstract concern for the cultural authority of science than by a quotidian pragmatic concern for producing a useful, valued product for an employer. Their "job description" is to provide an informed and objective context of expert knowledge for policy deliberations – and that requires the manifest *appearance* of a clean break between research and politics (even if edges are really rather sloppy). As a means to bolster its scientific credibility, the EPA created the Scientific Advisory Board (SAB) in 1974 to review studies

carried out by the agency's in-house Office of Research and Development. A challenge for the SAB was to "maintain its scientific authority
on the one hand" and on the other to avoid capture by the "partisan political interests" inherent in the EPA's mission (Jasanoff, 1990, p. 95).
The very name of the board attested to its concern for research rather
than policy, an image that was reinforced by the timing of SAB's re-
views; they typically occurred early in the deliberations, safely up-
stream from eventual policy initiatives. Of course, the socially
constructed distance between science and policy is helpful for maintaining the appearance of objectivity among science advisers, but even
that can go too far. Returning full circle to a point made at the start of
this section, Jasanoff (1990) notes that "the very success of the Board's
rhetorical strategy of distancing itself from policy creates a risk that its
advice will be seen as irrelevant to policy" (pp. 97–98) – far enough to
be objective and authoritative, close enough to be useful.

CONCLUSION

There is, of course, no single "boundary problem in STS." To appreciate the socially constructed, contingent, local, and episodic character
of cultural categories such as science opens up new perspectives on a
vast terrain of issues. Examination of how and why people do boundary-work – how they define "science" by attributing characteristics that
spatially segregate it from other territories in the culturescape – could
be the first step toward a cultural interpretation of historically changing allocations of power, authority, control, credibility, expertise, prestige, and material resources among groups and occupations.
Boundary-work here becomes an important feature of professionaliz-
ing projects of scientists, a rhetorical form well suited to the seizure,
monopolization, and protection of those goodies. From a very different
direction, exposing the contingent, flexible, pragmatic, and (to a degree) arbitrary borders-and-territories staked out for science is the start
of a critical evaluation of the consequences of such cultural classifications – not just for intellectual but real-world lives. Undermining the
givenness, naturalness, necessity, universalism, and essentialism in cartographic representations of science/non-science opens the possibility
of their reconfiguration.
   Whether studies of boundary-work are designed to interpret the
world or to change it, available methodological options are many and
varied. One could home in on the rhetorical tropes used by boundary

builders to distance science from non-science – with careful analytic attention to the particular and possibly unique discursive formulations in a local episode. Or one could move from the local and episodic to examine cultural categories in their obdurate, stable, and enduring forms, in order to understand cultural and social change over the long haul. If science need not be defined from scratch each time the question comes up, and if the question does not even come up because people tacitly agree that *this* is science (and *that* is not), then some analytic attention must be given to cultural spaces that transcend their momentary instantiation, to extant maps that get unfolded (not drawn fresh) in an episode of boundary-work.

This chapter carries the following prescription for STS. Get constructivism "out of the lab" to release its interpretative potency on claims and representations where the referent is not nature but culture. If science studies has now convinced everybody that scientific facts are only contingently credible and claims about nature are only as good as their local performance, the task remains to demonstrate the similarly constructed character of the cultural categories that people in society use to interpret and evaluate those facts and claims.

Getting constructivism out of the lab assumes that the actual practices of scientists in laboratories – and their "professional" inscriptions of nature – are surrounded by an interpretative flexibility that allows for multiple and variable accountings. Getting constructivism out of the lab moves science (the practices, claims, and instruments in need of mapping) closer to places where matters of power, control, and authority are settled. Constructivism is not just a stick for beating up old-fashioned epistemologies; it is as well a theoretical challenge to just-as-old sociologies – Marxism, functionalism, rational choice theory – seeking explanations for things at the top of that discipline's agenda: uneven distributions of authority, power, control, and material resources. Political economy is cultural, interests rest on shifting meanings, pursuits of power or wealth are carried out through interpretative categories – cultural maps – that people use to arrange their worlds.

NOTES

1. Steve Woolgar offered the concept of boundary-work in response to some half-baked ideas I discussed with him at a conference in 1981, and I first used it in Gieryn (1983). Since then, I have found too many others who

have used synonymous tags for the same idea, and one goal of this chapter is to bring these common but fragmented efforts together. Perhaps the earliest mention in science studies of the boundary problem as characterized here is Barnes (1974): "We should not seek to define science ourselves; we must seek to *discover* it as a segment of culture already defined by actors themselves... It may be of real sociological interest to know how actors conceive the boundary between science and the rest of culture, since they may treat inside and outside very differently' (p. 100). By 1982, when Steven Shapin (1982) reviewed sociologically informed work in the history of science, he was able to group 20 studies under a heading 'Interests and the Boundaries of the Scientific Community" (pp. 169–175). Shapin (1992) continues the review up to the minute.

2. There is irony in my lumping together Popper with those whose solution to the demarcation problem is *essentialist*, for he railed against essentialism: "I do not think that we can ever describe, by our universal laws, an ultimate essence of the world" (Popper, 1972, p. 196, see 1963, pp. 103 ff.). He was steadfast in his rejection of the essentialist idea that "if we can explain the behavior of a thing in terms of its essence – of its essential properties – then no further questions can be raised, and none need be raised" (1972, p. 194; see pp. 309–310). Evidently, Popper's anti-essentialism did not extend from "scientific explanations of the *world*" (i.e., of nature) to "philosophi-cal-cum-sociological explanations of *science*," for it is precisely my argument that falsifiability (bold conjecture, severe critique, and so on) constitutes the essence of science for Popper and, moreover, explains for him why science is superior to any alternative in advancing objective knowledge toward an ever closer approximation to reality. When talking *about* science, Popper's (1972) essentialism is inescapable: "For once we have been told... that the most satisfactory explanation will be the one that is most severely testable and most severely tested, we know all that we need to know *as methodologists*" (p. 203).

3. Amidst his study (with Trevor Pinch) of parapsychology, Collins explicitly denies the ability of falsifiability (tied as it is to replicability) to serve as a criterion demarcating science from non-science: "To a philosopher of science who follows Karl Popper's views the events must seem to reveal a classic case of unfalsifiability akin to the case of astrology. Proponents cannot agree on a procedure that will render falsifiable the existence of the phenomenon in which they are interested. Reasons or excuses are always provided to explain away failures of the subjects to perform in any set of circumstances. Yet, the scientists involved would not want to say that they were not doing science. It is difficult to see what they are doing if it is not science" (Collins and Pinch, 1982, p. 134; see Collins, 1975).

4. Surely more ethnographic studies of laboratory life and textual analyses of scientific discourse are needed to better understand the ambiguities of replication, the interpretation and deployment of social norms, and the construction and reconstruction of scientific consensus. For additional discussion of the contrast between essentialist and constructivist views of demarcation, see Woolgar (1988, pp. 11–21).

5. "Consequences" is smuggled in here to preempt an obvious (but wrong) explanation for the cultural authority of science: Scientific interpretations

are accorded credibility because they *work* when translated into technologies with demonstrated effectiveness in getting things done (airplanes fly thanks to aerodynamics and the rest of physics; polio rates are reduced by biomedical science). The error should be apparent to those familiar with the literature on the social construction of technology: the workability of a machine is itself a matter for negotiation, as is the attribution of technological success to science (as opposed to engineering, manufacturing, karma, or some other cultural realm). See Bijker, Pinch-and Hughes (1987) and Bijker and Law (1992).

6. This is a contentious issue, of course. The relationship between scientific practices or claims and subsequent reconstructions parallels exactly the relationship between nature and scientific knowledge as described in classically constructivist works (e.g., Pinch, 1986). Relativistic indeterminacy between referent and reference does not mean that "anything goes." Rather, it becomes an empirical question – of the first order in significance – to understand how *some* scientific claims assume the kind of obduracy that leads people to see them as real (i.e., given in nature). To shift now from construction of knowledge to construction of science, it is also an empirical question – of equal significance – to understand how some representations of scientific practice attain the same obduracy and perceived reality. Studies of this question are, sadly, in their infancy.

7. But a backdoor essentialism appears even among stalwart constructivists. Collins and Yearley's (1992) call for "social realism" verges dangerously toward *definitive* conclusions about the boundaries between science and other social institutions: "Close description of human activity makes science look like any other kind of practical work. Detailed description dissolves epistemological mystery and wonder. This makes science one with our other cultural endeavors" (pp. 308–309). And elsewhere, Collins (1992a) writes: "The sociology of scientific knowledge has only one thing to say about the institution of science: it is much like other social institutions" (p. 190). For sociological analysts to erase the cultural boundaries of science *is as essentialist* as Popper, Merton, and Kuhn's laying them down. It is not for analysts to decide (for example) that science is politics by other means, in some definitive sense. On occasion, people in society (scientists, for example) do build a Berlin Wall between science and politics that demarcates authority and human action every bit as effectively as the real one once did. Studies of boundary-work begin with constructivist findings that essentialist demarcation principles are poor descriptions of SSK-observed scientific knowledge and practice, but this line of inquiry would be stillborn if sociologists were to stop at Collins and Yearley's conclusion that science is much like other social institutions. That judgment rests with historical actors, who will make science into something "much like other social institutions" only if that cartographic representation serves their interests. Collins and Yearley (1992b) are on more fertile ground when they write: "None of this excludes an examination of the initial work of demarcating science and non-science" (p. 378). (Presumably, this "work" is done by someone other than sociologists, historians, or philosophers.)

8. The same point is made elsewhere in *Leviathan*: "The category of mechanical philosophy was an interpretative accomplishment; it was not some-

thing which resided as an essence in the texts" – or in their experimental practices (Shapin and Schaffer, 1985, p. 205).

9. Since 1986, not one but two biographies of Burt have been published (Fletcher, 1991; Joynson, 1989), each challenging Hearnshaw's guilty verdict and seeking some vindication of Burt's scientific image. Their appearance is evidence itself for the potential endlessness of debates surrounding the boundaries between good science and bad. Whatever Burt might have done or not done, his scientific practices are forever available for reconstruction as proper or not – and those reconstructions are best interpreted sociologically in terms of the contexts in which they appear than judged for accuracy against some elusive yardstick of what the man really did.

   Gieryn and Figert base their analysis primarily on exchanges among psychologists in the *Bulletin of the British Psychological Society.*

10. For a case study of how scientists seek to retain objectivity for their claims while still appearing useful for obviously partisan deliberations of acid rain regulation, see Zehr (1990).

REFERENCES

Barnes, B. (1974). *Scientific Knowledge and Sociological Theory,* London: Routledge & Kegan Paul.

—— (1982). *T.S. Kuhn and Social Science,* New York: Columbia University Press.

Barnes, B. and Edge, D. (eds.) (1982). *Science in Context: Readings in the Sociology of Science,* Milton Keynes: Open University Press/Cambridge, MA: MIT Press.

Bijker, W.E. and Law, J. (eds.) (1992). *Shaping Technology/Building Society,* Cambridge, MA: MIT Press.

Bijker, W.E., Hughes, T.P. and Pinch, T.J. (1987). *The Social Construction of Technological Systems. New Directions in the Sociology and History of Technology,* Cambridge, MA: MIT Press.

Collins, H.M. (1975). "The Seven Sexes: A study in the sociology of a phenomenon, or the replication of experiments in physics", *Sociology,* 9, pp. 205–224.

—— (1985). *Changing Order: Reputation and Induction in Scientific Practice,* London: Sage.

—— (1992). *Changing Order: Replication and Induction in Scientific Practice- (revised edn),* Chicago: The University of Chicago Press.

Collins, H.M. and Pinch, T.J. (1982). *Frames of Meaning: The Social Construction of Extraordinary Science,* London: Routledge & Kegan Paul.

Collins, H.M. and Yearley, S. (1992a). "Epistemological chicken", in A. Pickering (ed.) *Science as Practice and Culture,* Chicago: The University of Chicago Press, pp. 301–326.

—— (1992b). "Journey into space", in A. Pickering (ed.), *Science as Culture and Practice,* Chicago: The University of Chicago Press, pp. 369–389.

Darnton, R. (1984). "Philosophers trim the tree of knowledge: The epistemological strategy of the *Encyclopédie*", in R. Darnton, *The Great Cat Massacre,* New York: Basic Books, pp. 191–213.

Fletcher, R. (1991). *Science, Ideology and the Media: The Cyril Burt Scandal*, New Brunswick, NJ: Transaction Books.

Gieryn, T. F. (1983). "Boundary work and the demarcation of science from non-science: Strains and interests of professional ideologies of scientists", *American Sociological Review*, 48, pp. 781–795.

Gieryn, T. and Figert, A. (1986). "Scientists protect their cognitive authority: The status degradation ceremony of Sir Cyril Burt", in G. Bohme and N. Stehr (eds.), *The Knowledge Society*, Dordrecht: Reidel, pp. 67–86.

—— (1990). "Ingredients for the theory of science in society: O-rings, ice water, c-clamp, Richard Feynman and the press", in S. E. Cozzens and T. Gieryn (eds.), *Theories of Science in Society*, Bloomington: Indiana University Press, pp. 67–97.

Gilbert, G. N. and Mulkay, M. (1984). *Opening Pandora's Box: A Sociological Analysis of Scientists' Discourse*, Cambridge: Cambridge University Press.

Hollinger, D. (1983). "The Defense of Democracy and Robert K. Merton's Formulation of the Scientific Ethos", *Knowledge and Society*, 4, pp. 1–5.

Jasanoff, S. (1990). *The Fifth Branch: Science Advisors as Policy Makers*, Cambridge, MA: Harvard University Press.

Joynson, R.B. (1989). *The Burt Affair*, New York: Routledge & Kegan Paul.

Kuhn, T. S. (1962). *The Structure of Scientific Revolutions*, Chicago: University of Chicago Press.

—— (1970). "Logic of discovery or psychology of research?", in I. Lakatos and A. Musgrave (eds.), *Criticism and the Growth of Knowledge*, Cambridge: Cambridge University Press, pp. 1–23.

—— (1977). *The Essential Tension*, Chicago: University of Chicago Press.

Laudan, L. (1983). "The demise of the demarcation problem", in R. Laudan (ed.), *The Demarcation between Science and Pseudo-Science*, Blacksburg: Virginia Tech, Center for the Study of Science in Society, pp. 7–35.

Masterman, M. (1970). "The nature of a paradigm", in I. Lakatos and A. Musgrave (eds.), *Criticism and the Growth of Knowledge*, Cambridge: Cambridge University Press, pp. 59–90.

Merton, R. K. (1942). "Science and technology in a democratic order", *Journal of Legal and Political Sociology*, 1, pp. 15–26.

—— (1973). *The Sociology of Science: Theoretical and Empirical Investigations*, N.W. Sorter (ed.), Chicago: University of Chicago Press.

—— (1977). "The sociology of science: An episodic memoir", in R. Merton and J. Gaston (eds.), *The Sociology of Science in Europe*, Carbondale: Southern Illinois University Press, pp. 3–141.

Mitroff, I. I. (1974). "Norms and counter-norms in a select group of Apollo moon scientists", *American Sociological Review*, 39, pp. 579–595.

Mukerjii, C. (1989). *A Fragile Power: Scientists and the State*, Princeton: Princeton University Press.

Mulkay, M. (1980). "Interpretation and the use of rules: The case of the norms of science", in T. Gieryn (ed.), *Science and Social Structure: A Festschrift for Robert K. Merton*, New York: New York Academy of Sciences, pp. 11–125.

Pinch, T. (1982). "Kuhn: The conservative and radical interpretations", *4S Newsletter*, 7, pp. 10–25.

Popper, K. R. (1959). *The Logic of Scientific Discovery,* New York: Harper (originally published in 1934).

—— (1963). *Conjectures and Refutations,* New York: Harper.

—— (1970). "Normal science and its dangers", in I. Lakatos and A. Musgrave (eds.), *Criticism and the Growth of Knowledge,* Cambridge: Cambridge University Press, pp. 51–58.

—— (1972). *Objective Knowledge: An Evolutionary Approach,* Oxford: Clarendon.

Shapin, S. (1982). "History of Science and its Sociological Reconstructions", *History of Science,* 20, pp. 157–211.

—— (1992). "Discipline and Bounding. The history and sociology of science as seen through the externalism–internalism debate", *History of Science* 30, pp. 333–369.

Shapin, S. and Schaffer, S. (1985). *Leviathan and the Air Pump: Hobbes, Boyle and the Experimental Life,* Princeton: Princeton University Press.

Staw, P. (1982). *The Social Transformation of American Medicine,* New York: Basic Books.

Woolgar, S. (1976). "Writing an Intellectual History of Scientific Development: The use of discovery accounts", *Social Studies of Science,* 6, pp. 395–422.

—— (1988). *Science: The Very Idea,* London: Tavistock.

Zehr, S.C. (1990). *Acid Rain as a Social, Political, and Scientific Controversy,* unpublished doctoral dissertation, Bloomington: Indiana University.

Zuckerman, H. (1977). "Deviant behavior and social control in science", E. Sagarin (ed.), *Deviance and Social Change,* Beverly Hills, CA: Sage, pp. 87–138.

Part V
Science and Values

12 Ethical Problems in Using Science in the Regulatory Process*

Nicholas A. Ashford and Karin A. Gregory

This article provides a framework for considering ethical issues in the use of science in the regulatory process. The science in question includes both the assessment of technological risk – from chemicals, consumer products, energy sources, transportation technology, etc. – and the assessment of technological options to reduce those risks, such as hazard control technology, product substitution, and industrial process redesign.

The focus of our inquiry is on the role of the scientist-engineer as analyst–assessor and not as the designer of new products or processes, although the latter activity also presents serious ethical questions for scientists and engineers. Similarly, instead of analyzing the use of science by lawyers and other players in the legal system, we adopt the premise that some criticisms of the adversary process in resolving science-based disputes can be better addressed by examining the role of scientists or engineers who participate in that process.

For the purposes of this article, participation in the regulatory process includes a variety of activities from the undertaking of scientific and technological inquiry relevant to regulation to advocacy of particular actions within the regulatory process. The regulatory process is broadly defined to include the notification of interested parties of technological risks and options, control of those risks, and compensation for harm resulting from technology. Government at the federal or state level intervenes in all three sets of activities through regulatory agencies and the courts. But governmental decisions at each stage of the regulatory process critically depend on scientific and technological information. The question we consider is the appropriate role of the

* From *Natural Resources and the Environment*, American Bar Association, 2(2) (Fall 1986), pp. 13–16, 55–57.

scientist-engineer to ensure both a fair process and a fair outcome for science-based disputes in which he or she participates.

SCIENCE AND VALUES: CAN THEY BE SEPARATED?

While scientific inquiry often claims to be value-neutral (i.e., nonnormative), the same cannot be said for the uses of scientific information in decisions concerning the control of science and technology. It is therefore important to ask whether the conduct of policy-relevant scientific inquiry, such as risk assessment, can ever really be value-neutral. The practice of science has been described as reductionist, that is, science teases out the most likely correct truth in an uncertain world by using simplifying assumptions and theories. The traditions and conventions adopted by science in order to establish "truth" are traditions and conventions to deal with uncertainties in both scientific theory and data. In the evolution of a scientific paradigm or methodology, for example, science often establishes clearly visible standards which must be achieved for something to be considered true. The things that are considered true according to these standards are called facts. When we are certain about scientific explanations, we call these explanations law. When we are less certain, we call them theories. To change a scientific theory into a scientific law, we need both confirmation of the theory by existing data and acceptable explanations of data that appear to deviate from the predictions of the theory. Science recognizes that such confirmation or explanations cannot be 100 percent certain. Scientific tradition and conventions establish the minimum *scientifically* acceptable probability of being correct and the maximum *scientifically* acceptable probability of being wrong in reaching a conclusion.

Legal actions seek to be fair and to encourage the correct outcome in societal activities, including the applications of science and technology. In prescribing or prohibiting a given activity, the law, like science, is sensitive to the probability that a certain view of the world might be right or might be wrong. What the law calls a fact – sometimes called a legal fact – is based on a set of data that is certain enough to justify a given directive or conclusion. But the law also seeks to encourage the correct result in the normative sense, that is, what Rawls (1971) would call the just thing. The law and policy process recognize that something must be true enough to justify an action, but absolute truth is not required as a prerequisite to arriving at a just outcome. Law thus seemingly creates a paradox whereby things can be regarded as true for some purposes but not for others. Science, on the other hand, insists that things are either true or untrue, and by marshalling established scientific conventions as the tests, encourages us to believe that no value judgment ever attends the establishment of truth.

It is, however, clear that those who undertake scientific inquiry today, in fact, hold values concerning the use of their science. Within a given framework of scientific traditions and conventions, there are many ways to analyze and present data. There are also many ways to frame the scientific question and choose which data to collect. A scientist's choice among these possibilities is shaped by values. By either speaking out about those values or by remaining silent, scientists exercise a value judgment about the way science is used in regulation. Accordingly, as Professor Mark Rushefsky of Southwest Missouri State University has observed, "Ostensible disputes over the science are, in reality, over the values inherent in the assumptions" (1985, p. 31).

If science is not value-free, then how can it best inform the public policy debate? Many would address that question by requiring agencies to establish a two-step process for dealing with risk: risk assessment and risk management. The former is expected to be a nonnormative scientific determination, and the latter a value-laden political decision to control a given hazard. However, the key question for policy makers is this: At what level of proof does a showing of risk or danger trigger a requirement for regulatory action? What is considered sufficient proof is a social policy determination, requiring a judgment about the consequences of both regulating and not regulating a possibly hazardous activity. Science can inform, but should not necessarily dictate, the results of that analysis.

Such judgments can be in error because of uncertainties with regard to the nature and extent of the risk or the economic and technological feasibility of regulatory controls. Type I errors are committed when, because of insufficient evidence, society fails to regulate an activity which turns out to be harmful. Type II errors are committed when society regulates an activity which turns out not to be harmful and resources are needlessly expended. Aversion to making Type I and Type II errors reflect differing value decisions about (1) the nature of the mistakes made and (2) the extent, prevalence, or magnitude of the mistakes. A bias against Type II errors underlies the often- expressed plea that we "move the regulatory process toward better science." In some cases this may simply be a request to be more permissive in controlling technological activities.

The interplay of facts, or science, and values can be illuminated by three general scenarios concerning carcinogen regulation:

1. If a causal relationship is shown which satisfies the accepted scientific conventions for establishing that a chemical causes cancer, then a *scientific determination* has been made which can inform the public policy process. (An example is the overwhelming evidence that asbestos exposure causes mesothelioma.) Then, the decision to notify, regulate, or compensate is essentially a social or public policy decision.
2. If a sizable majority of the relevant and respected scientific community believes that a substance is probably carcinogenic (perhaps more likely than not), although causality has not been proven at the conventional (high) level of statistical significance or with sufficient strength of association, then a *science policy determination* has been made that justifies treating the substance as if it were carcinogenic. The scientists who reach such a consensus have similar values regarding the use to which the scientific data will be put. Specifically, the decision to view the substance as a probable carcinogen in this scenario reflects a bias against Type I errors, that is, erring on the side of caution. Their science policy decision can then inform the social policy decision taken by the regulatory agency.
3. If the scientific community reaches no consensus about labeling a substance as carcinogenic (for regulatory purposes), then there is no scientific or science policy determination to inform regulatory decision making. However, it may still be sound *social policy* to control that substance. A decision to regulate under these circumstances would merely mean that the regulator's preference for making Type I versus Type II errors is different from that of the scientists who reviewed the evidence.

These three scenarios, of course, represent points on what is really a continuum of scientific uncertainty. They merely illustrate the varying relationship of science and values encountered in the regulatory process. Under conditions of uncertainty, the nature of the scientific consensus, or science policy determination, may depend on the use to which data and information will be put. Consensus on the minimum evidence required for action will, and probably should, differ according to whether the purpose is notification, regulation, or compensation.

Thus, while a uniform intellectual *approach* to the question of risk assessment and risk management is theoretically desirable, *uniform* conventions, such as statistical significance or the rejection or acceptance of negative studies, are not advisable in deciding what level of proof is acceptable for policy purposes. Value judgments clearly attend decisions whether to lean in favor of Type I or Type II errors in specific cases. This is because the cost of being wrong in one instance may be vastly different from the cost of being wrong in another. For example, banning a chemical which is essential to a beneficial societal activity (such as the use of radionuclides in medicine) has potentially more drastic consequences than banning a non-essential chemical for which there is a close, cost-comparable substitute. It may be perfectly appropriate to rely on most likely estimates of risk in the first case and on worst case analysis in the second. This approach illustrates not only a preference for making Type II rather than Type I errors, but also illustrates the dependence of that preference on the size of the Type II error.

In each of the three scenarios described above, both a fair process and a fair outcome are desirable. A fair *process* has its origins in the legal tradition of due process, but more generally means that a procedure for the determinations of both science and policy has provided adequate opportunity for presentation and discussion of the data, their relevance for society, and the underlying values and preferences of the participants regarding the use of the data or findings. Whether a process is fair or not can usually be determined objectively by any observer without deciding questions of fact or policy one way or another. In contrast, a fair *outcome* has as its reference or basis a particular observer's view of the same issues. People who would make different decisions concerning a fact or policy might not call the outcome fair, although they might agree that the process leading to it was fair.

In Scenario I scientists can contribute their work product (or publish their findings) and hope that the facts will speak for themselves when considered by the decision makers. Of course, there may be vigorous dissent about the scientific studies themselves, and this may require open discussion of the science. But this process can usually be handled with fairness by the scientists themselves in an informal way through peer review and other avenues for open exchange and criticism. A more formal process targeted toward elucidating facts and values will then be required for the subsequent policy decision to ensure that the data are put to an appropriate use for regulatory purposes.

In Scenario 2 a fair process is needed not only for the risk management decision, but also during risk assessment in order that the *science policy* determinations are properly arrived at. Such a process is required to illuminate the values that may be hidden behind science policy conclusions.

In Scenario 3 we are unsure about the correctness of the outcome, that is, Type I and/or Type II errors are large. The best we can do is to provide a fair process for the resolution of competing values, since the final social policy decision turns on value choices concerning Type I and Type II errors.

The remainder of the article deals with the ethical problems encountered by scientists participating in the regulatory process and the problems of ensuring fairness in their contributions to policy decisions. Where the consideration of facts and values cannot be institutionally separated, as they arguably can be in Scenario I, the trier of fact or decision maker must insist on a clarification of the values which influence the determinations of the scientist-participant. Only then can a decision maker fairly make the decision which society has mandated.

PARTICIPATION IN THE REGULATORY PROCESS AND ATTENDANT ETHICAL DILEMMAS

As noted earlier, the major components of the regulatory process are (1) notification of interested individuals or groups about technological risks; (2) mandatory action for control of those risks; and (3) compensation in the form of workers' compensation, insurance for the general public, and tort (product liability) suits. Each type of regulatory action relies on scientific data and theories, but each seeks to fulfill a different societal purpose.

It is helpful to expand these major types of regulatory activity into some subordinate activities that may create ethical dilemmas for the scientist-participant:

1. Alerting a regulatory agency or an industry to a problem or an opportunity to improve public health and safety. The institution thus alerted is then free to determine how serious the problem is and how it can be corrected. The regulatory agency or industry may, of course, control the hazard and/or notify people at risk.

2. Advising a regulatory agency or industry of the seriousness of a problem, that is, performing a risk assessment and/or identifying a substitute technology or industrial practice which might reduce the problem.

3. Participating in the regulatory standard-setting process.

4. Contesting a regulatory violation.

5. Participating in a proceeding wherein a victim seeks compensation for injury allegedly caused by exposure to a particular hazard.

As scientific data become available, a scientist may conclude that the results suggest action. Should the scientist alert a regulatory agency or an industry to his or her findings? Scientists work more or less independently on projects that interest them. While they may have personal autonomy on a daily basis, they are constrained in their research by available resources, including external funding. Scientists may be asked to investigate the carcinogenicity of a chemical substance by a regulatory agency or by industry. Should they alert both if they discover a problem or only the body that funded the research? If the scientist alerts the agency or industry to the problem, is there a further duty to advise them both of the seriousness of the problem?

Another way scientists can participate is in the standard-setting process. Participation can take several forms, including serving on agency advisory committees, petitioning the agency to act, providing expert testimony in regulatory hearings, cross-examining other witnesses commenting on another expert's testimony, and participating in a challenge to the regulatory standard. Scientists operate with varying degrees of independence in these activities, participating to bring attention to their research to critique a scientific principle or policy recommendation, and on behalf of an interested party.

Scientists may run risks in being too outspoken in an advisory committee, since they may not be re-appointed or asked to join others. Agencies differ among themselves and over time as to how much critical comment they want to hear. Similarly, scientists may be reluctant to press for regulatory action because of the threat of loss of jobs, status, or research funds. "The field is littered with the bones of scientists who spoke out bravely about controversial issues and were then forced to be at the mercy of a totally unresponsive and uncharitable society," claims Ralph Nader (1974). Because of the effective grapevine that exists throughout industry and regulatory agencies, scientists who aggressively push agencies to act can be labeled troublemakers and find themselves jobless. Yet, Nader argues than "scientists should be the initiators, the interpreters, and the advocates in many public policy issues."

How much information should scientists give when they testify? Should scientists volunteer facts unfavourable to the party for whom they are testifying? Should scientists assist in actively cross-examining other points of view. Should they only answer questions which are put to them? Is there a distinction between the duty of a scientist who testifies for an interested party and one that works as a special master for the court? These questions present ethical and professional dilemmas.

Once the regulatory hearing process is completed, the agency can use the data, testimony, and recommendations from the various parties in setting a standard or taking other regulatory action. Here, too, dilemmas exist. Should scientists volunteer facts and interpretations unfavorable to their own argument? Scientists can assist industry, public interest groups, or labor in challenging the agency's policy decision. Once a standard is in place, scientists can further involve themselves in enforcement activities by helping to defend or challenge a citation for noncompliance.

Finally, scientific evidence is vital in adjudicating claims for compensation. What level of proof is sufficient to establish causation for purposes of compensation? Should a scientist's judgment be weighed heavily in this determination? The scientist, as much as any sworn witness, promises to tell the truth, the whole truth, and nothing but the truth. If a scientist tells the whole truth, he should volunteer all the strengths and weaknesses of the scientific theory, data, and reasoning relevant to the case at hand. The whole truth includes an exposition of the weaknesses of one's own testimony. Otherwise, the scientist is indeed no different from any other well-informed advocate who seeks to advance advantageous truth and to minimize or hide disadvantageous truth.

When a scientist gives testimony on behalf of an interested party, his likely biases are recognized and accepted. His testimony is discredited only if it is clearly wrong. It is not discredited if he merely omits to disclose a fact or viewpoint upon which he was not asked to comment directly. Yet, if a scientist were to present a scientific paper with such omissions, it probably would not pass a review of his peers. He would not be allowed to ignore a contradictory theory or unfavorable data, or avoid answering an embarrassing question by a reviewer. Yet, some scientists who provide incomplete or partial testimony in regulatory proceedings maintain that they are complying with the standards of their profession.

Even when a scientist is not formally affiliated with an interested party, influences exist that can bias his views. One important influence is industry's funding of academic research. Industry can "coopt the academic experts," and control the direction and content of academic research (Owen and Braeutigam, 1978). Industry funding for research at many universities has increased many times more than federal support and has raised questions about the independence of academic scientists.

Numerous motives can underlie an industry decision to fund a particular academic scientist. A manufacturer of a chemical that is already in production and whose safety is now questioned would prefer to fund a scientist whose research will not confirm the problem. It is not necessary to find a scientist who will lie, only one who customarily uses a relatively insensitive method of experimental or statistical analysis, or evaluates data by methods that fail to provide conclusive results. On the other hand, a firm may adopt the opposite strategy when it wants to determine whether a new compound is carcinogenic prior to marketing it and making a large capital investment in it. Here, all the incentives work in the direction of discovering carcinogenicity, if it exists. The firm wants to avoid a later and potentially more costly discovery of toxicity.

It is important to consider whether toxicologists asked by a regulatory agency to evaluate a chemical may not also be influenced by the sponsor. The difference, however, is that those performing government-funded research, whether through mission agencies or basic research institutes like the National Institutes of Health, have to maintain their long-term credibility and must contend with public oversight and advisory committees. Public funding comes under more public scrutiny. Decisions to fund certain types of research over others are made in a context sensitive to social demands, often directly by a peer group of scientists as well as indirectly by politicians who budget funds for projects viewed as important to society.

Scientists direct their research toward areas for which public or private funding is available. But public interest or labor organizations that want scientific data to support their position are not likely to be able to fund such research even if they can find a scientist sympathetic to their cause. In a real sense, therefore, scientific expertise is likely to be biased against the interests of such organizations.

Regulation can confer considerable costs and benefits on the players. The anticipated consequences influence not only the political process, but also the conduct and use of science because, as we have argued, the practice of science is seldom interest-free. It is important to be sensitive to the allegiances or biases of the scientist-participant in order to uncover possible distortions of scientific data and conclusions.

SORTING OUT SUBJECTIVE VALUES FROM OBJECTIVE VIEWS

To achieve society's purposes, the use of science in the regulatory process should be objective, balanced, and attended by a fair process. A fair process is valuable in itself, independent of the outcome it produces. What constitutes a fair process may be different for the different uses to which science is put and in the different components in the regulatory process.

It is important to distinguish balance from objectivity. If a sufficiently diverse group of scientists is brought together to address a problem or interpret a set of data, that group's determination may be called balanced. That determination may or may not represent a consensus. The group may simply express a wide diversity of legitimate views. In either case, we should avoid the use of the term objective to describe the determination. It is quite another matter to call an individual scientist's testimony balanced. An individual's views are balanced only if they are objective. As we have discussed, however, a variety of factors make it difficult for any individual scientist to achieve true objectivity. It is also difficult for the external observer to determine whether an individual has given balanced consideration to the facts before him. Since there are so many assumptions, traditions, and conventions which he may use in the face of uncertainty, it is well nigh impossible to delve into the reasons for all of his choices. Sometimes the mechanism of cross-examination can be used to discredit a particularly biased or uninformed analysis. For a cross-examination to be effective, the examiner should demand disclosure of all those factors which could, at their worst, result in distortions of fact, omissions of data, or failures to consider alternative views and theories.

The Federal Advisory Committee Act (FACA) requires that advisory committees be balanced among differing points of view. In formulating an operational definition of balance to meet the FACA requirements, three criteria should be considered: competence, discipline and bias or allegiance. Ensuring balance in the effectiveness of technical argument surely requires equivalent competence among antagonists, but experts as a group may need to be tempered by nonexpert members in order to achieve a fully balanced perspective in other ways. Each discipline carries its own paradigmatic bias. To guard against this, both nonexperts and multiple disciplines should be represented, the mixture depending on the committee's particular agenda or purpose. Finally, political, institutional, ethnic, or sexual bias or allegiance should be adequately balanced.

Voluntary disclosure of bias and identification of interests is imperative to a fair process for the resolution of science-informed disputes. If balance is to be achieved, whether in the context of a committee or in a hearing, the decision maker must be made aware of bias and interests. Within an advisory committee, disclosure of current professional affiliations and past or present consulting arrangements will let the public know what influences may be affecting the views of members. Disclosure is also important in the context of individual testimony. In a hearing, an expert can be cross-examined to test not only his competence, but the objectivity of his views, and timely disclosure of bias or allegiance can influence the cross-examining process.

The decision maker, whether it is an administrative law judge, a regulatory official, or a court, needs to be alerted to both bias *and* values of the participants in order to evaluate the evidence and put it to proper use. Science claims to be guided by the search for and the explication of truth. In regulation, however, the central question is not what is true but at what level of proof does the evidence trigger a requirement for action. This question creates a tension for scientists. It requires experts to balance the desire to follow scientific conventions for finding truth against the imperative of furthering a just social policy through the use of science. Scientists may be uncomfortable with and unaccustomed to making such judgments or trade-offs. It is safer for most experts to follow an accepted professional tradition rather than to argue over values. As we have noted, however, social policy determinations about risk are directly related to the aversion of the scientific observer or interpreter of evidence to commit Type I and Type II errors. Though scientists frequently make such value judgments, they are no more expert at making them than other informed decision makers.

What makes scientists different is that their intellectual contribution to regulation – science – is the starting point for the entire decision-making process. Distortions of the scientific issues at the beginning can have serious consequences later on. Thus, one can argue that because of the powerful position that science and scientists occupy in the regulatory process, they have a special responsibility to guard against distortions or misuse of evidence.

TOWARD A SOLUTION

As experts in their fields, scientists have tremendous power over the outcome of the process. Mistakes or tricks that misguide a decision maker lessen the chances for a fair process, let alone a fair outcome.

Many commentators on expert witnesses blame lawyers for failing to ask the right questions of experts. Scientists could help here. Lawyers are also blamed for using any available technique to win their case. But each side in a regulatory proceeding finds experts claiming different objective views of the data. "The obvious objective of the courts in respect to expert testimony is to optimize the search for the truth," says Paul Meier of the University of Chicago (Meier, 1983). Judges and lawyers may not have the expertise to extract truths from manipulated data, but experts do.

The misuse of science serves no one's ultimate purpose. A desired outcome may be obtained in a particular hearing or trial in spite of mistakes or an unfair process, but the credibility of both science and law will be damaged. Because of the uncertainties of science, it is critical for an attorney to marshall the opinions favoring the side he is representing, but neither dishonest interpretation nor omission of crucial facts allows for a fair process. Lawyers strive to explain scientific and technological information to the decision maker using cross- examination to discredit data that have been manipulated or conclusions that have no factual basis. Only experts can provide the remaining elements that ensure both a fair process and a fair outcome. These are best achieved when experts voluntarily clarify their values and make explicit their conventions, assumptions, and especially the *limitations* of their own findings and conclusions.

REFERENCES

Meier, P. (1986). "Damned Liars and Expert Witnesses", *Journal of the American Statistical Association*, 81, pp. 269–276.

Nader, R. (1974) "Obligations of Scientists to Respond to Society's Needs", in W.A. Thomas (ed.), *Scientist in the Legal System*, Ann Arbor: Ann Arbor Science Publishers.

Owen, B.M. and Braeutigam, R. (1978). *The Regulation Game*, Cambridge, MA: Ballinger.

Rawls, J. (1971). *A Theory of Justice*, Cambridge, MA: The Belknap Press of Harvard University.

Rushefsky, M.E. (1985): "Assuming the Conclusions: Risk Assessment in the Development of Cancer Policy", *Politics and the Life Sciences*, 4; pp. 31–44.

13 Ethics and Science*

Robert S. Cohen

Speaking about science and humanistic education, which is to speak about the social relations of science, it is useful to address ourselves to the purest of social questions: not only such practical questions as arise in technological change and economic development, in allocation of our finite resources, and in mastering the runaway accumulation of knowledge and of people, but also the most ancient and simplest of questions, how shall a man live with his fellows and with himself? In every civilization we know, Asian, African, European, there have been those men who have said with Socrates that the unexamined life is not worth living. Indeed, the history of those who have been called 'wise' is a history of such explicit examinations of life.

How men shall live with one another is the subject matter of ethics. And here I mean to include social ethics, as another name for politics, along with personal ethics. From the beginning of Western thought, certainly in Greece, and perhaps in Egypt and the Mesopotamian Valley, the habits and customs and myths, which constitute the morality of daily life, have been subjected to a professional scrutiny and criticism. In this specialized part of the division of intellectual labor, the priests and the philosophers, the poets and playwrights, tried to establish grounds for reinforcing, or overthrowing, or plainly establishing beliefs about what is good and what is right. The very fact of inquiry into right and wrong was a tremendous advance, for such thinking recognized that there were alternatives from which you might choose, or at any rate your teachers and priests might choose. Freed from the narrow moral vision of a single clan or tribe, men began to think as widely as they travelled. It is likely that this widening of vision was a by-product of international trade, and especially of those remarkable Ionian Greeks of 600 BC who sold their goods throughout the Near Eastern lands of Persia and Egypt, who learned techniques and medicines and religions, and then compared and contrasted them for reasonableness, for evidence, for moral satisfaction, and for happiness.

* From R.S. Cohen, J.J. Stachel and M.W. Wartofsky (eds.), *For Dirk Struik. Scientific, Historical and Political Essays in Honor of Dirk J. Struik* (Dordrecht: D. Reidel Publishing Co., 1974), pp. 307–323.

The moral vision in European thought had another source, and apparently quite separate, in the Hebrew tribes whose conception of their Gods changed over a relatively short number of generations. From the usual agricultural and nomadic idolatries, and the pantheons of tribal gods, they turned to the idea of One God. More importantly, they went further to conceive two even more striking changes. First, the Hebrews' One God was thought to be God of all the world, and not because this God would lead them to conquest over all men, and Himself to conquest over all other tribal gods. Rather, He is the God of all by virtue of the inherently reasonable notion that the universe as a whole deserved an explanation which the separate gods could only partially offer and then only with incoherence or with unexplained contingent conflicts. The chaos of gods now became a unified (still personified) system; and the system was to be rational. So, for the Hebrew mind a rational center of responsibility for the whole world was established: here is proto-science and early religion.

The second striking Hebrew characteristic was the nonpictorial nature of the new One God. Even at the beginning, with the story of Moses and the burning bush, God was to be a god of ethical qualities, not so much an anthropomorphic material being *with* behavioral qualities of justice and goodness and anger, as He was the being of the ethical qualities themselves, joined together in a mysterious personality to be sure, since only a person could be understood as 'responsible' for the World.

In both these sources, Greek and Hebrew, there is therefore a tendency toward reason combined with morality, of justice dispensed in the light of evidence, of a searching for explanation not by acts of arbitrary powers but either by reasoned godly actions or within patterns of natural sequences. In Greece particularly, then, thought about the actions of men and nature and gods turned self-critical, towards the preliminary questions: how to think, how to be sure, how to recognize truths, or, briefly, how (and how much) can we know? Here in this question of how to establish knowledge, in this long search for no magic key but for a common method to truth, is the first link of science with ethics.

The search was difficult, and the philosophers of many outlooks. There have been idealists like Plato who offered theories of absolute insightful and *a priori* ethics tied closely to their brilliantly fruitful insightful mathematical approach to absolute knowledge of Nature; there were Aristotelian materialists who were successful taxonomic and developmental empiricists as much in their social and political ethics as in their biologically modelled approach to understanding Nature; there were the humanist empiricists like Hippocrates who offered experimental evidence and other lessons from experience for their reasonable scientific estimates of the probabilities or the impossibilities of diagnosis and cure, along with humane and rational advice on the conduct of life in this uncertain, only partly known, but wholly natural world; and among many others, there were the skeptics like Protagoras and Diogenes and Sextus Empiricus who logically criticized the sufficiency of the evidence for every belief about Nature, indeed who demonstrated the untrustworthiness of our sense-perceptions and our rationality alike, and who likewise were cynical about any transhuman moral standards whatsoever for human conduct. Theories of knowledge and theories of ethics were not only analogous; they were closely linked.

Later analogues to these tendencies are evident enough. We have had our Platonists in science seeking logical and mathematical assurances and with these, the ethical certainties of almost mathematically absolute rigor, of Descartes, and Leibniz; and later the Kantian parallel *a priori* of natural and moral laws. We have had the pure empiricists, the experimenters and observers of science, and with them Baconian ethics; and, later, equally empirical observations of moral differences throughout the world with the simple conclusion of ethical relativity. How appropriate for empiricism. And we have those utter skeptics about a reasonable basis for human knowledge, Montaigne and Hume, with their equally skeptical abandonment of any reasonable basis for ethics too. Science and ethics have seemed to march together, throughout the history of civilization.

Traditionally then, philosophy linked knowledge of nature with knowledge of morality through epistemology. And since epistemology arose and evolved primarily by its reciprocal dependence upon science, so ethics was dominated by science. This dominant influence took several forms: conceptions of how to think at all; philosophical conceptions of what constitutes an explanation; practical scientific ideas of what is possible and what is impossible in this world; scientifically known (or allegedly known) facts of human nature, social nature and the world at large; often, too, a scientific restatement of theology so that at several crucial stages the laws of nature were thought to be divine legislation just as the laws of politics and personal ethics. Much of the historical interaction of science and ethics reflects just this ambiguity in the concept of law, in the slow development of the distinction between the laws of men (which may be violated and which then may demand just and presumably equal punishment) and the laws of Nature, which cannot be violated. The idea of causality, so central in the scientific world-view and however subtle it has become recently, came from the ethical notion of retribution,[1] of a punishment fitting a crime and hence of effects equaling their causes with respect to some essential quality.

Yet we must also admit that there have traditionally been major gaps, even conflicts, between science and ethics. For many European thinkers, this was the great gap between nature and the supernatural, between body and spirit, or – more subtle – perhaps between responsible private men and irresponsible and unreasoning public society.[2] Even within the non-religious outlook of naturalists or atheists, the ancient observation of a gap between fact and ideal was disturbing. It was simple enough to see that knowledge is not necessarily liberating; indeed, in one formulation, what may be used for good by one hand may be used for evil by another. Science deals with facts, we often say, and ethics deals with values, and the gap becomes unbridgeable. Most succinctly, it is said that an *ought* cannot be derived from an *is*, no imperative from a declarative. Shall we formulate this as a contrast of knowledges: knowledge of facts versus knowledge of values?

The matter is clear, even within the elementary ethics of common sense. If we know what we want, then knowledge of facts will help us, either to achieve our goal or to tell us that it cannot be achieved. So science, if conceived as knowledge of the relevant facts, will inform us about the *means* to our ends, but it does not thereby shed light on the wisdom of those ends. And the grim fact remains today that men not only may differ about their ends, from man to man, from class to class, from nation to nation, but also that their ends frequently have been incompatible. It is better, I believe, not to claim too much for science at this time. Not many centuries ago, science was thought to be *inherently* good. Phrasings vary: the truth shall make you free; the unity of the True, the Good, and the Beautiful; if you know the good, you will do it. Indeed, the revolutions of politics and science of the sixteenth and seventeenth centuries incorporated into their various ideologies the maxims of both Socrates and Bacon; knowledge is virtue, knowledge is power. Science is knowledge, and science is good.

But the full truth is bitter. Science is no longer the wholly enlightening ally of human progress that it once seemed to be: and humane men will look warily at any model of a scientifically 'rationalized' social order, with too strict a devotion to facts, with concentrated focus of intellectual resources upon those technical fields which have made possible the mechanization of human life and culture. Whether by a Marxist critique of human relations in present advanced industrial society, or by an existentialist critique of the private individual's isolated situation in modern mass rootless societies, or by religious despair at the prevailing absence of love and of genuine comradeship: by any of these we come to realize again that science alone and by itself is morally neutral and painfully so. It is not automatically, is not spontaneously, a force for good; and hence neither individual nor society can depend upon such a neutral social institution as science. Being itself uncommitted, it is dangerous as well as powerful, irresponsible. Without responsibility, there is no ethics.

Furthermore, the extension of rational science beyond mechanics to the study of society and humanity, to history, is no guarantee of a rational humane commitment within the scientific community, no guarantee of moral wisdom within scientific knowledge. If nuclear bombs were the twentieth century's greatest destructive advance in man's masterly manipulative knowledge of nature, then the extremely rapid achievement of unsurpassed barbarism in Germany in the early 1930s, that is in one of the most civilized nations of Europe, is an achievement of man's masterly manipulative knowledge of human nature. And yet, so many humane thinkers in so many different traditions have known what is wrong. From Jeremiah to Jesus, from Erasmus and Comenius to Diderot and Jefferson and Kant, from Saint-Simon and Marx and John Stuart Mill to Veblen and Dewey and, I believe, from Lenin and Wilson and Gandhi to Franklin Roosevelt and John XXIII and Bertrand Russell, Brecht, Camus and Sartre, all have seen that the problem of the good society is to know how to treat each man and woman as an end, not only as a tool or as an instrument; not as an object alone but as a subject too. The grave difficulty thus far in the history of mankind has been that the material basis for life remains limited and finite, with attendant and apparently unavoidable exploitation of many men and women by a powerful and fortunate few. The ethic of have and have-not peoples remains, the gap continues and astonishingly it widens with technical advances, steadily. But then perhaps more science will help at last?

Will nuclear energy set us free? Free from poverty and hence free from greed and jealousy and even from war and exploitation of men? Will widespread science make us free from idolatry and other illusions, free for truth and curiosity, play and art and love? Indeed, will more and better science, a more completed science of nature, end the ghastly monstrosities which have plagued European civilization with its technology and its science?

These, and connected questions, have stimulated a rejuvenation of thought about ethics in recent decades, by philosophers in many countries and with different viewpoints. And in their several ways they have repeatedly sought to discover how science and ethics are related.

The role of science in ethics is complex. First, scientific *methods* may provide methods of ethical thinking and moral discovery. Second, scientific *investigations* of societies may provide what Spencer called 'the data of ethics.' Third, scientifically plausible *theories* may explain by comparative analysis the historically situated differing moral systems by using psychology and history in the broadest senses. Fourth, scientific *achievements* may determine the scope and limits of responsible moral choice and decision. Fifth, since science is a human enterprise, the *life of science* may itself provide a moral lesson of how men may live with one another, that is, the study of the history of science may show us a scientific ethic. Sixth, science brings us to new choices, new problems, and new circumstances for old problems. These modes of relevance between science and ethics may be labelled as: first, analytic; second, descriptive; third, comparative and causal; fourth, instrumental; fifth, behavioral; and sixth, problematic. Let us examine these briefly in turn.

But you may ask, what is the first of these, 'analytic', intended to mean? Surely not what is meant by 'mathematical analysis' nor again what is meant by 'experimental analysis' in a laboratory or in the field. It means 'philosophical analysis'; and this phrase presupposes, first, that philosophy is possible at all (which tough-minded scientists often deny, at least until, in their elder years, they retire to write books of philosophy), and further, it presupposes that philosophy has a distinctive method which differentiates it from science. The claim has been made that science supplants philosophy, and the historical record does affirm that the description of the world of nature, and of human society, is no longer to be the work of contemplative philosophers but rather that of patient, observing, and *active* scientific investigators. Physics and astronomy came from natural philosophy; psychology from the philosophy of mind; history (which I believe is a science in its own way) from the philosophy of culture; and sociology, scientific anthropology, economics and the rest of the social sciences have almost entirely supplanted the old speculative assertions of metaphysics.

What is left? What is the function of philosophy aside from the study of its own history, taken as an instructive and at times amusing recounting of prescientific illusions and prejudices? But such irony is doubly wrong. For one thing, our modern ways of scientific knowing should be scrutinized for their own metaphysical presuppositions, gaps, illusions, and prejudices, and in this their linkage with the past will be certain. And for another, what in the past was speculation about metaphysical abstractions often turns out to have been sharply reasoned analysis of human situations.

We have a profound, if simple, example in the history of that fundamental term of European metaphysics: Reason. Ultimately, all the philosophical idealists sought with *a priori* argument, to establish a belief in a rational world-order; and the materialists, both those who followed a dialectic and those who worked in the empiricist or pragmatic traditions, found the social content of metaphysical "Reason" by means of an elementary transformation which has revolutionary consequences: namely, if we first see that a demand for genuine individual happiness is the content of the metaphysical demand for a rational world-order, then we may comprehend that "the realization of reason is not a fact but a task."[3] The task is to materialize values.

If to make the world reasonable is our task, then, the philosopher must turn from philosophy to the material and causal sciences of physical and human nature on the one hand, and to critical social theory and practical concrete activity on the other. Not idealist utopian analysis, but concrete knowledge is needed for this task; not pure thought but scientific practice and thoughtful criticism. And so, as with Reason, the other metaphysical categories and speculations of the prescientific philosophers and poets, while set forth in strange, abstract, and even irritating languages, deserve respectful study and then reconstruction and reinterpretation.

After such sociological reconstruction of metaphysics (into humane social sciences), what *then* remains to philosophy? Perhaps only logic, but not mathematical logic, which has passed to the mathematicians; it would be a critical logic which remains, and indeed philosophy has only this critical role to play. Never really creative or discovering, philosophy is, as Hegel gloomily reminded himself, the wise owl who flies only at night, when the bright creative life of day has passed. And thus conceived, philosophy is ancillary both to science and to social realities, a Socratic gadfly still, a conceptual irritant, seeking to clarify meanings and techniques, to pose overlooked questions, to relate hitherto unrelated fields, to transcend egocentric and sociocentric predicament. Max Horkheimer was right, at the depth of Nazi savagery: More than logic, philosophy still has its old tasks, to explicate human knowledge and to criticize the human situation. And, at least in the recent English-speaking countries, the technique is called 'philosophical analysis.'

One of the philosophical analysts, John Wisdom, put the matter this way: "Philosophical progress does not consist in acquiring knowledge of new facts but in acquiring new knowledge of facts – a passage via inspection from poor insight to good insight."[4]

The rigorous analytic techniques of scientific philosophizing have been used to clarify the major issues and terms of ethical discussion too. A slogan for analysis, and its ally logical empiricism, has been one of 'rational reconstruction' of scientific theories and ethical doctrines alike. In ethics we must distinguish means from ends, and know when achieved ends become means to higher ends-in-view; we distinguish purposes, wishes, exhortations, commands, which have easily been confused. And above all, with Charles Stevenson, the meaning of ethical terms has been seen to involve two distinct components, that which implicitly asserts matters-of-fact (whether true, false, or probable in some degree) and that which expresses an emotional significance.[5] There is an 'emotive meaning' as well as a cognitive meaning. In the specialized division of philosophical labor, such analysis of ethics may hope to provide an answer to the puzzling and important question, "What sort of reasons can be given for normative conclusions, i.e., for conclusions about how men ought to behave under specific circumstances?"

Such analysis is scientific in spirit but *limited* to logical and linguistic techniques and to neutral abstractions. With *descriptive* ethics the scientific scope of ethical investigations widens greatly. The history of practical moralities and of ethical theories, and the investigations by ethnologists and sociologists, have by now provided an astonishing range to the phenomenon of moral consciousness. Not merely the breadth but also the subtlety of these moral phenomena provide scope to artists too, for beyond the questionnaires and surveys of spontaneous awareness, beyond social science, we can also recognize moral situations, moral responses, and moral responsibilities through novels and drama. Penetrating description of the distinctively human situation is the scientific function of literature, for it is a cognitive function: to reveal how the world feels to this man in this place. Through a union of subject and object in art, this feeling of a real world by a real subject (however fantastic the fiction may be, its truth remains genuine), we learn of ethical phenomena scientifically, through literature as through anthropology. As a final descriptive factor the existentialist and phenomenological philosophers have stressed the psychological phenomena of individual responsibility as against sociological or historical accounts of moral behavior.

A further stage which uses all the resources of social and psychological science is *comparative*. Examine the moralities of different times and places, different peoples, different generations within one people, differing economic and racial and linguistic circumstances within one society; explore the historical development and cross-cultural diffusion of moral ideals; and then certain conclusions may be drawn by such comparisons, if drawn scientifically rather than romantically and impressionistically. Thus, in the most direct though elementary way, we might be freed from ethnocentrism, that constricting vision which sees our Western ethical rules as a universal standard. We will be assisted in a more valid search for ethical invariants throughout the variety of conditions of life on earth without any *a priori* commitment to a specific set, or even to the existence of such. Further, by setting forth the "different ways in which human beings have tried to do similar jobs" we can be aided in a practical evolution of criteria for success. In MacBeath's phrase, we examine "experiments in living." Comparative studies offer the soundest basis for understanding the relation between ethical behavior and historical culture, and for appreciating the extraordinary inventive and creative possibilities for stable societies, indeed for appreciating the fine title of Gordon Childe's history of early societies, *Man Makes Himself.* We see that what might be legitimated as goals will be not only deeply rooted in the immediacy of daily life and recent events but also may transcend any local situation; a scientific insight which describes what may be called, with Paul Meadows, a situational dialectic.[6]

It is easy to see social and historical content in ethical theories, for they usually state and guide the interests of a class or church or other group. In this sense, comparative analysis reveals that ethical theories themselves are not just about values; they *are* values, since they *consist* of the goals and interests of certain groups.

Does a given ethical theory rationally justify a pattern of behavior? Is it true? Whether true or false, any operative pattern simply is the values of its adherents; and it provokes us to try causal-historical analysis, to a sociology of ethical ideas. Indeed the social functioning of ideas and the social determining of ideas are the crisscross intersection of science-as-truth and ethical values-as-interests, both investigated by scientific comparative studies.

Probably Hume and Marx in Europe and Veblen in America were the clearest voices to call for prompt social-scientific investigation into the causes of moral sentiments. What were the origins, the variables of function and displacement, the effects and complexities of truths and distortions in human goals? Remarkably, both Marx and Veblen combined a passion for empirical observation and hypothetical theory, that is, a scientific base for criticism, with an equally passionate moral critique of their own times. And when Marx said that the criticism of religion was the foundation of all criticism, he was calling direct attention to the hypothesis, as Reinhold Niebuhr phrased it, that every "claim of the absolute is used as a screen for particular competitive historical interests."

Now this hypothesis can only be established by empirical investigation. Indeed, there is an instructive example of the interaction of causal social science with ethics in the Marxist critique of religion: for, as the French priest Henri Chambre remarked aptly, Marx was not battling gods but idols and fetishes: "Marxist atheism does not believe in God but in men... it disputes the supremacy of things (idols) over men."[7] Hence we might say, the central *scientific* problem of the theory of knowledge, a problem whose solution is the basis of liberated thinking, is the nature of idol-producing thought, whether it is thought about nature or about values. And so once again, the relevance of science to values is established.

Just as scientists may study ethical codes, so they may study the specific ethical code of *scientists*. Let us idealize, but not the biographies of our teachers, friends and colleagues, nor even of great thinkers and experimenters. If we were to do that, I suppose the evidence would only show the same moral behavior that other men and women have had: superstitions, opportunism, wishful thinking, bandwagon fashionable enthusiasm, prejudice against the new or the foreign, all these and more can be found by studying the behavior of scientists with a slightly jaundiced eye; and of course we probably have our share of merit also. But what I mean to do is to recall the ideal which scientists do know about. The life of science, the ways we live with each other when we are true to the unimpeded knowledge-seeking goal of science, is characterized by an ethic with notable positive features. We form a democracy whose citizens decide what shall be the policy, what accepted as truth for guiding the commonwealth. The citizens who are the voters have an educational test, a special literacy test, perhaps the Ph.D. degree or another form of introduction to participation in the literature and discussion within the forum of ideas and techniques. In this situation, which is political despite its lack of formal structure or national boundaries, we have leaders but we choose them ourselves. And each of the leaders is ever open to challenge and then to replacement. Shall I say the leading ideas are replaced rather than the leaders?

But the matter ought to be put a bit more strongly. Science does have a unique fusion of obligation and rebellion. We cannot have science without realizing our obligation to past workers and current colleagues; and we know there will be no science if we do not honor our obligation to future scientists; we may choose not to have science, but insofar as we are scientists, we act with devotion to social cooperation. And we also seek social criticism since the very obligations to others which ensure the exsistence of science also ensure that mutual criticism occurs. Indeed the cumulative nature of science includes a recurring theme of revolt, since at the crisis points of investigation and interpretation, we do honor to our masters, teachers and heroes precisely by coming to reject them, learning to repudiate them. In its inner character, science is both objective and tentative, a temporary knowledge about persisting realities. And the political scheme is simple enough. It is a democracy, almost naively so in spirit, a democracy of the sort described by Thomas Jefferson in the early years of the United States. We scientists do not have formal elections, much less regularly scheduled ones, but we do have that plausibility of a true democratic practice: we give an idea, or a theory, or a technique, a test; we choose some men and their proposals and let them run the affairs which are on our agenda and after a while, we test them against our experience, and decide whether they are right or wrong, wise or foolish, the best likely or the least likely to succeed. And usually we replace them. And while Einstein replacing Newton does so with the greatest respect for Newton, there is yet psychological interest in the manner by which science utilizes the revolt of young generations against the old, for positive ends.

This scientific democracy has an additional quality which should demand respect: social collaboration is combined with extraordinary respect for individual work. If ever the conflicting claims of classic bourgeois individualism and classic socialism will be reconciled in a fully healthy society, it will reflect this beautiful legitimacy of *independence and inter-dependence* within science.

Along with this synthesis there are several other contrasts. In science, we combine subjective attitudes with objective demands, for example, esthetic delight with a demanding reasonableness. We combine beauty with utility, pride with modesty. We combine authority and leadership with private judgment and constant individual criticism. And we should treat each other with respect. Despite violation by pride and other weaknesses, the ethic of the international community of scientists is persistently known. *And the ethic of science is the democratic ethic of a cooperative republic.* Insofar as its own character may be a model for other human enterprises science has therefore a noble relation to ethical behavior. Indeed, if science teachers would bring to their students a conscious attention to these factors in the history and current practices of science, those students would be morally educated as well as technically.

Nevertheless praise of science must be limited. This internal ethic has not always characterized the external social relations of science. In fact, science is in society as a subordinate instrument of power. If we are to understand the place of science in society, we must turn to the history of the several social functions of science. Here milieu and context are decisive. Moreover, just as we have seen how glorious or terrifying has been, and yet may be, the impact of science on society, so also we must assess the fact that society has its decisive impact upon science. It provides resources, of course, material and human, but in addition science derives from society problems to be solved, ideas and metaphors to be used, techniques for inquiry, and, not least, standards of explanation (the notion of what will be accepted as an explanation in one time is so different from that which motivates thinking at another). Understanding the history of these external relations of science is essential to understanding our own science. And the interpretations of that history are not always favorable. They link science with society in such a way that science amplifies the social signals which stimulate it, and even exaggerates the worst of them.

One theorist, Husserl, understood modern science as a human "project," an attempt to treat Nature as an instrument, as an idealized extension of the utilization of prescientific craft tools. And this mathematical projection upon Nature, not only in Husserl's eyes, has evidently been successful. One might claim that the metaphor of Nature study as craft became literally true, for the human environment has been quite objectively changed. Nature can be, and has been, amenable to this projection. The mechanical may not be the only way to truth about Nature, but it has succeeded beyond expectation, beyond science in classic Greek society wherein there was curiosity and high intelligence but no drive for mastery over Nature and its processes.

To other thinkers, science shares the competitive and calculating spirit of modern society because it has been conceptualized in full parallel with the technological demands of industry. If so, science in its recent ways of thought and achievement will probably be judged less than universal in its scope and technique, less apt for those tasks of mankind which lie *outside* of, or which arise *after* full industrialization. But, how can one claim that science is restrictively characterized by the local epoch in which we now live? Briefly we can simply list the criteria for the empirical and theoretical success of science and see that they may be stated as criteria for a working, successful, well-engineered mass-production factory: *precision; simplicity; analysis of parts and components; impersonal, objective* and completely standardized workers and supplies; *economy* of thought, tools and materials; *efficiency* of administration and labor; *unified,* consistent, harmonious and *complete* development from raw materials to finished product; and finally *determinate* relations between input and output. Any theory for man today, any philosophy of science, would reflect upon these emphasized criteria in terms which in turn reflect the industrial foundations of our times.

This relation of industrial society to its technological culture, and in particular to the scientific ways of thinking, may be seen by yet another comparison. Thus, Simmel and Schumpeter have investigated how the modern money economy, as it developed out of feudal society, *reflected* and, more importantly, *promoted* a scientific manner of thought as efficiently as did the developing technology; indeed I think the economic preceded the technological in this relationship by nearly three centuries. A money economy encourages such scientific thought by providing a social context wherein the standards are *abstract, impersonal, objective, quantitative, rational,* and where money itself is *conserved* and as empty of intrinsic perceived properties as physical mass in the particles and structures of Newtonian mechanics was sometimes thought to be.

The question of the social connections of thought is subtle, to be sure, and we cannot be certain about much in the history and sociology of ideas. If science amplified and was amplified by its society as I have just suggested, even receiving part of its rational methodology from that society, it must also be plausible that our science received whatever may be irrational in this society. However, what is irrational in modern technology and in a money economy has not often been agreed upon. The complex configuration of technology–money–science, taken as the material base and background for Western history, produced a technological civilization which has always been partly dehumanized. And within this partially dehumanized society, we see a situation where working and living are generally divorced, where the culture and enlightenment of individual persons is often and easily manipulated and, in the extreme, socially coordinated, where the concept and consciousness of happiness is itself regulated.

But I believe an optimistic note may be sounded. We have, as yet, a society of only *incomplete* technology. *Complete* technology, the total use of precise and automatic scientific achievements, would be a precondition not necessarily, as the gloomy humanists often fear, of an extrapolation of the present human situation, namely a totally dehumanized society, but rather of its opposite: thorough and complete utilization of science and technology would be the necessary *precondition* of a humanized society founded upon dehumanized production. Necessary, of course; surely not sufficient. But such completed science will compel man to make certain moral and political decisions, for it will be true that freedom from necessary and dehumanizing labor, if taken as the character of a civilization and not just of a leisure class, poses fundamental problems of knowledge and morality afresh. If our science cannot actually lead us to new ethical 'experiments in living' it will be the new tradition from which the possibility of such experiments will grow. We are not faced with the single road to an end to a period of history, a new decline and fall, but rather with another option, a transition which is fashioning its own way.

Unfortunately, it is plausible to expect that our science, here and now, will continue to function as an ideology as well as a servant of technology, aside from its success as conqueror of greater ranges of Nature. Long connected with elites, and now accustomed to and requiring support by political powers, science alone can scarcely be asked to infuse our socialist, capitalist, or mixed societies with a humanistic spirit. Weak as the model may be, we are fortunate enough that its internal ethic is a humane one. Only as the habits of thought which arose in the millenia of necessary work begin to be supplanted in social and personal life, might we also expect that new patterns of posing problems, new ways of thought, new ways of curiosity, new feelings for working and playing, might also arise within a new way of science which is open to all men. The leisure we cherish, and yet sometimes we fear, will bring, I suspect, this new way of knowing. For we will have to pass, in the next social transformation, from the essential prerequisite for all previous struggles for life, a *science of work*, to another prerequisite, a *science of pleasure*. And if this phrase is too puzzling, if it seems to demand that esthetics become science, and science learn much from art, at least it suggests the quality of the transition which a peaceful world would have ahead.

We do not know under what conditions the social impact of science would be wholly humane. But we can say that science provides one great quality of the enlightened and humane life: objectivity. And why objectivity? Surely we have met those who criticize science as coldly objective? But if we wish to have a society of citizens without illusion, self-governing, self-evaluating, judging for themselves what is possible, what is impossible, what is probable, and how to choose the most likely course in the light of inadequate evidence – and that life of probabilities and uncertainties is the scientific life – then we must see objectivity taught throughout our culture. Objectivity, after all, is the ability to see the facts; it never need entail an inability to assess their human significance.

And in the realm of ethics, science can contribute this utterly valuable core of objectivity in a number of ways:

1. logical necessity which should be inherent in the procedures of validating hypotheses.
2. logical consistency among the ethical norms of a proposed system.
3. factual objectivity of the characterization given of the empirical features of human attitudes and conduct, which are the subject-matter for moral appraisal.
4. factual objectivity of statements about the relations of means to ends, and of conditions to consequences.
5. factual objectivity of statements about human needs, interests, and ideals as they have arisen in social context.
6. conformity of the proposed norms of a system with the basic biological, social, and psychological nature of man, the preservation of his existence, the facts of growth, development, and evolution.
7. factual objectivity of the claims of universality which certain moral norms may embody, as shown in the conscience of men in certain groups, and perhaps invariantly across all groups.

In summary, developments in science relate to ethics as follows: first, scientific discoveries may force ethical decisions; second, scientific discoveries may make certain ethical decisions possible; third, scientific methods may help men to rational control and hence ethical planning, of their lives and societies; fourth, science may offer a model of democratic living for those who may be persuaded to see.

But, last, science should be modest. It is part of our own times and also part of our own weaknesses, not an escape from them. And it must not be a decision-making device for us. When choices are difficult, it is rational to choose in the light of probability estimates, rational to avoid the intellectual cowardice of Hamlet in his demand for factual and ethical certainty. As Reichenbach reminded us in his analysis of

Hamlet's great soliloquy,[8] it takes more courage than *a priori* metaphysics knows in order to live by scientific criteria, that is, by empirical *and* rational standards. But we must never think that science gives us that moral courage. On the contrary, it is a moral choice to be courageous, an ethical act to be scientific.

NOTES

This lecture was first prepared for a conference on "Science and Technology and Their Impact on Modern Society", Herceg Novi, Yugoslavia (September 14–23, 1964).

1. See the pioneer study by Hans Kelsen, *Kausalität und Vergeltung* (translated into English as *Society and Nature*).
2. As recently as forty years ago, Reinhold Niebuhr entitled his somber and radical critique of modern social order *Moral Man and Immoral Society.*
3. See, Herbert Marcuse, *Reason and Revolution: Hegel and the Rise of Social Theory* (Humanities Press, Inc., New York), especially pp. 16–28 and 253–257: also R.S. Cohen, "Dialectical Materialism and Carnap's Logical Empiricism", in *The Philosophy of Rudolf Carnap* (ed. by P.A. Schilpp), (Open Court Publishing Co., La Salle, Ill.), pp. 99–158.
4. John Wisdom, "Logical Constructions", *Mind*, **42** (1933), 195.
5. C.L. Stevenson, *Facts and Values* (Yale University Press, New Haven, 1963) and earlier *Ethics and Language* (Yale University Press, New Haven, 1944).
6. Much of this and other paragraphs is derived from Abraham Edel's stimulating *Method in Ethical Theory* (Bobbs-Merrill, Indianapolis, 1963) and *Science and the Structure of Ethics* (University of Chicago Press, Chicago, 1961).
7. Henri Chambre, *Le Marxisme en Union Soviétique*, p. 334.
8. See Hans Reichenbach, *The Rise of Scientific Philosophy* (University of California Press, Berkeley, 1951), Chapter 15.

14 Beyond the Fact/Value Dichotomy*

Hilary Putnam

Several years ago I was a guest at a dinner party at which the hostess made a remark that stuck in my mind. It was just after the taking of the American embassy in Iran, and we were all rather upset and worried about the fate of the hostages. After a while, my hostess said something to the effect that she envied, or almost envied, the consolation that their intense faith in Islam must give the Iranian people, and that *we* are in a disconsolate position because "science has taught us that the universe is an uncaring machine."

Science has taught us that the universe is an uncaring machine: the tragic Weltanschauung of Nietzsche prefaced with "science has taught us." Not since Matthew Arnold talked so confidently of "the best that has been thought and known" has anyone been quite so confident; and Arnold did not think that science was all, or even the most important part, of "the best that has been thought and known." Those who know me at all will surmise correctly that I did not let this claim about what "science has taught us" go unargued-against, and a far-ranging discussion ensued. But the remark stayed with me past that almost eighteenth-century dinner conversation.

Some months later I repeated this story to Rogers Albritton, and he characterized my hostess's remark as "a religious remark." He was, of course, quite right: it *was* a religious remark, if religion embraces one's ultimate view of the universe as a whole in its moral aspect; and what my hostess was claiming was that science has delivered a new, if depressing, revelation.

One popular view of what is wrong with my hostess's remark was beautifully expressed by Ramsey, who closed a celebrated lecture with these words:

* Chapter 9 in J. Conant (ed.), *Realism with a Human Face* (Cambridge, MA: Harvard University Press, 1990), pp. 135–141. © 1982 Universidad Nacional Artonoma de Mexico.

My picture of the world is drawn in perspective, and not like a model to scale. The foreground is occupied by human beings and the stars are all as small as threepenny bits. I don't really believe in astronomy, except as a complicated description of part of the course of human and possibly animal sensation. I apply my perspective not merely to space but also to time. In time the world will cool and everything will die; but that is a long time off still, and its present value at compound discount is almost nothing. Nor is the present less valuable because the future will be blank. Humanity, which fills the foreground of my picture, I find interesting and on the whole admirable. I find, just now at least, the world a pleasant and exciting place. You may find it depressing; I am sorry for you, and you despise me. But I have reason and you have none; you would only have a reason for despising me if your feeling corresponded to the fact in a way mine didn't. But neither can correspond to the fact. The fact is not in itself good or bad; it is just that it thrills me but depresses you. On the other hand, I pity you with reason, because it is pleasanter to be thrilled than to be depressed, and not merely pleasanter but better for all one's activities.[1]

If one has seen a little more of life than the 22-year-old Ramsey who delivered this lecture, and if one has faced the beastliness of the world (not just the wars and the mass starvation and the totalitarianism – how different our world is from Ramsey's England of 1925! – but the beastliness that sensitive novelists remind us of, and that even upper-middle-class life cannot avoid), one *is* more likely to be depressed than "thrilled." Also, that Ramsey himself died when he was only 27 depresses *me*.

But notice – I think it comes out even in the bit of Ramsey's lecture that I quoted, and it certainly comes out in the phrase "science has taught us" – notice how sure we are that we are *right*. Our modern revelation may be a depressing revelation, but at least it is a *demythologizing* revelation. If the world is terrible, at least we *know* that our fathers were fools to think otherwise, and that everything they believed and cherished was a lie, or at best superstition.

This certainly flatters our vanity. The traditional view said that the nature of God was a mystery, that His purposes were mysterious, and that His creation – Nature – was also largely mysterious. The new view admits that our knowledge is, indeed, not *final*; that in many ways our picture will in the future be changed; that it can everywhere be superseded by new scientific discoveries; but that in broad outlines we *know* what's what. "The universe is an uncaring machine," and we are, so to speak, a chance by-product. Values are just *feelings*. As Ramsey put it elsewhere in the same lecture, "– Most of us would agree that the objectivity of good was a thing we had settled and dismissed with the existence of God. Theology and Absolute Ethics are two famous subjects which we have realized to have no real objects."

I think that this consolation to our vanity cannot be overestimated. Narcissism is often a more powerful force in human life than self-preservation or the desire for a productive, loving, fulfilling life, as psychologists have come to realize: I think that, if someone could *show* that Ramsey's view is wrong, that objective values are *not* mythology, that the "uncaring machine" may be all there is to the worlds of physics and chemistry and biology, but that the worlds of physics and chemistry and biology are not the only worlds we inhabit, we would welcome this, *provided* the new view gave us the same intellectual confidence, the same idea that we have a superior method, the same sense of being on top of the facts, that the scientistic view gives us. If the new view were to threaten our intellectual pride, if it were to say that there is much with respect to which we are unlikely to have more than our fathers had – our fallible capacity for plausible reasoning, with all its uncertainty, all its tendency to be too easily seduced by emotion and corrupted by power or selfish interest – then, I suspect, many of us would reject it as "unscientific," "vague," lacking in "criteria for deciding," and so on. In fact, I suspect many of us will stick with the scientistic view even if it, at any rate, can be *shown* to be inconsistent or incoherent. In short, we shall prefer to go on being depressed to losing our status as sophisticated persons.

Such a new view is what I try to sketch and defend in my book *Reason, Truth, and History*.[2] I only sketch it, because it is intrinsic to the view itself that there isn't much more one *can* do than sketch it. A *textbook* entitled "Informal Non-Scientific Knowledge" would be a bit ridiculous. But I feel sure that it is, in its main outline, more on the right track than the depressing view that has been regarded as the best that is thought and known by the leaders of modern opinion since the latter part of the nineteenth century. This essay is, then, a short sketch of something that is itself a sketch.

W. V. Quine has pointed out that the idea that science proceeds by anything like a formal syntactic method is a myth. When theory conflicts with what is taken to be fact, we sometimes give up the theory and sometimes give up the "fact"; when theory conflicts with theory, the decision cannot always be made on the basis of the known observational facts (Einstein's theory of gravitation was accepted and White-head's alternative theory was rejected fifty years before anyone thought of an experiment to decide between the two). Sometimes the decision must be based on such desiderata as simplicity (Einstein's theory seemed a "simpler" way to move from Special Relativity to an account of gravitation than Whitehead's), sometimes on conservativism (momentum was redefined by Einstein so that the Law of the Conservation of Momentum could be conserved in elastic collisions); and "simplicity" and "conservativism" themselves are words for complex phenomena which vary from situation to situation. When apparent observational data conflict with the demands of theory, or when simplicity and conservativism tug in opposite directions, trade-offs must be made, and there is no formal rule or method for making such trade-offs. The decisions we make are, "where rational, pragmatic," as Quine put it.

Part of my case is that *coherence* and *simplicity* and the like are themselves *values*. To suppose that "coherent" and "simple" are themselves just emotive words – words which express a "pro attitude" toward a theory, but which do not ascribe any definite properties to the theory – would be to regard *justification* as an entirely subjective matter. On the other hand, to suppose that "coherent" and "simple" name *neutral* properties – properties toward which people may have a "pro attitude," but there is no objective rightness in doing so – runs into difficulties at once. Like the paradigm value terms (such as "courageous," "kind," "honest," or "good"), "coherent" and "simple" are used as terms of praise. Indeed, they are *action guiding* terms: to describe a theory as "coherent, simple, explanatory" is, in the right setting, to say that acceptance of the theory is *justified*; and to say that acceptance of a statement is (completely) justified is to say that one ought to accept the statement or theory. If *action guiding* predicates are "ontologically queer," as John Mackie urged, they are nonetheless indispensable in epistemology. Moreover, every argument that has ever been offered for noncognitivism in ethics applies immediately and *without the slightest change* to these epistemological predicates: there are disagreements between cultures (and within one culture) over what is or is not coherent or simple (or "justified" or "plausible," and so forth). These controversies are no more settleable than are controversies over the nature of justice. Our views on the nature of coherence and simplicity are historically conditioned, just as our views on the nature of justice or goodness are. There is no neutral conception of rationality to which one can appeal when the nature of rationality is itself what is at issue.

Richard Rorty[3] might suggest that "justified relative to the standards of culture *A*" is one property and "justified relative to the standards of culture *B*" is a different property. But, if we say that it is a *fact* that acceptance of a given statement or theory is "justified relative to the standards of culture *A*," then we are treating "being the standard of a culture" and "according with the standard of a culture" as something *objective*, something itself *not* relative to the standards of this-or-that culture. Or we had better be: for otherwise, we fall at once into the self-refuting relativism of Protagoras. Like Protagoras, we abandon all distinction between *being right* and *thinking one is right*. Even the notion of a culture crumbles (does every person have his or her own "idioculture," just as every person has his or her own idiolect? How many "cultures" are there in any one country in the world today?).

The fact is that the notions of "being a standard of a culture" and "being in accord with the standards of a culture" are as difficult notions (epistemically speaking) as we possess. To treat these sorts of facts as the ground floor to which all talk of objectivity and relativity is to be reduced is a strange disease (a sort of scientism which comes from the social sciences as opposed to the sort of scientism which comes from physics). As I put it in *Reason, Truth, and History*, without the cognitive values of coherence, simplicity, and instrumental efficacy we have no world and no facts, not even facts about what is so *relative to* what. And these cognitive values, I claim, are simply a part of our holistic conception of human flourishing. Bereft of the old realist idea of truth as "correspondence" and of the positivist idea of justification as fixed by public "criteria," we are left with the necessity of seeing our search for better conceptions of rationality as an intentional activity which, like every activity that rises above the mere following of inclination or obsession, is guided by our idea of the good.

Can coherence and simplicity be restricted to contexts in which we are choosing between *predictive* theories, however? Logical positivism maintained that nothing can have cognitive significance unless it contributes, however indirectly, to predicting the sensory stimulations that are our ultimate epistemological starting point (in empiricist philosophy). *I* say that *that* statement itself does not contribute, even indirectly, to improving our capacity to predict anything. Not even when conjoined to boundary conditions, or to scientific laws, or to appropriate mathematics, or to all of these at once, does positivist philosophy or any other philosophy imply an observation sentence. In short, positivism is self-refuting. Moreover, I see the idea that the only purpose or function of reason itself is prediction (or prediction plus "simplicity") as a prejudice – a prejudice whose unreasonableness is exposed by the very fact that *arguing for it* presupposes intellectual interests unrelated to prediction as such. (That relativism and positivism – the two most influential philosophies of science of our generation – are *both* self- refuting is argued in one of the chapters of my book, the one titled "Two Conceptions of Rationality," by the way.)

If coherence and simplicity are values, and if we cannot deny without falling into total self-refuting subjectivism that *they* are objective (notwithstanding their "softness," the lack of well-defined "criteria," and so forth), then the classic argument against the objectivity of ethical values is *totally* undercut. For that argument turned on precisely the "softness" of ethical values – the lack of a noncontroversial "method," and so on – and on the alleged "queerness" of the very notion of an *action guiding fact.* But *all* values are in this boat; if *those* arguments show that ethical values are totally subjective, then cognitive values are totally subjective as well.

Where are we then? On the one hand, the idea that science (in the sense of exact science) exhausts rationality is seen to be a self-stultifying error. The very activity of arguing about the nature of rationality presupposes a conception of rationality wider than that of laboratory testability. If there is no fact of the matter about what cannot be tested by deriving *predictions*, then there is no fact of the matter about any philosophical statement, including *that* one. On the other hand, any conception of rationality broad enough to embrace philosophy – not to mention linguistics, mentalistic psychology, history, clinical psychology, and so on – must embrace much that is vague, ill-defined, no more capable of being "scientized" than was the knowledge of our forefathers. The horror of what cannot be "methodized" is nothing but method fetishism; it is time we got over it. Getting over it would reduce the intellectual *hubris* that I talked about at the beginning of this essay. We might even recover our sense of *mystery*; who knows?

I am fond of arguing that popular philosophical views are incoherent or worse. In *Reason, Truth, and History* I also try to show that the two most influential theories of truth – the empiricist theory (it's all a matter of getting the "sense data" right – note that Ramsey endorsed that one in the quote I gave earlier) and the correspondence theory (there is some special "correspondence" between words and objects, and *that* is what explains the existence of reference and truth) – are either unexplanatory or unintelligible.

So far, what I have said could be summarized by saying that if "values" seem a bit suspect from a narrowly scientific point of view, they have, at the very least, a lot of "companions in the guilt": justification, coherence, simplicity, reference, truth, and so on, all exhibit the *same* problems that goodness and kindness do, from an epistemological point of view. None of them is reducible to physical notions; none of them is governed by syntactically precise rules. Rather than give up all of them (which would be to abandon the ideas of thinking and talking), and rather than do what we are doing, which is to reject some – the ones which do not fit in with a narrow instrumentalist conception of rationality which itself lacks all intellectual justification – we should recognize that *all* values, including the cognitive ones, derive their authority from our idea of human flourishing and our idea of reason. These two ideas are interconnected: our image of an ideal theoretical intelligence is simply a *part* of our ideal of total human flourishing, and makes no sense wrenched out of the total ideal, as Plato and Aristotle saw.

In sum, I don't doubt that the universe of *physics* is, in some respects, a "machine," and that it is not "caring" (although describing it as "uncaring" is more than a little misleading). But – as Kant saw – what the universe of physics leaves out is the very thing that makes that universe possible for us, or that makes it possible for us to construct that universe from our "sensory stimulations" – the intentional, valuational, referential work of "synthesis." I claim, in short, that without *values* we would not have a *world*. Instrumentalism, although it denies it, is itself a value system, albeit a sick one.

15 Darwin and Philosophy*

Marjorie Grene

The theme of Darwin's influence on philosophy has been a recurrent one, notably in John Dewey's lecture in the semi-centennial year and again in J.H. Randall, Jr.'s defense of Dewey in the centennial year of 1959.[1] Randall was, in effect, defending Dewey against the charge of another centennial essayist, J.S. Fulton, who writes:

> An essay on the philosophy of evolution in the century since the publication of Darwin's *Origin of Species* can be written in two sentences. By the end of the first fifty years, everybody in the educated world took evolution for granted, but the idea was still intellectually exciting and its philosophical exploitation was entering upon its period of full maturity. By the end of the next fifty years, evolution belongs to 'common sense' almost as thoroughly as the Copernican hypothesis and other early landmarks of the scientific revolution; but the idea is no longer exciting, and evolutionary philosophy is out of fashion.[2]

Now it seems obvious, at first glance at any rate, that Randall is wrong and Fulton right. In general, we don't talk evolution in philosophy these days. Why not? Let us ask once more, what has been the destiny of Darwinian theory in connection with philosophy, both professional and popular? This entails three questions. First, what is the fundamental move of Darwinian explanation? Second, how does it fit into the history of nineteenth and twentieth century thought? Third, what is its relation to present-day problems in philosophy?

First, what Darwin did was both grand and simple. He showed how the myriad features of living things which fit them to cope with their environments, their ways of being *adapted*, could be explained, not as designed *for* this purpose, but as produced by ordinary cause-and-effect relations. Thus what *looks* purposive becomes explicable in a perfectly humdrum causal fashion, and life on earth takes its place, as the heavens had done earlier, in a plain naturalistic cosmology. There had

* Chapter XI in R. Cohen and M. Wartofsky (eds.), *The Understanding of Nature: Essays in the Philosophy of Biology* (Dordrecht: D. Reidel, 1974), pp. 189–200.

of course been anticipations of Darwin's theory, not to mention its concurrent formulation by Wallace; there is no time to go into that question here – nor is it pertinent, for it was the *Origin* that put the theory on a new and imposing scale – backed by all the experience of a great naturalist – and, as C.F.A. Pantin has demonstrated, put it in *deductive* form.[3] Given the obvious facts of heredity and variation and the Malthusian parameter of population increase, a 'struggle for existence' (metaphorically, not literally, understood) must follow, and from this in turn must follow the survival of the fitter (not the fitt*est*), that is, natural selection as the process by which relatively superior adaptations arise and relatively inferior ones disappear. What looks purposive is explained causally, and by a logically compelling argument.

Moreover, the situation is still basically the same when you build in first genetics and then biochemistry. When, after three decades of opposition or neglect by eminent geneticists (roughly 1900 to 1930), the theory of Natural Selection was reconciled with Mendelism in what has come to be called the synthetic theory, it was genetics that was assimilated to the theory of Natural Selection as the underpinning it had previously lacked. For the measure of changing gene frequencies in interbreeding Mendelian populations *is* the measure of Natural Selection in the sense of such causal organism-environment interactions as produce heritable results (beginning of course with the internal environment within the organism, indeed, within the genome itself). The biochemical revolution, moreover, has only served to confirm once more the power of Darwinian thinking. There has been some talk of 'neutral mutations' (by adherents of the King Crowe school), but the random occurrence of single base pair substitutions in DNA is easily subordinated by orthodox evolutionists to the framework of an account of changing selective pressures. Of course mutations are random with respect to selection and so may be 'neutral' as well as 'deleterious' or 'useful'. But there are always plenty of them available, and which ones persist is dictated by the causal processes summarized as Natural Selection. However precise biochemical analysis becomes, in other words, the Darwinian explanation (of the way adaptive structures are favored and maintained) is confirmed rather than undermined, or even modified, by this increased precision.

One could find countless examples of modern, yet classically Darwinian, work, the whole literature of population genetics, for example. Take as one instance of this kind of pure Darwinian thinking, for instance, E. O. Wilson's application of ergonomic theory to the problem of the evolution of sociality in insects.[5] Darwin and others after him had considered the phenomena of altruistic behavior in the social insects stubbornly resistant to explanation in terms of natural selection. But today, with advancing knowledge of insect behavior, its interpretation in population-genetical terms, and the application of a benefit/cost calculus analogous to that of game theory, these same phenomena can not only be assimilated to a classically Darwinian framework of explication; they provide a striking instantiation of its explanatory power. At the level of the colony, the 'altruism' of the worker bee or the slave ant turns out to be the highly probable outcome of selective processes. There is still plenty of controversy in this area, but even the arguments are purely Darwinian. Nobody considers that anything *but* selection could be the controlling concept in an evolutionary debate.

One other example: on a more massive scale, and assimilating a vast range of recent data from such fields as biochemical genetics, biochemical evolution and Precambrian paleontology, Lynn Margulis, in her *Origin of Eukaryotic Cells*, has presented an impressive argument for the endosymbiotic theory of eukaryote evolution. She supports, with modifications, Whittaker's five-kingdom taxonomy, from Monera, then Protists (the earliest eukaryotes, which in this classification excludes bacteria and blue green algae), both of these originating in the Precambrian, and then the three great 'modern' kingdoms, Plants, Fungi and Animals.[6] Dr. Margulis's reasoning is indeed 'speculative', a suspect quality to most modern biologists, and revolutionary for our concepts of taxonomy and phylogenesis. But my point here is that her argument uses a very wide range of biochemical and genetic knowledge in the service of straightforward, classically Darwinian, concepts: heredity (whose origin one can now also hypothesize, as Darwin could not), variation and multiplication, resulting in the persistence of certain forms of life rather than others in appropriate environments.

So much for my first point: the structure of Darwinian explanation. Now let us look, equally sketchily, at the major trends of philosophical thought in the nineteenth and early twentieth centuries and see where Darwin's achievement was influential in shaping (or at least giving a special expression to) the philosophical tradition. (Perhaps I should say here parenthetically what I shall have to stress in the last part of my argument, that scientific discoveries, however important, do not in themselves 'solve' philosophical problems. They sometimes generate meta-problems by their own conceptual confusions or paradoxes, and they sometimes – and that is the case that chiefly interests us here – put constraints on philosophical reflection by favoring certain very general metaphysical or cosmological ideas. But back to philosophy!)

The history of philosophy in the nineteenth century and longer –
shall we say till 1927 and *Sein und Zeit* on the continent of Europe and
until the influence of late Wittgenstein in England in the late 1930s and
thereafter (we English speakers are notoriously slow in catching on, or
catching up) – the history of philosophy in this period is the history of
the destiny of Kant's critical system. Kant had limited our theoretical
powers to appearances, to the phenomenal only. According to him, we
can organize *a priori*, and therefore *know*, only the appearances of
sense, inner as well as outer. Practically, however, he held, we know, in
some non-theoretical sense of 'know', that we are free beings with a
duty to follow autonomously a moral law. The tension between these
two aspects of our nature is painful, even for Kant; but they nevertheless harmonize for him in the light of God's creation and of our fallen
state.[7] The secularization of thought in the nineteenth century, therefore – which of course also set the stage for Darwin – loosened the tie
between the theoretical and practical as Kant had conceived them. It
left us on the one hand with a 'scientific' or better 'scientistic' tradition
that was both mechanistic and phenomenalist, that is, believing that
causal explanation is universal but that we can never penetrate through
phenomenal cause and effect relations to the reality behind appearance.
And on the other hand, *via* Fichte and Romanticism, it left us with a
voluntarist tradition that stressed, in one form or another, the unique
reality, however dialectically expressed, of Will or Act. These two
demi-traditions culminated, one, in the philosophy of the Vienna Circle, and the other in existentialism. The world came to be seen not, as
Schopenhauer would have it, as Will and Idea, but alternatively as Will
or Machine. Marxists, among others, I dare say, will object to this dichotomy, for Marx supposedly reconciled science and dialectic in the
shape of dialectical materialism. In reply I would have to argue that all
dialectical philosophies are romantic, or at least that the attempted
gap-bridging failed. But even supposing I could succeed in this peril-
ous venture, I cannot attempt it here. For the sake of my present argument let me, with a *small* apology, permit my dichotomy to stand:
either machine or will, a necessary nexus of mere phenomena, or the
upsurge of reality as pure act. The question here is: where did the impact of Darwin fit into this story? And the answer is: on both sides.
From, say, about 1870 to 1930, in both styles of philosophizing, different
though they were, evolutionary thinking reigned supreme. Not, indeed,
that Darwin was the *only* influence here: on both sides at least two
other factors were important: the development of historical method
and nineteenth century progressivism. The latter in its more 'scientific'

version celebrated the triumphs of technology and of liberal politics; but in the Romantic tradition as well the doctrine of progress had its influence: see for instance Victor Hugo's poetic praise of progress.[8] But the acceptance of organic evolution was certainly a major factor in this development. On the science-oriented side, there was in England the popular developmental theory of Spencer, which both antedated and supported the acceptance of the *Origin*. In Germany there was the mechanistic monism of Haeckel, whose simple doctrine was certainly swallowed whole, along with Spencer *and* Darwin, by the educated middle class in America (I can attest to this from my own early recollections of popular evolutionism). Then, a little later, on the whole, but equally influential, was the Romantic version: the emergence theories of Bergson, Alexander, Whitehead and Collingwood, which were still *just* in vogue when I was a graduate student in the early thirties. In fact in 1933–34 I heard Whitehead's lectures in cosmology, which represent the grand culmination of this tradition.

Yet Fulton is right: all that is over. So I want to ask, thirdly, and this is my major question: what remains of evolutionary theory today? What is Darwin's present influence on our views of nature and of man? In general, of course, it is correct to say that evolutionary thinking has been assimilated to our common sense view of nature, and of man in relation to nature, so much so that it is, because wholly uncontroversial, philosophically uninteresting. On three of four counts, any philosophical argument on any problem has to take the outcome of the Darwinian revolution as among its unquestioned premises.

First, Darwinism (again along with other factors, historical and anthropological relativism, for example) has forced us to recognize mutability, the omnipresence of change. Not, as some people have naively argued, that change is good in itself, but simply that it *is*. The acceptance of organic evolution eliminates an Aristotelian theory of nature, dependent as such a theory must be on the existence of permanent natural kinds. Aristotle had of course been banished from physics by the first scientific revolution, but religious sanctions combined with the plain facts of organic diversity had so far kept the new naturalism from incorporating biology. There seemed no way to bring the staggering variety of organic forms and the apparent teleology of organized beings under simple physical principles. So Kant predicted there could be no 'Newton of a blade of grass'. But Darwin not only unified living nature through transformist principles – many others had tried that too – he did it, as we have seen, by the simple but sweeping move of rendering the seeming purposiveness of organic structures and functions susceptible of causal explanation. Indeed, it was *through* that move, through the concept of natural selection as explanatory of organic change, that the *Origin* persuaded so many of the fact of evolution, precisely because it brought what had before seemed simply mysterious within the purview of scientific investigation. Yet in so doing, paradoxically, the Darwinian revolution undermined not only Aristotelian species, but the simple eternity of the Newtonian world itself. It put process before permanence, development before structure. True, this emphasis is not now, explicitly at least, current in philosophical discussion, at least in the English-speaking world. The slogan that we should 'take time seriously' has now a quaintly old-fashioned ring. But if, without acknowledging it, we *have* a metaphysic (and we always do), it is a metaphysic of mutability.

Secondly, with the permanence of Aristotelian nature we have also abandoned metaphysical necessity in favor of a recognition of fundamental contingency – and here too evolution, along with historiography as well as history itself, has played a role. Everything there is, we must acknowledge, might have been otherwise. This again is paradoxical: for the fact that we can seek a causal explanation of any natural event (and so everything is in some sense necessary) seems tied to the insight that everything is contingent, and so, in another sense, nothing is necessary. At each stage of evolution there are chances of success and risks of failure. Each new population appears as the heir of past successes and the replacement of past failures. But successes and failures are always the effects of processes that *could* have gone another way. Neither God's Providence nor the all-embracing Natura Naturans of Spinoza, let alone Hegelian dialectic, which attempts to synthesize the two, can stand against this basic principle. Again, it is not talked of in philosophy – though some discussions of determinism perhaps reflect it – but surely we have it in our bones.

The firmest lesson of Darwinism for metaphysics, however – thirdly – is of course the lesson of our own animal nature, our demotion from supernatural support to a place in nature comparable to that of any other living thing. By now, indeed, not only the doctrine of the 'descent of man', but many lines of research in many biological fields serve to confirm this change in our view of ourselves. Indeed, so pervasive is this change that it is often misapplied, even by distinguished scientists, let alone by pure popularizers. (For a discussion of these misuses, see Leon Eisenberg's article in *Science* on 'The *Human* Nature of Human Nature'.[9]) The most intimate influence of biological research on philosophy at present, I think, comes from neurophysiology. Herbert

Fingarette, for example, in presenting his theory of self-deception which supplements philosophical argument with empirical support from work on commisural patients.[10] Or more sweepingly, Richard Rorty's argument for identity theory as against functionalism in the philosophy of mind seems to rest on a general confidence in the explanatory power of physiology or physiological cybernetics.[11] But be that as it may, certainly the attempt to overcome Cartesian dualism, which still remains, alas, the major philosophic task of the waning twentieth century, found its first massive support in the Darwinian theory. No doubt about it: whatever kinds of strange fish we are, we are organic beings, not half bodies and half immortal souls.

There is a fourth point, however, on which the lesson of Darwinism is more ambiguous. In some contexts, as we have seen, evolutionary theory and the nineteenth century belief in progress were mutually reinforcing. But *is* evolution progressive? From the beginning the views of scientists have been conflicting. Darwin is said to have written in the margin of this copy of Chambers' *Vestiges*, 'Never speak of higher and lower in evolution' and T.H. Huxley's friend Kingsley wrote in *The Water Babies* about a strange land where evolution went backwards – from man to amoeba, so to speak, as in terms of environmental pressures interacting with variable heritable structures, it very well might. Huxley's grandson, however, thinks otherwise, and so do many leading evolutionists today: Simpson, Dobzhansky, Stebbins, for example, not to mention others like Kimura or Thoday: but all in terms of differing criteria of 'progress'. (The various views are well summarized, as well as criticized, in George Williams' *Adaptation and Natural Selection*.)[12] For the Romantic evolutionists, of course, the rise to higher levels, somehow propelled by tendencies in life itself, was the very heart of evolutionary truth. But what is 'progress' on a cosmic scale? Can evolution be both 'opportunistic', as in Darwin's terms, it is, and inherently progressive? Or is 'progress' what we call it when *we* look back and see the billions of years of living history leading to – ourselves? There are, indeed, in the neobiosphere, innumerable populations of organisms structured in terms of a greater number of levels of self-regulation than were the Monera of the pre-Cambrian or even much later forms. This is clearly true of multi-cellular eukaryotes as against protists, again of organisms possessing a central nervous system as against their forerunners, and so on. Stebbins gives a table of such levels of organization, for example, in his *Basis of Progressive Evolution*; others may wish to enumerate such levels somewhat differently.[13] There *is* progress, somehow, from blue-green algae and bacteria to mice and monkeys, quite aside from men. But what in such contexts does the concept of progress mean exactly, and how do we measure what it represents? On this question, it seems to me, the message of Darwinian thinking is much less clear. I am inclined, indeed, at the moment, to think it safer to equate evolution with the theory of organic change, and to reserve 'higher' and 'lower' or their analogues for the analysis of systems once they exist, not for the causal explanation of their origin. Living systems *are* hierarchically organized, and hierarchically systems can be richer or poorer in their hierarchical structure. But these systems-theoretical considerations perhaps serve chiefly to confuse when we try to inject them into the evolutionary account itself. Once life originates (and that's another story), we have at every period already *some* self-regulating systems, mutations which permit and environmental conditions which determine the rise and spread of new such systems. Sometimes a new level of homeostasis results, sometimes not, but a new level of organization demands no new ground rules for the historical account. (On this point Slobodkin's essay on 'The Strategy of Evolution' seems to me definitive.)[14]

So far I have been talking about the influence of Darwinian thought on what may roughly be called contemporary metaphysics. There are other areas in philosophy, however, where we find, not so much the assimilation as unspoken presuppositions of the implications of evolutionary theory, as the explicit application of the theory itself to philosophical problems. Let me look briefly in conclusion at three of these areas: evolutionary ethics, functionalism in the theory of the social sciences, and evolutionary epistemology. All three, I'm afraid, are extensions of a powerful scientific theory to areas where it is misplaced, where its application serves to obscure the problem and where, therefore, instead of explaining, it explains away.

First, evolutionary ethics. Unless we go back to Dewey this is in any case an evolutionist's, not a philosopher's theory, and from a philosophical point of view hard to take seriously. What it *is* to be an ethical animal, as Waddington puts it, is not explained, as he tries to explain it, by the fact that we came into being, as animals, through mutation and natural selection, but with this strange propensity to moralize.[15] Nor is it explained by analogy with 'evolutionary progress', itself, as we have seen, a muddled concept. The problem of ethics is not how apparent norms (or ends) arose, nor how biological processes run parallel (if they do) to processes of ethical judgment. The problem of ethics is: what real norms (or ends) are: norms that are human artifacts, but real to us, as real as anything in our lives can be. Ethics, in other words, is a critical inquiry about (1) the general character of norms of conduct and goals of intentional action, and (2), and more fundamentally, about what it *is* to be an ethical norm or an end of intentional action at all. We all distinguish somehow right from wrong and good from evil. What kinds of claims are we making when we do this? Evolution does set constraints on our answers: we are not, evolution tells us, or ought not to be, appealing to supernatural revelation for our judgments of 'ought' and 'good'. But apart from depriving us of this simple if superstitious account, the story of our evolution, the explanation of how we originated from earlier anthropoids, cannot in itself clarify critically, as philosophical ethics has to do, the *nature* of moral values or of responsible choice. We are animals, but culture-dependent animals, tradition-dwelling animals, and it is inherent in culture, in the human variety of tradition, to work by self-imposed, historically received, but individually authorized, standards. What this means, is the problem of ethics, not how it all began – the problem of human evolution. Evolutionary studies, like Portmann's comparative study of early mammalian development, may indeed have an interesting bearing on this question, but in themselves they do not even put the ethical question, let alone answer it. Only if we decide, as Hobbes did, that adaptation for survival, and absolutely nothing else, is the sole standard of rational choice, can we equate the answer to these two questions. But that is already to *have* an ethical theory, which *then* turns out to be convergent with the theory of natural selection. As an ethical theory, it needs philosophical grounds, in terms of a theory of the nature of moral standards, not of the origin of the creatures who adopt, and follow, them.

A similar confusion infects functionalism in the social sciences. Here there is an apparent homology with the evolutionary situation which leads to a parallel explanation that begs the very questions which social philosophy needs to raise. The claws of male isopods, let's say, are adapted for mating. Their origin can be causally explained in terms of Darwinian theory as making more probable for the organisms bearing the appropriate genes the production of greater numbers of organisms bearing the same genes. Survival is here the only criterion of 'success' – that's what evolution is about. Is it the same with social institutions? It might seem so. For example, Max Gluckman has described rebellion in some African societies as a device to decrease dissatisfaction and so keep the society going.[16] Thus, as with the rereading of apparent 'purposiveness' in causal terms by Darwinian biology, what looks like intentional behavior is turned back to receive a purely causal explanation (though the continuity here is cultural rather than genetic – already a substantive difference). Moreover, in these terms, again, as in every evolutionary context, survival is the sole criterion of excellence. But will this really do? Don't we want to be able to *compare* societies, to say that sometimes the superior society succumbs to its inferior? As in the case of evolutionary ethics, so here, the problem of standards, the question on what grounds we judge what makes one functioning structure (in this case social structure) better than another – the philosophical question is unseen or prejudged before it has even been raised.

What, finally, of evolutionary epistemology?[17] Isn't natural selection here a proper paradigm for the growth of knowledge? Don't those theories survive which predict successfully and those go to the wall that don't? So don't we judge claims to knowledge in terms of something analogous to the algebra of gene-frequencies: a greater probability of leaving descendants, that is, in this case, scientists who believe the theory, in the future? But remember that the algebra of gene-frequencies, like the biochemistry of DNA, is the *carrier* of Darwinian explanation, the one its formalization, the other its material; neither *is* the explanation itself. As an explanatory principle natural selection is simply shorthand for a vast nexus of cause and effect relations between organism and environment. But a causal explanation, a when-then explanation, of cognitive achievements, omits, and must omit, any account of their claims as *knowledge*. Hence Dewey's confusion in calling mathematics an experimental science because children have to learn it. Hence also, I must say with all due respect, the confusion inherent, for all its great achievements, in the very concept of 'genetic epistemology'.

That is not to say that there is not a real and significant continuity between cognitive activities and other behaviors on the part of living things. All knowledge is orientation of some kind (but even that is more an ecological than an evolutionary statement). What makes us claim, however, that some sorts of orientation in our world constitute *knowledge* is not simply their success in guiding us in our surroundings, as the plankton is oriented in the streaming ocean or the robin to the red breast of its mate. Illusions, if systematic, can also guide successfully. When we say we *know* something, we are not saying it is true simply because it will survive, but contrariwise, it will survive because it is true. We are claiming, not just that we have made a beautiful theory, a theory so elegant that it will continue to fool people for a long time to come, like advertisers crying their wares, we are saying that we have found, in some limited respect and within the canons of some discipline in which we have competence, the way things work: we are claiming, sometimes in anticipation of our theory's empirical consequences, sometimes by virtue of its very elegance, that we are in contact with reality. *A scientist*, Norman Campbell said, is a man who passionately believes that nature will conform to his intellectual desires.[18] This claim to truth, this gamble on being in contact with reality – of course it is a gamble and we may always be mistaken: another lesson of Darwinism if you like: fallibility follows from the acceptance of metaphysical contingency – this risky claim, then, is precisely what epistemology has critically to examine. By what standards, of accuracy, of objectivity or disinterestedness, of systematic relevance, do we judge statements to be true or false? On what grounds are such standards supported? Again, it is this critical normative discipline that is the work of philosophical reflection. To try to make a scientific theory, even, or especially, a seemingly comprehensive theory like the Darwinian, do the job for us, is to blind ourselves to the reflective task inherent in being the kind of self-questioning animals we are. Granted, we are not as simply unique, even as animals, as we used to think. Chimps, too, Jane Goodall has taught us, make and use tools; chimps too recognize, at least in some laboratory situations, their own mirror images. Each easy cutoff between men and other animals becomes more delicate, it seems, with advancing research. Possibly even language. Yet it is still true that we live more massively than other animals *in* artifacts, *in* culture, *in* language. We live therefore within our own norms for making, needing to accept those we are taught in infancy and youth, yet needing also to remake them. It is these strange unnaturally natural, or naturally unnatural, normative structures that we need to examine in every philosophical field, in epistemology the structures, for instance, that we call sciences, or theories, or empirical laws. To confuse their organization, their axiology, with the biological and psychological roots of their origin is to forget, in Eisenberg's phrase, the *human* nature of human nature, to live that unexamined life which, as Socrates told us long ago, it is not worthwhile for a man to live.

NOTES

1. J.H. Randall, Jr., 'The Changing Impact of Darwin on Philosophy', *Journal of the History of Ideas* **22** (1961), 435–462.
2. J.S. Fulton, 'Philosophical Adventures of the Idea of Evolution, 1859–1959', *Rice Institute Pamphlets* **46** (1959), 1.

3. C.F.A. Pantin, 'The Origin of Species', in *The History of Science*, London 1951, p. 129ff.
4. See e.g., T. Dobzhansky, *The Genetics of the Evolutionary Process*, New York 1970.
5. E.O. Wilson, *The Insect Societies*, Cambridge, Mass., 1971.
6. Lynn Margulis, *The Origin of Eukaryotic Cells*, New Haven 1970.
7. See G. Krueger, *Philosophie und Moral in der Kantischen Kritik*, Tübingen 1931.
8. See Victor Hugo, *Poésie*, Collection l'Intégrale, Paris 1972, Vol. 2, pp. 560–61; Vol. 3, p. 663.
9. L. Eisenberg, 'The *Human* Nature of Human Nature', *Science* **176** (1972), 123–28.
10. Herbert Fingarette, *Self-Deception*, London 1969.
11. Richard Rorty, 'Functionalism, Machines and Incorrigibility', *Journal of Philosophy* **69** (1972), 203–220.
12. George Williams, *Adaptation and Natural Selection*, Princeton 1966.
13. G.L. Stebbins, *The Basis of Progressive Evolution*, Chapel Hill, North Carolina, 1969.
14. Lawrence B. Slobodkin, 'The Strategy of Evolution', *American Science* **52** (1964), 342–357.
15. C.H. Waddington, *The Ethical Animal*, London 1960.
16. Max Gluckman, *Custom and Conflict in Africa*, Glencoe, Ill., 1955.
17. See for instance Sir Karl Popper's *Objective Knowledge*, Oxford 1973, p. 67.
18. Norman Campbell, *What is Science?*, New York 1952.

16 Science as a Vocation*
Max Weber

... Scientific work is chained to the course of progress; whereas in the realm of art there is no progress in the same sense. It is not true that the work of art of a period that has worked out new technical means, or, for instance, the laws of perspective, stands therefore artistically higher than a work of art devoid of all knowledge of those means and laws – if its form does justice to the material, that is, if its object has been chosen and formed so that it could be artistically mastered without applying those conditions and means. A work of art which is genuine "fulfilment" is never surpassed; it will never be antiquated. Individuals may differ in appreciating the personal significance of works of art, but no one will ever be able to say of such a work that it is 'outstripped' by another work which is also "fulfilment."

In science, each of us knows that what he has accomplished will be antiquated in ten, twenty, fifty years. That is the fate to which science is subjected; it is the very *meaning* of scientific work, to which it is devoted in a quite specific sense, as compared with other spheres of culture for which in general the same holds. Every scientific "fulfilment" raises new "questions"; it *asks* to be "surpassed" and outdated. Whoever wishes to serve science has to resign himself to this fact. Scientific works certainly can last as "gratifications" because of their artistic quality, or they may remain important as a means of training. Yet they will be surpassed scientifically – let that be repeated – for it is our common fate and, more, our common goal. We cannot work without hoping that others will advance further than we have. In principle, this progress goes on *ad infinitum*. And with this we come to inquire into the *meaning* of science. For, after all, it is not self-evident that something subordinate to such a law is sensible and meaningful in itself. Why does one engage in doing something that in reality never comes, and never can come, to an end?

One does it, first, for purely practical, in the broader sense of the word, for technical, purposes: in order to be able to orient our practical

* Excerpted essay in *From Max Weber: Essays in Sociology*, trans. and ed. H.H. Gerth and C.W. Mills (New York: Oxford University Press, 1946), pp. 137–144; 150–156.

activities to the expectations that scientific experience places at our disposal. Good. Yet this has meaning only to practitioners. What is the attitude of the academic man towards his vocation – that is, if he is at all in quest of such a personal attitude? He maintains that he engages in "science for science's sake" and not merely because others, by exploiting science, bring about commercial or technical success and can better feed, dress, illuminate, and govern. But what does he who allows himself to be integrated into this specialized organization, running on *ad infinitum*, hope to accomplish that is significant in these productions that are always destined to be outdated? This question requires a few general considerations.

Scientific progress is a fraction, the most important fraction, of the process of intellectualization which we have been undergoing for thousands of years and which nowadays is usually judged in such an extremely negative way. Let us first clarify what this intellectualist rationalization, created by science and by scientifically oriented technology, means practically.

Does it mean that we, today, for instance, everyone sitting in this hall, have a greater knowledge of the conditions of life under which we exist than has an American Indian or a Hottentot? Hardly. Unless he is a physicist, one who rides on the streetcar has no idea how the car happened to get into motion. And he does not need to know. He is satisfied that he may "count" on the behavior of the streetcar, and he orients his conduct according to this expectation: but he knows nothing about what it takes to produce such a car so that it can move. The savage knows incomparably more about his tools. When we spend money today I bet that even if there are colleagues of political economy here in the hall, almost every one of them will hold a different answer in readiness to the question: How does it happen that one can buy something for money – sometimes more and sometimes less? The savage knows what he does in order to get his daily food and which institutions serve him in this pursuit. The increasing intellectualization and rationalization do *not*, therefore, indicate an increased and general knowledge of the conditions under which one lives.

It means something else, namely, the knowledge or belief that if one but wished one *could* learn it at any time. Hence, it means that principally there are no mysterious incalculable forces that come into play, but rather that one can, in principle, master all things by calculation. This means that the world is disenchanted. One need no longer have recourse to magical means in order to master or implore the spirits, as did the savage, for whom such mysterious powers existed. Technical means and calculations perform the service. This above all is what intellectualization means.

Now, this process of disenchantment, which has continued to exist in Occidental culture for millennia, and, in general, this "progress," to which science belongs as a link and motive force, do they have any meanings that go beyond the purely practical and technical? You will find this question raised in the most principled form in the works of Leo Tolstoi. He came to raise the question in a peculiar way. All his broodings increasingly revolved around the problem of whether or not death is a meaningful phenomenon. And his answer was: for civilized man death has no meaning. It has none because the individual life of civilized man, placed into an infinite "progress," according to its own imminent meaning should never come to an end; for there is always a further step ahead of one who stands in the march of progress. And no man who comes to die stands upon the peak which lies in infinity. Abraham, or some peasant of the past, died "old and satiated with life" because he stood in the organic cycle of life: because his life, in terms of its meaning and on the eve of his days, had given to him what life had to offer; because for him there remained no puzzles he might wish to solve; and therefore he could have had "enough" of life. Whereas civilized man, placed in the midst of the continuous enrichment of culture by ideas, knowledge, and problems, may become "tired of life" but not "satiated with life." He catches only the most minute part of what the life of the spirit brings forth ever anew, and what he siezes is always something provisional and not definitive, and therefore death for him is a meaningless occurrence. And because death is meaningless, civilized life as such is meaningless; by its very "progressiveness" it gives death the imprint of meaninglessness. Throughout his late novels one meets with this thought as the keynote of the Tolstoyan art.

What stand should one take? Has "progress" as such a recognizable meaning that goes beyond the technical, so that to serve it is a meaningful vocation? The question must be raised. But this is no longer merely the question of man's calling *for* science, hence, the problem of what science as a vocation means to its devoted disciples. To raise this question is to ask for the vocation of science within the total life of humanity. What is the value of science?

Here the contrast between the past and the present is tremendous. You will recall the wonderful image at the beginning of the seventh book of Plato's *Republic:* those enchained cavemen whose faces are turned toward the stone wall before them. Behind them lies the source of the light which they cannot see. They are concerned only with the shadowy images that this light throws upon the wall, and they seek to fathom their interrelations. Finally one of them succeeds in shattering his fetters, turns around, and sees the sun. Blinded, he gropes about and stammers of what he saw. The others say he is raving. But gradually he learns to behold the light, and then his task is to descend to the cavemen and to lead them to the light. He is the philosopher; the sun, however, is the truth of science, which alone seizes not upon illusions and shadows but upon the true being.

Well, who today views science in such a manner? Today youth feels rather the reverse: the intellectual constructions of science constitute an unreal realm of artificial abstractions, which with their bony hands seek to grasp the blood-and-the-sap of true life without ever catching up with it. But here in life, in what for Plato was the play of shadows on the walls of the cave, genuine reality is pulsating; and the rest are derivatives of life, lifeless ghosts, and nothing else. How did this change come about?

Plato's passionate enthusiasm in *The Republic* must, in the last analysis, be explained by the fact that for the first time the *concept*, one of the great tools of all scientific knowledge, had been consciously discovered. Socrates had discovered it in its bearing. He was not the only man in the world to discover it. In India one finds the beginnings of a logic that is quite similar to that of Aristotle's. But nowhere else do we find this realization of the significance of the concept. In Greece, for the first time, appeared a handy means by which one could put the logical screws upon somebody so that he could not come out without admitting either that he knew nothing or that this and nothing else was truth, the *eternal* truth that never would vanish as the doings of the blind men vanish. That was the tremendous experience which dawned upon the disciples of Socrates. And from this it seemed to follow that if one only found the right concept of the beautiful, the good, or, for instance, of bravery, of the soul – or whatever – that then one could also grasp its true being. And this, in turn, seemed to open the way for knowing and for teaching how to act rightly in life and, above all, how to act as a citizen of the state; for this question was everything to the Hellenic man, whose thinking was political throughout. And for these reasons one engaged in science.

The second great tool of scientific work, the rational experiment, made its appearance at the side of this discovery of the Hellenic spirit during the Renaissance period. The experiment is a means of reliably controlling experience. Without it, present-day empirical science would be impossible. There were experiments earlier; for instance, in India physiological experiments were made in the service of ascetic yoga technique; in Hellenic antiquity, mathematical experiments were made for purposes of war technology; and in the Middle Ages, for purposes of mining. But to raise the experiment to a principle of research was the achievement of the Renaissance. They were the great innovators in *art*, who were the pioneers of experiment. Leonardo and his like and, above all, the sixteenth-century experimenters in music with their experimental pianos were characteristic. From these circles the experiment entered science, especially through Galileo, and it entered theory through Bacon; and then it was taken over by the various exact disciplines of the continental universities, first of all those of Italy and then those of the Netherlands.

What did science mean to these men who stood at the threshold of modern times? To artistic experimenters of the type of Leonardo and the musical innovators, science meant the path to *true* art, and that meant for them the path to true *nature*. Art was to be raised to the rank of a science, and this meant at the same time and above all to raise the artist to the rank of the doctor, socially and with reference to the meaning of his life. This is the ambition on which, for instance, Leonardo's sketch book was based. And today? "Science as the way to nature" would sound like blasphemy to youth. Today, youth proclaims the opposite: redemption from the intellectualism of science in order to return to one's own nature and therewith to nature in general. Science as a way to art? Here no criticism is even needed.

But during the period of the rise of the exact sciences one expected a great deal more. If you recall Swammerdam's statement, "Here I bring you the proof of God's providence in the anatomy of a louse," you will see what the scientific worker, influenced (indirectly) by Protestantism and Puritanism, conceived to be his task: to show the path to God. People no longer found this path among the philosophers, with their concepts and deductions. All pietist theology of the time, above all Spener, knew that God was not to be found along the road by which the Middle Ages had sought him. God is hidden, His ways are not our ways, His thoughts are not our thoughts. In the exact sciences, however, where one could physically grasp His works, one hoped to come upon the traces of what He planned for the world. And today? Who – aside from certain big children who are indeed found in the natural sciences – still believes that the findings of astronomy, biology, physics, or chemistry could teach us anything about the *meaning* of the world? If there is any such "meaning," along what road could one come upon its tracks? If these natural sciences lead to anything in this way, they are apt to make the belief that there is such a thing as the 'meaning' of the universe die out at its very roots.

And finally, science as a way "to God"? Science, this specifically irreligious power? That science today is irreligious no one will doubt in his innermost being, even if he will not admit it to himself. Redemption from the rationalism and intellectualism of science is the fundamental presupposition of living in union with the divine. This, or something similar in meaning, is one of the fundamental watchwords one hears among German youth, whose feelings are attuned to religion or who crave religious experiences. They crave not only religious experience but experience as such. The only thing that is strange is the method that is now followed: the spheres of the irrational, the only spheres that intellectualism has not yet touched, are now raised into consciousness and put under its lens. For in practice this is where the modern intellectualist form of romantic irrationalism leads. This method of emancipation from intellectualism may well bring about the very opposite of what those who take to it conceive as its goal.

After Nietzsche's devastating criticism of those "last men" who "invented happiness," I may leave aside altogether the naive optimism in which science – that is, the technique of mastering life which rests upon science – has been celebrated as the way to happiness. Who believes in this? – aside from a few big children in university chairs or editorial offices. Let us resume our argument.

Under these internal presuppositions, what is the meaning of science as a vocation, now after all these former illusions, the "way to true being," the "way to true art," the "way to true nature," the "way to true God," the "way to true happiness," have been dispelled? Tolstoi has given the simplest answer, with the words: "Science is meaningless because it gives no answer to our question, the only question important for us: 'What shall we do and how shall we live?'" That science does not give an answer to this is indisputable. The only question that remains is the sense in which science gives "no" answer, and whether or not science might yet be of some use to the one who puts the question correctly.

Today one usually speaks of science as "free from presuppositions." Is there such a thing? It depends upon what one understands thereby. All scientific work presupposes that the rules of logic and method are valid; these are the general foundations of our orientation in the world; and, at least for our special question, these presuppositions are the least problematic aspect of science. Science further presupposes that what is yielded by scientific work is important in the sense that it is "worth being known." In this, obviously, are contained all our problems. For this presupposition cannot be proved by scientific means. It can only be *interpreted* with reference to its ultimate meaning, which we must reject or accept according to our ultimate position towards life.

Furthermore, the nature of the relationship of scientific work and its presuppositions varies widely according to their structure. The natural sciences, for instance, physics, chemistry, and astronomy, presuppose as self-evident that it is worth while to know the ultimate laws of cosmic events as far as science can construe them. This is the case not only because with such knowledge one can attain technical results but for its own sake, if the quest for such knowledge is to be a "vocation." Yet this presupposition can by no means be proved. And still less can it be proved that the existence of the world which these sciences describe is worth while, that it has any "meaning," or that it makes sense to live in such a world. Science does not ask for the answers to such questions.

Consider modern medicine, a practical technology which is highly developed scientifically. The general "presupposition" of the medical enterprise is stated trivially in the assertion that medical science has the task of maintaining life as such and of diminishing suffering as such to the greatest possible degree. Yet this is problematical. By his means the medical man preserves the life of the mortally ill man, even if the patient implores us to relieve him of life, even if his relatives, to whom his life is worthless and to whom the costs of maintaining his worthless life grow unbearable, grant his redemption from suffering. Perhaps a poor lunatic is involved, whose relatives, whether they admit it or not, wish and must wish for his death. Yet the presuppositions of medicine, and the penal code, prevent the physician from relinquishing his therapeutic efforts. Whether life is worth while living and when – this question is not asked by medicine. Natural science gives us an answer to the question of what we must do if we wish to master life technically. It leaves quite aside, or assumes for its purposes, whether we should and do wish to master life technically and whether it ultimately makes sense to do so...

Finally, you will put the question: "If this is so, what then does science actually and positively contribute to practical and personal 'life'?" Therewith we are back again at the problem of science as a 'vocation."

First, of course, science contributes to the technology of controlling life by calculating external objects as well as man's activities. Well, you will say, that, after all, amounts to no more than the greengrocer of the American boy.[1] I fully agree.

Second, science can contribute something that the greengrocer cannot: methods of thinking, the tools and the training for thought. Perhaps you will say: well, that is no vegetable, but it amounts to no more than the means for procuring vegetables. Well and good, let us leave it at that for today.

Fortunately, however, the contribution of science does not reach its limit with this. We are in a position to help you to a third objective: to gain *clarity*. Of course, it is presupposed that we ourselves possess clarity. As far as this is the case, we can make clear to you the following:

In practice, you can take this or that position when concerned with a problem of value – for simplicity's sake, please think of social phenomena as examples. *If* you take such and such a stand, then, according to scientific experience, you have to use such and such a *means* in order to carry out your conviction practically. Now, these means are perhaps such that you believe you must reject them. Then you simply must choose between the end and the inevitable means. Does the end "justify" the means? Or does it not? The teacher can confront you with the necessity of this choice. He cannot do more, so long as he wishes to remain a teacher and not to become a demagogue. He can, of course, also tell you that if you want such and such an end, then you must take into the bargain the subsidiary consequences which according to all experience will occur. Again we find ourselves in the same situation as before. These are still problems that can also emerge for the technician, who in numerous instances has to make decisions according to the principle of the lesser evil or of the relatively best. Only to him one thing, the main thing, is usually given, namely, the end. But as soon as truly 'ultimate' problems are at stake for us this is not the case. With this, at long last, we come to the final service that science as such can render to the aim of clarity, and at the same time we come to the limits of science.

Besides we can and we should state: In terms of its meaning, such and such a practical stand can be derived with inner consistency, and hence integrity, from this or that ultimate *weltanschauliche* position. Perhaps it can only be derived from one such fundamental position, or maybe from several, but it cannot be derived from these or those other positions. Figuratively speaking, you serve this god and you offend the other god when you decide to adhere to this position. And if you remain faithful to yourself, you will necessarily come to certain final conclusions that subjectively make sense. This much, in principle at least, can be accomplished. Philosophy, as a special discipline, and the essentially philosophical discussions of principles in the other sciences attempt to achieve this. Thus, if we are competent in our pursuit (which must be presupposed here) we can force the individual, or at least we can help him, to give himself an *account of the ultimate meaning of his own conduct*. This appears to me as not so trifling a thing to do, even for one's own personal life. Again, I am tempted to say of a teacher who succeeds in this: he stands in the service of 'moral' forces; he fulfils the duty of bringing about self-clarification and a sense of responsibility. And I believe he will be the more able to accomplish this, the more conscientiously he avoids the desire personally to impose upon or suggest to his audience his own stand.

This proposition, which I present here, always takes its point of departure from the one fundamental fact, that so long as life remains immanent and is interpreted in its own terms, it knows only of an unceasing struggle of these gods with one another. Or speaking directly, the ultimately possible attitudes toward life are irreconcilable, and hence their struggle can never be brought to a final conclusion. Thus it is necessary to make a decisive choice. Whether, under such conditions, science is a worth while "vocation" for somebody, and whether science itself has an objectively valuable 'vocation' are again value judgments about which nothing can be said in the lecture-room. To affirm the value of science is a presupposition for teaching there. I personally by my very work answer in the affirmative, and I also do so from precisely the standpoint that hates intellectualism as the worst devil, as youth does today, or usually only fancies it does. In that case the word holds for these youths: "Mind you, the devil is old; grow old to understand him." This does not mean age in the sense of the birth certificate. It means that if one wishes to settle with this devil, one must not take to flight before him as so many like to do nowadays. First of all, one has to see the devil's ways to the end in order to realize his power and his limitations.

Science today is a "vocation" organized in special disciplines in the service of self-clarification and knowledge of interrelated facts. It is not the gift of grace of seers and prophets dispensing sacred values and revelations, nor does it partake of the contemplation of sages and philosophers about the meaning of the universe. This, to be sure, is the inescapable condition of our historical situation. We cannot evade it so long as we remain true to ourselves. And if Tolstoi's question recurs to you: as science does not, who is to answer the question: "What shall we do, and, how shall we arrange our lives?" or, in the words used here tonight: "Which of the warring gods should we serve? Or should we serve perhaps an entirely different god, and who is he?" then one can say that only a prophet or a savior can give the answers. If there is no such man, or if his message is no longer believed in, then you will certainly not compel him to appear on this earth by having thousands of professors, as privileged hirelings of the state, attempt as petty prophets in their lecture-rooms to take over his role. All they will accomplish is to show that they are unaware of the decisive state of affairs: the prophet for whom so many of our younger generation yearn simply does not exist. But this knowledge in its forceful significance has never become vital for them. The inward interest of a truly religiously 'musical' man can never be served by veiling to him and to others the fundamental fact that he is destined to live in a godless and prophetless time by giving him the *ersatz* of armchair prophecy. The integrity of his religious organ, it seems to me, must rebel against this.

Now you will be inclined to say: Which stand does one take towards the factual existence of "theology" and its claims to be a "science"? Let us not flinch and evade the answer. To be sure, "theology" and "dogmas" do not exist universally, but neither do they exist for Christianity alone. Rather (going backward in time), they exist in highly developed form also in Islam, in Manicheanism, in Gnosticism, in Orphism, in Parsism, in Buddhism, in the Hindu sects, in Taoism, and in the Upanishads, and, of course, in Judaism. To be sure their systematic development varies greatly. It is no accident that Occidental Christianity – in contrast to the theological possessions of Jewry – has expanded and elaborated theology more systematically, or strives to do so. In the Occident the development of theology has had by far the greatest historical significance. This is the product of the Hellenic spirit, and all theology of the West goes back to it, as (obviously) all theology of the East goes back to Indian thought. All theology represents an intellectual *rationalization* of the possession of sacred values. No science is absolutely free from presuppositions, and no science can prove its fundamental value to the man who rejects these presuppositions. Every theology, however, adds a few specific presuppositions for its work and thus for the justification of its existence. Their meaning and scope vary. Every theology, including for instance Hinduist theology, presupposes that the world must have a *meaning*, and the question is how to interpret this meaning so that it is intellectually conceivable.

It is the same as with Kant's epistemology. He took for his point of departure the presupposition: "Scientific truth exists and it is valid," and then asked: "Under which presuppositions of thought is truth possible and meaningful?" The modern aestheticians (actually or expressly, as for instance, G. v. Lukacs) proceed from the presupposition that "works of art exist," and then ask: "How is their existence meaningful and possible?"

As a rule, theologies, however, do not content themselves with this (essentially religious and philosophical) presupposition. They regularly proceed from the further presupposition that certain "revelations" are facts relevant for salvation and as such make possible a meaningful conduct of life. Hence, these revelations must be believed in. Moreover, theologies presuppose that certain subjective states and acts possess the quality of holiness, that is, they constitute a way of life, or at least elements of one, that is religiously meaningful. Then the question of theology is: How can these presuppositions, which must simply be accepted be meaningfully interpreted in a view of the universe? For theology, these presuppositions as such lie beyond the limits of "science." They do not represent "knowledge," in the usual sense, but rather a "possession." Whoever does not "possess" faith, or the other holy states, cannot have theology as a substitute for them, least of all any other science. On the contrary, in every "positive" theology, the devout reaches the point where the Augustinian sentence holds: *credo non quod, sed quia absurdum est.*

The capacity for the accomplishment of religious virtuosos – the "intellectual sacrifice" – is the decisive characteristic of the positively religious man. That this is so is shown by the fact that in spite (or rather in consequence) of theology (which unveils it) the tension between the valuespheres of "science" and the sphere of "the holy" is unbridgeable. Legitimately, only the disciple offers the 'intellectual sacrifice' to the prophet, the believer to the church. Never as yet has a new prophecy emerged (and I repeat here deliberately this image which has offended some) by way of the need of some modern intellectuals to furnish their souls with, so to speak, guaranteed genuine antiques. In doing so, they happen to remember that religion has belonged among such antiques, and of all things religion is what they do not possess. By way of substitute, however, they play at decorating a sort of domestic chapel with small sacred images from all over the world, or they produce surrogates through all sorts of psychic experiences to which they ascribe the dignity of mystic holiness, which they peddle in the book market. This is plain humbug or self-deception. It is, however, no humbug but rather something very sincere and genuine if some of the youth groups who during recent years have quietly grown together give their human community the interpretation of a religious, cosmic, or mystical relation, although occasionally perhaps such interpretation rests on misunderstanding of self. True as it is that every act of genuine brotherliness may be linked with the awareness that it contributes something imperishable to a super-personal realm, it seems to me dubious whether the dignity of purely human and communal relations is enhanced by these religious interpretations. But that is no longer our theme.

The fate of our times is characterized by rationalization and intellectualization and, above all, by the "disenchantment of the world." Precisely the ultimate and most sublime values have retreated from public life either into the transcendental realm of mystic life or into the brotherliness of direct and personal human relations. It is not accidental that our greatest art is intimate and not monumental, nor is it accidental that today only within the smallest and intimate circles, in personal human situations, in *pianissimo*, that something is pulsating that corresponds to the prophetic *pneuma*, which in former times swept through the great communities like a firebrand, welding them together. If we attempt to force and to "invent" a monumental style in art, such miserable monstrosities are produced as the many monuments of the last twenty years. If one tries intellectually to construe new religions without a new and genuine prophecy, then, in an inner sense, something similar will result, but with still worse effects. And academic prophecy, finally, will create only fanatical sects but never a genuine community.

To the person who cannot bear the fate of the times like a man, one must say: may he rather return silently, without the usual publicity build-up of renegades, but simply and plainly. The arms of the old churches are opened widely and compassionately for him. After all, they do not make it hard for him. One way or another he has to bring his "intellectual sacrifice" – that is inevitable. If he can really do it, we shall not rebuke him. For such an intellectual sacrifice in favor of an unconditional religious devotion is ethically quite a different matter than the evasion of the plain duty of intellectual integrity, which sets in if one lacks the courage to clarify one's own ultimate standpoint and rather facilitates this duty by feeble relative judgments. In my eyes, such religious return stands higher than the academic prophecy, which does not clearly realize that in the lecture-rooms of the university no other virtue holds but plain intellectual integrity. Integrity, however, compels us to state that for the many who today tarry for new prophets and saviors, the situation is the same as resounds in the beautiful Edomite watchman's song of the period of exile that has been included among Isaiah's oracles:

He calleth to me out of Seir, Watchman, what of the night? The watchman said, The morning cometh, and also the night: if ye will enquire, enquire ye: return, come.

The people to whom this was said has enquired and tarried for more than two millennia, and we are shaken when we realize its fate. From this we want to draw the lesson that nothing is gained by yearning and tarrying alone, and we shall act differently. We shall set to work and meet the "demands of the day," in human relations as well as in our vocation. This, however, is plain and simple, if each finds and obeys the demon who holds the fibers of his very life.

EDITOR'S NOTE

1. Weber is referring to an earlier remark (omitted in this selection): "The American's conception of the teacher who faces him is: he sells me his knowledge and his methods for my father's money, just as the greengrocer sells my mother cabbage. And that is all ... And no young American would think of having the teacher sell him a *Weltanschauung* or a code of conduct."

Epilogue
Alfred I. Tauber

THE PROBLEM OF TWO CULTURES

This anthology has considered science as a metaphysical world view and a system of epistemological inquiry. We have contemplated the nature of its evolution, as well as attempted to chart the sociological boundaries and modes of scientific discourse embedded in its complex cultural context; finally, the values governing science and the implications of scientific ideals for society-at-large were also entertained. Left for short comment is perhaps the most elusive of topics, namely the deeper significance of how science is meaningful in its human dimension, and it is here that I would like to offer an addendum to Max Weber's essay in Chapter 16. It is appropriate in considering the role of science in society to ponder its purpose and cultural meaning. Weber makes a powerful statement in characterizing science as "disenchanting," and the moral import of that judgement reverberates throughout our age. On that view, *meaning* is sought outside science, an issue that deserves, at the very least – to borrow from Kierkegaard – a "concluding unscientific postscript."

The issue arises because Science has come to be regarded as a separate activity within culture, something outside normal human discourse – a technical endeavor, enagaged in by a few stalwart minds for the benefit of society. Perhaps we might designate Science as analogous to a colony, with valuable resources that society harvests as the technological and practical applications of discovery and technical mastery. There are, to be sure, intrinsically valuable epistemological conquests, where we perceive an ever-expanding domain of nature as better understood, but Francis Bacon was politically astute when he sold state-supported science on the basis of the power that it offered. The most obvious benefits are better tools for living in a frequently hostile world, improved utilization of resources that increase our wealth and quality of life, and enhanced health and longevity. But for the non-scientist there remains something enigmatic, if not alien, about the world of science, for while it is obviously important to general interests, the actual work and thought of science reside outside our intimate experience. The sheer technicality of discourse, the daunting intellectual requirements to grasp its factual and conceptual details, are for many a distant kingdom of foreign knowledge and therefore highly mysterious, and often frightening.

Fundamentally, knowledge has been fragmented. This disjunction between different forms of knowledge became blatantly apparent in the nineteenth century when, in a series of complex cultural and intellectual developments, the natural philosophers became scientists and the moral philosophers, humanists. A widening schism over the next century evolved into what C.P. Snow called the "Two Cultures" (Snow, 1959), which distinguished the domains of two distinct intellectual modalities. What Snow crystalized as an intransigence between the worlds of the humanities and of science/technology, echoes a broad cultural conflict, when the analytical, mechanical, and abstract qualities of science displaced the elements of the primary encounter that may be characterized in terms of the personal, emotional, or aesthetic. When the poet communes with nature, the artist does so almost always in rejection of the scientific stance. Our culture has been saturated with the schism between the Two Cultures, and this partition remains deeply problematic as the attempts of different philosophers and philosophical schools to bridge (or for some to maintain) the divide attest.

Whatever the basis for the moat separating advanced Western society from its fabulously successful "colony," I disagree with Weber that science is solely pragmatic. More radically, I believe that science remains deeply "enchanting" and maintain that the 'colonists' seek meaning in their work through the aesthetic dimension of their experience. From that perspective I will argue that the scientific attitude, beyond its stance of an objectification toward nature, also extends to encompass dimensions regarded as more elusive – the emotional, the subjective. Consideration of the aesthetic facet of science allows a way to focus upon these nebulous dimensions of experience. So beyond the epistemological challenges of science as a system of objective knowledge, this other constitutive element of scientific practice and meaning, namely how science is aesthetic, deserves attention: I seek to link the impersonal positivist activity of science with a dimension which, while widely shared and often acknowledged as central by scientists, is truncated from official science as a matter "merely" of personal, private experience. But precisely because it is highly personal and pervasively appreciated, the aesthetic dimension may reveal the road back from an apparently intractable cultural divergence. To be explicit, I suspect that if science is appreciated in its full humane potential, we might recognize how closely the 'colony' lies to us.

My interest is to tie together our major strands with the rope of the scientist's phenomenological experience. To do so I briefly revisit some of our topics from a somewhat different perspective, hoping that this brief exercise will accentuate a psychological theme that has too often been submerged beneath the cloak of an idealized objectivity, where, as Evelyn Fox Keller observed, the scientist is subsumed, if not replaced, by her machine. There is resistance to this positivist stance, for in its drive toward objectivity and logical analysis, Science is too often seen as distorting our existential being in the world. On their view, always to scrutinize objectively is to divorce ourselves from personal meaning (Polanyi, 1962). This was a basic facts is Heidegger critique (Chapter 2) and this neo-Romantic complaint persist in many quarters. My strategy in these brief remarks is to show how science draws upon both the private and public personae of the scientist, and how the putative duality of the self dispassionately observing its world is an epistemological conceit. The issue here is not how the scientist attempts to be objective, but rather how knowledge of the world becomes personally meaningful. Beyond the matter of psychological integration, the import of that synthesis concerns the very nature of science itself.

TWO EPISTEMOLOGICAL DOMAINS – PUBLIC AND PRIVATE

The conflict between the objectified world of scientific facts and theory, and the private domain of personalized experience of those facts dates from the very origins of science, which aspired to discover facts as divorced from a subjective projection of the mind upon nature. Descartes epitomized this philosophical stance of a newly defined science: separation of mind and body simply formalized a split between the "I" and the "world," which was required to objectify nature. A dissociated self was demanded in order to study natural phenomena – separated, or better, distanced from intimate human involvement. From an objective perspective, projection of the self contaminated the process of attaining scientific knowledge. The unleashing of Descartes' mind–body dualism bequeathed the dilemma of how to render whole that which is broken in the division between self and world. The Cartesian method imparts a tension, for while dissecting the world into parts it offers no ready means for those elements to become reintegrated.[1] The Cartesian split resurfaces in the public and private scientific experience of "fact," the product of an ostensibly "cold-eyed" scrutiny of nature.

Who "knows" facts, how are facts used, and what do they mean? Answers to these questions are required in any epistemological system seeking the basis of how and what the subject knows, the nature of the known object, and the relationship between the knower and the known. Although the discovery or, more precisely, the construction of a fact is intimately linked to its creator, the dynamics of the fact can hardly be limited to the private domain of the observer's experience; others have a claim to a fact, which is often shared in the narrow proprietary sense, but *always* as the expected outcome of the scientific process. A scientific fact is fundamentally public, for it must be universalized by the scientific community at large. A hidden fact is useless to that community, because its placement outside discourse forbids its scrutiny. Scientific objectivity is focused upon the discovery or creation of facts and the public debates surrounding them. Scientific facts acquire the status of public entities as they become objectified and widely circulated, identified increasingly less with the subjective, private report of the scientist. Critical to the development of modern science was precisely this process by which *shared* experience was universalized among scientific practitioners. This is the domain in which Objectivity is attained.

Yet there remains a second, private sphere of the fact, which arises from the scientist's identity as an autonomous epistemological agent. This idea of autonomy was recognized at the crest of Newton's epochal discoveries in the philosophy of John Locke, who in a deep sense translated the Cartesian scientific ideal into the political and moral domains. Locke's philosophy hinged upon arguing for the ability of the individual to detach from the world, and himself, and observe each objectively. The individual then becomes an independent consciousness, which allows for the possibility of idiosyncratic or original epistemological strategies of knowing the world.[2] Autonomy is key in the political and normative universe of modern science, and whereas the process of scientific discourse and verification occur in public (facts require confirmation), the integrity of the scientist as a private, knowing agent remains an implicit and critical characteristic of scientific activity. To know the world remains a fundamental individual aspiration in the age of the Self (Tauber, 1994, pp. 141 ff.), and while we emphasize the social aspects of science as a cultural activity, the scientist remains committed to that Cartesian agent who experiences the world as an independent agent. Thus scientific knowledge has strong commitments to its dual character, namely a universalized corpus of fact and theory, which arises as the product of individual private experience. We are left with a complex dialectic between the observer's "personal" relations to those facts as the product of his autonomous personhood and the need for entering that experience into the public sector. We see here a convergence of epistemological, political, and moral ideals in the birth of modern science, and the scientist becomes the personification of these deeply embedded cultural values. I am not referring here to the role of Prometheus, to conquer nature and all of its richness, but rather to that of Janus, who must resolve the tension of peering to the public and private domains, simultaneously.

I have used the "fact" as our discursive vehicle, because at first glance a fact represents "what is out *there*." From the positivist perspective, this independence of the known 'fact' is based on the correspondence of a reality that an objective science might know. In the Introduction, I discussed this realist stance, and here I would only note that while perhaps an esteemable goal, the positivist project becomes problematic in light of the perspectival, contextualized, subjective observer, who cannot adhere to the rigid identification of the "facts" based on an idealized separation of the knower and the known. As Goethe recognized:

> my thinking is not separate from objects; that the elements of the object, the perceptions of the object, flow into my thinking and are fully permeated by it; that my perception itself is a thinking, and my thinking a perception. ([1823] 1988, p. 39).

Goethe's realization of a cognitive confluence between subject and object was formalized and developed in twentieth-century philosophy as an epistemological problem of science, and yet his "solution" – namely, that the aesthetic experience may serve to integrate self and the world – was essentially ignored (Tauber, 1993; 1996a).

Phenomenological philosophers dealt most directly with the challenge presented by Goethe, the epistemologist. They invoked the "gaze" as the privileged vehicle of the subject's relation to the world, in the sense that consciousness and meaning were understood to depend quite literally on how we see things (Husserl, [1935], 1970). As Morrissey comments, for phenomenologists,

> the objects that surround us function less "as they are" than "as they mean," and objects only mean for someone...To see implies seeing meaningfully. (1988, p. xx)

The inextricability of subject and object that Goethe pronounced and phenomenology has reaffirmed, clashes with the ideal of the scientist as independent from the world – as austere observer, collector of data uncontaminated by projected personal prejudice. How then do the crucial and variable elements of creative intuition, deduction, observation, replicable method, and assembly of disparate information create "objective" reality? This has been the question informing most discussion in the philosophy of science during this century, and we have endeavored to address it in Part II of this anthology. Here we turn our attention to the concern of relating scientific objectivity, the facts of scientific discourse, to the personal dimension, where much of our understanding and appreciation of the scientific experience rests on a different, perhaps less formalizable intelligence in which "events are not counted but weighed, and past events not explained but interpreted" (Heisenberg, 1979, p. 68).

THE CRISIS

In the general acceptance of a scientific inductive and mechanical ideal, the personal has been rejected. Descartes' separation of mind and body simply formalized a split between the "I" and the "world" – even between the "self" and the "body" in which it lives (Leder, 1990). Our predicament may be traced to this duality, originating with Cartesian skepticism, where a knowing subject, torn between objective and subjective modes of understanding the world became more and more self-reflective, and thereby disengaged from the plenum of existence. Science was based on this disinterested observer, relying solely on an empiricism molded by a positivist rationality. By exploring the consequences of an imperialistic rational science, we discern that other modes of experience are cut off from the rationalism invoked by "objectivity" and left adrift to find their own philosophical foundations. At the heart, then, of the elusive synthesis of "personal" and "objective" knowledge is the quest for their common philosophical foundation. The problem was posed dramatically by Edmund Husserl in *The Crisis of European Sciences:*

> Merely fact-minded sciences make merely fact-minded people... - Scientific, objective truth is exclusively a matter of establishing what the world, the physical as well as the spiritual world, is in fact. But can the world and human existence in it, truthfully have a meaning

if the sciences recognize as true only what is objectively established in this fashion...? (Husserl, [1935] 1970, pp. 6–7)

Husserl highlighted the distinction between science, one of the products of reason, and reason in its existential and metaphysical role that serves as the vital source of scientific inquiry: "The point is not to secure objectivity, but to understand it" (*ibid.* p. 189).[3]

Husserl offered a powerful reflection on the perils of scientific fruitfulness, which obfuscates that science is genuine–rationally grounded–only as long as it remains conscious of its philosophical basis, which is its starting point and foundation (Patoika, 1989, p. 226; Harvey, 1989). The crisis of the schism between science and its philosophical roots profoundly challenges the meaning of the sciences.

> [U]ltimately, all modern sciences drifted into a peculiar, increasingly puzzling crisis with regard to the meaning of their original founding as branches of philosophy, a meaning which they continued to bear within themselves. This is a crisis which does not encroach upon the theoretical and practical successes of the special sciences: yet it shakes to the foundations the whole meaning of their truth. (Husserl, [1935] 1970, p. 12)

A central consequence of the crisis Husserl identifies is a profound skepticism in defining a unifying metaphysics. Scientific reason is assigned to govern one domain of knowledge, while a different kind of reason is applied to matters of value and ethics. As a result of this division, we have witnessed the collapse of belief in absolute reason (*ibid.*, p. 13). Modern man is now truly torn between a naive faith in reason and the skepticism which negates or repudiates it in empiricist fashion.

> Unremittingly, skepticism insists on the validity of the factually experienced world, that of actual experience, and finds in it nothing of reason or its ideas. Reason itself and its [object], "that which is", become more and more enigmatic... [W]e find ourselves in the greatest danger of drowning in the skeptical deluge and thereby losing our hold on our own truth. (*Ibid.*, pp. 13–14)

Husserl recognized the need to synthesize experience – the scientific and the personal. The "crisis" was precisely a reflection of the deeply divisive nature of personal and objective knowledge, and he sought to bring both spheres of experience within a common philosophical domain.

The objectification of nature, for the purposes of studying it and using it, separated and distanced man from intimate involvement.

From an objective perspective, projection of the Self was regarded as contaminating the process of knowledge. Goethe clearly identified the problem in the early period of the Romantic era, and Nietzsche proclaimed an irredeemable gulf between the possibility of objective knowledge and personal meaning (Tauber, 1996a). Goethe's was a lonely, doomed heroic synthetic effort, while Nietzsche's argument, although the obverse of the positivists' position, also rejected a composite, or holistic ideal that might attempt to incorporate personal passion and objective science.

> [P]recisely the most superficial and external aspect of existence...- would be grasped first, and might even be the only thing that allowed itself to be grasped. A "scientific" interpretation of the world... might therefore be one of the *most stupid* of all interpretations of the world, meaning that it would be one of the poorest in meaning. (Nietzsche, [1882, 1887] 1974, p. 335) (emphasis in original)

In separating personal experience from what he viewed as a despotic rationality, Nietzsche advocated the rejection of an ascetic science, and sided with the primacy of personal, self-creative experience. In this sense, he articulated the Two Culture crevasse, but placed the disjunction within the individual (Tauber, 1992; 1994).

THE AESTHETIC DIMENSION

To our modern sensibility, Nietzsche's disclaimers have become familiar and sympathetically received as a legitimate response to positivism's attempted hegemony. Correspondingly, the aestheticization of the postmodern experience taps into a romantic desire for reunion of a self, integrated with its world, and within.

> When one views an object aesthetically, one lives in the object in the sense that one allows oneself to be entirely swayed by the laws of the object without any opposition upon one's own part...The aesthetic attitude may be likened to rowing downstream with the current and following all of its windings. One is here active in that one moves with the stream, but passive in that one opposes no resistance to the force which is carrying one on. The attitude is lost when one attempts to push upstream or off on the side eddy of one's choice. One is reminded of an illustration by Fidus of children in a boat paddling with the stream who think they are pushing while in reality they are drifting with the current. (Langfeld, 1920, pp. 59–60)

This sense of arriving at an easy integration, of the Self becoming fully embedded in its engagements, where subject–object dichotomies are dissolved, reflects a nostalgia for a world made whole again, but now based on a passionate rationality.[4] I am not advocating a return to Romantic ideals in the form of an imposition of the subjective into scientific methodology. Yet the attempt to separate completely the subject from its object – a crucial aspect of the self-understanding of Western science – has led to the exclusion of elements that are operative in science, but deemed inappropriate to a methodology governed only by rigorous logic and impersonal objectivity. However to displace the personal from scientific experience is to deny a large measure of the human dimension of the venture. With the truncation of the knowing subject, the scope of the entire enterprise is reduced and impoverished as a personally meaningful activity.

This enterprise may well be regarded as a peculiarly contemporary project, one that addresses the current need to find integration in our fragmented postmodern condition. Luc Ferry characterizes the contemporary aesthetic as no more than the

> expression of individuality... which does not see itself in any way as a mirror of the world but the creation of the world, a world... - which doesn't impose itself on us as an a priori common universe. (1993, p. 8)

This need to create a world of meaning was clearly articulated by Nietzsche, a critical architect of contrasting the rationality faculty, and science in particular, against the quest for personal meaning, which he espoused as fundamentally aesthetic (see n. 4).[5] I am not advocating a 'choice' between rationality and emotion, an either/or predicament, but rather a self-conscious admission of the need to acknowledge subjective experience in a world increasingly objectified. To see science as also partaking in that fundamental aesthetic experience of "personally meaningful," we humanize what might otherwise escape as sterilized 'objectivity' and reintegrate science with other cultural endeavors. Certainly those epistemological insights offer us technical mastery, but there are even greater stakes to consider. Making Science more humane is critical in making our culture more humane. Science must not be regarded as solely a paragon for objectifying experience, but should be recognized as drawing upon our full human faculties. Ultimately, the deepest value of science seems to me to be this moral lesson. To har-vest that ethos, it is imperative to reverse the vector driving Science further into a mode of autonomous colonial aloofness, and ultimately combine the splintered Two Cultures, a mirror of our own personal fragmentation.

How does one address the problem of celebrating a science that integrates all facets of human experience? It is insufficient merely to call upon such notions as "key insight," "beautiful experiments," and "elegant theory," as glosses for the "extra-scientific" aspects of the experiences of a few scientists of titanic creativity. The very fabric of everyday science, beyond its drudgery and frustration, must embody recognition and realization of a personalized ideal which governs the undertaking as much as impersonal standards of objectivity. While this position may be supported by a variety of strategies, it is most effectively advanced in the recognition that science, being essentially a creative project, must acknowledge that component of the personal which we call the aesthetic.[6] Almost two centuries ago, Goethe keenly understood our modern predicament, embracing the notion that the poet's eye might serve science in seeking Nature's true design:

Nowhere would anyone grant that science and poetry can be united. They forgot that science arose from poetry, and did not see that when times change the two can meet again on a higher level as friends. (Goethe, [1817] 1989)

The aesthetic in this sense may serve as a crucial faculty of re-integrating experience. In short, the aesthetic dimension may be the bridge that unifies the objective, *qua* scientific, with the subjective, *qua* personal. My thesis, quite simply, is that the scientist, just as the poet, draws upon the same aesthetic resources as a primary component of his experience (Tauber, 1996a).

Indeed a useful way of illustrating the elusive synthesis of scientific experience may be seen in the way the aesthetic has often served to span, in scientists' own accounts, the deep metaphysical schism (reviewed in Tauber, 1996b). For instance, when the physicist Paul Dirac said, it is more important that a theory be beautiful than that it be true,[7] he did not proclaim qualitative equivalence as did John Keats (Beauty is truth, truth beauty), but offered the sense of the beautiful as paramount. Dirac's proclamation jolts as it challenges the usual perception of scientific inquiry, and yet it is neither a novel assessment, nor a radical position. Instead, it reverberates as only an apparent contradiction might, for beneath our initial incredulity, we sense our responsive recognition.[8]

We have become self-conscious of the beautiful or the true as we intellectually ponder our experience. A purported tension between them is enmeshed in the very roots of Western thinking. Pretechnical cultures allow for little separation of subject–object encounters. The mythic reality fuses the subjects' dimensions into a continuous universe of affect and effect. But in our intellectualizations, our separation of the self from its world, where we become detached observers in our strategy to claim mastery, we define truth in terms separate from other dimensions of our experience. To discern truth is to comprehend a reality beyond ourselves, knowable and, to be sure, understood in our terms, but to discern beauty is to experience some reality emotionally, to participate directly in a communal sharing between the object and the *experiencing* subject. It lies closer to that mythic unity of subject with its object.[9]

It is this recognition of the self, the personal I, which is expanded in its encounter with its object that qualifies and defines aesthetic experience. This experience is – or perhaps, better, ought to be acknowledged as – inseparable from the aims of science. Science often appears most driven by its quest for technical mastery, but its aspirations for explanation draw upon a deep aesthetic reservoir, one steeped in the metaphysical thirst for meaning.[10] The dissection of the world has yielded a kind of knowledge which beckons to be coordinated in our full human experience. The scientific object may reside seemingly separate – out there – the focus of an inquiry of what it is – in itself – (ignoring the philosophical difficulties of that expectation), but the issue is to integrate that object to our full experience, rational *and* emotional. The search for this common ground is the elusive synthesis of our very selves in a world ever more objectified from us – a beguiling reminder of the lingering fault of our very identity. To the extent that we appreciate that our two cultured world reflects a disjunction of that integration, we gain insight into a metaphysical chasm that may still be mended.

NOTES

1. Holism, as a philosophical construct, grew out of seventeenth-century debate over the metaphysical structure of Nature. In response to the dualistic construction of mind and body proposed by Descartes, Spinozean pantheism was the direct antecedent to the Romantic notion of Nature's unity (McFarland, 1969).
2. This view also has profound ethical ramifications, for the mode of objective disengagement becomes a moral requirement in scrutinizing not only the world but also the self. Autonomy becomes a value, limited only to the extent that an individual's freedom infringes upon the freedom of others.

Entwined in Locke's epistemological definition, we find his legal foundation, for the individual so defined becomes the unit of government, divided between its freedom and the rights of the majority. "Self" becomes a forensic term to which the law is applicable, and "possessive individualism" (MacPherson, 1962) is thus celebrated and moreover assured as established by the epistemological system from which an independent ethical unity consistently arises. We are heirs of this seventeenth-century liberalism, of which the essence of man being the proprietor of his or her person and property is the obvious result.

3. What began as Descartes' Dream, a philosophy that seeks to encompass in the unity of a theoretical system all meaningful questions in a rigorous scientific manner, has left science as a residual concept (*ibid.*, p. 9). By this, Husserl notes that metaphysical or philosophical problems that should still be broadly linked to science under the rubric of rational inquiry are separated over the criterion of fact. In a powerful sense, positivism... decapitates philosophy (*ibid.*) by legitimizing one form of knowledge at the expense of another. For Husserl, the crisis was not limited to science or philosophy, but reflected a fundamental challenge to European cultural life, its total *Existenz*, the very collapse of a universal philosophy.

4. Heidegger eloquently described the process in Nietzsche's aesthetic:

> By having a feeling for beauty the subject has already come out of himself, he is no longer subjective, no longer a subject. On the other side, beauty is not something at hand like an object of sheer representation. As an attuning, it thoroughly determines that state of man. Beauty breaks through the confinement of the object placed at a distance, standing on its own, and brings it into essential and original correlation to the "subject." Beauty is no longer objective, no longer an object. (Heidegger, 1979, p. 123)

5. Much of the postmodernist attitude alluded to here may be traced to Nietzsche, and certainly the pluralism of knowledge and the centrality of the individual's search (and definition) of meaning was his anthem:

> "There are no states of fact as such" but "only interpretations," not a world, but an "infinity of worlds," themselves only the perspectives of the living individual. "The question 'what is it' is a way to create meaning... It is, at bottom, always the question 'what is it *for me*.'" (Ferry, 1993, p. 12)

(Ferry's quotation is a rather free combination of several lines from Nietzsche's *Will to Power*. In the section, "The Will to Power as Knowledge," Nietzsche has many aphorisms directed against the "tyranny of reason." Ferry has seemingly combined elements from aphorism no. 481 (Nietzsche, [1904] 1967, p. 267) and no. 556 (*ibid.*, p. 301).) Ferry's project, analogous to my own, is to seek in the aesthetic a reconciliation of the subjective, individual experience with the demands related to the political problems of democratic society.

6. When discussing science and aesthetics, it is difficult to draw the line between psychology and philosophy. It is particularly vexing because the inter-

section of the discursive languages are incomplete, and even at cross purposes. I have chosen not to dwell on the issues of how we recognize natural beauty; geometric form and other visual metaphors generally fulfill criteria of form that we "perceive" as beautiful. But whether the appreciation of a phenomenon or form as beautiful is learned (i.e. culturally derived), or in fact fulfills some resonant cognitive function remains a vexing question (Rentschler *et al.*, 1988). The literature concerned only with defining the beautiful *in* science is vast; see Tauber (1996b) for a partial listing.

7. Quoted by Charles Hartshorne as heard in a lecture (Hartshorne, 1982). Dirac clearly emphasized a mathematical–aesthetic method at the expense of inductive empiricism:

> A theory of mathematical beauty is more likely to be correct than an ugly one that fits some experimental data [and] there are occasions when mathematical beauty should take priority over agreement with experiment. (Kragh, 1990, quoting Dirac, p. 284)

Or again, "It is more important to have beauty in one's equations than to have them fit experiment" (Dirac, 1963). This so-called Dirac–Weyl doctrine in fact can be traced in modern physics to Hermann Minkowski, but perhaps of more influence for Dirac was Einstein, who was guided by principles of simplicity and exhibited legendary confidence in his equations of gravitation theory. Dirac and many other physicists of his time regarded Einstein's gravitation theory as created virtually without empirical reasoning, although Einstein himself was more circumspect in his trust in aesthetic parameters (Kragh, 1990, pp. 286–287; see also McAllister, 1990). The entire issue of the subjectivity and changing standards of aesthetic criteria are of course beyond our concern, but have focused much discussion on this issue. (See Renscher, 1990; Tauber, 1996b for introductions.)

8. Dirac's pronouncement falls prey to the disjunction of the rational scientific from the emotive beautiful. In the very separation of beauty from truth we perhaps might be satisfied with Keats' assignment of equality, or at least complementarity, in as much as "truth" fulfills certain necessary criteria and "beauty", others. Dirac delineates them as different and hierarchical. In a most profound sense, by separating truth and beauty, we again admit a potentially debilitating dichotomy.

9. The merging of the subject and object in mythical and magical cultures has been exhaustively studied. Whether viewed from a neo-Kantian (Cassirer, 1955), structuralist (Lévi-Strauss, 1966), psychoanalytic (Neumann, 1954) or aesthetic perspective (Gauguin, 1921), the crucial elements of the mythical universe are a psychic–religious belief system and cognitive world view that collapse the subject–object dicotomy. This is the origin of the self-conscious espousal of a Nietzschean return to the Dionysian, which is obviously contrasted with the analytic posture characterized by a subject scrutinizing its object, and in the same gesture striving to define itself. Not trusting its context from which only indistinct boundaries might arise, modern man is denied the ability to simply identify, and thus merge, with a magical world. A particularly powerful example is the contrasting role of the artist. The self-consciousness of the artist in our culture, in contrast to the head-hunting Asmet of New Guinea, is a particularly poignant case in point. A cogent

observation was made by Michael L. Rockefeller on his visit to acquire museum artefacts:

> The Asmet [New Guinea] culture offers the artist a specific language in form. This is a language which every artist can interpret and use according to his genius, and a language which has symbolic meaning for the entire culture. Our culture offers the artist no such language. The result is that each painter or sculptor must discover his own means of communicating in form. Only the greatest genuises are able to invent an expression which has meaning for a nation or people. (Letter of November 16, 1961, quoted at an exhibit at the Metropolitan Museum of Art, New York)

10. See Henri Atlan's discussion of this issue ([1986] 1993, p. 193), already alluded to in the Introduction, where he draws upon Gaston Bachelard's original insights concerning the psychological motivations of scientists (Bachelard, [1934] 1984).

ACKNOWLEDGEMENTS

Much of the foregoing appears in somewhat different format elsewhere (Tauber, 1996a), where a more detailed historical and philosophical account is offered. I am most grateful to my colleagues at the Boston University Center for Philosophy and History of Science, Robert S. Cohen and Eileen Crist, for their critical comments and encouragement.

REFERENCES

Atlan, H. ([1986] 1993) *Enlightenment to Enlightenment. Intercritique of Science and Myth*, trans. L.J. Schramm (Albany: State University of New York).

Bachelard, G. ([1934] (1984) *The New Scientific Spirit*, trans. A. Goldhammer (Boston: Beacon Press).

Cassirer, E. (1955) *The Philosophy of Symbolic Forms. Vol. 2, Mythical Thought*, trans. R. Manheim (New Haven: Yale University Press).

Dirac, P. A. M., (1963) "The evolution of the physicist's picture of nature," *Scientific American*, 208: 45–53.

Ferry, L. (1993) *Homo Aestheticus. The Invention of Taste in the Democratic Age*, trans. R. de Loaiza (Chicago: The University of Chicago Press).

Gauguin, P. (1921) *Intimate Journals* (New York: Liveright).

Goethe, J.W. ([1817] 1989) "History of the printed brochure," in B. Mueller (ed.), *Goethe's Botanical Writings* (Woodbridge: Oxbow Press) pp. 170–176.

—— ([1823] 1988) "Significant help given by an ingenious turn of phrase," in D. Miller (ed.), *Scientific Studies* (New York: Suhrkamp Publishers), pp. 39–41.

Hartshorne, C., (1982) "Science as the search for the hidden beauty of the world," in D. W. Curtin (ed.), *The Aesthetic Dimension of Science, 1980 Nobel Conference* (New York: Philosophical Library), pp. 95–106.

Harvey, C. W. (1989) *Husserl's Phenomenology and the Foundations of Natural Science* (Athens: Ohio University Press).

Heidegger, M. (1979) *Nietzsche Vol. 1*, trans. D. P. Krell, (San Francisco: Harper).

Heisenberg, W. (1979) "The teachings of Goethe and Newton on colour in the light of modern physics," in *Philosophical Problems of Quantum Physics* (Woodbridge: Oxbow Press), pp. 60–76.

Husserl, E. ([1935] 1970) *The Crisis of European Sciences and Transcendental Phenomenology*, trans. D. Carr (Evanston: Northwestern University Press).

Kragh H., (1990) *Dirac, A Scientific Biography* (Cambridge: Cambridge University Press).

Langfeld, H. S. (1920) *The Aesthetic Attitude* (New York: Harcourt, Brace & Howe).

Leder, D. (1990) *The Absent Body* (Chicago: The University of Chicago Press).

Lévi-Strauss, C. (1966) *The Savage Mind* (Chicago: The University of Chicago Press).

McAllister, J. W. (1990) "Dirac and the aesthetic evaluation of theories," *Methodology and Science*, 23: 87–102.

MacPherson, C. B. (1962) *The Political Theory of Possessive Individualism: Hobbes to Locke* (Oxford: Clarendon Press).

McFarland, T. (1969) *Coleridge and the Pantheist Tradition* (Oxford: Clarendon Press).

Morrissey, R. J. (1988) "Introduction. Jean Starobinski and Otherness," in *Jean-Jacques Rousseau. Transparency and Obstruction*, trans. A. Goldhammer (Chicago: The University of Chicago Press).

Neumann, E. (1954) *The Origins and History of Consciousness* (Princeton: Princeton University Press).

Nietzsche, F. ([1882] 1974) *The Gay Science*, trans. W. Kaufman (New York: Vintage).

—— ([1904] 1967) *The Will to Power*, trans. W. Kaufmann and R. J. Hollingdale (New York: Vintage).

Patoika, J. (1989) *Philosophy and Selected Writings*, ed. and trans. E. Kohak (Chicago: The University of Chicago Press), pp. 223–238.

Polanyi, M. (1962) *Personal Knowledge. Towards a Post-critical Philosophy* (Chicago: The University of Chicago Press).

Rentschler, L., Herzberger, B. and Epstein, D. (eds.) (1988) *Beauty and the Brain. Biological Aspects of Aesthetics* (Basle: Birkhauser Verlag).

Rescher, N. (1990) *Aesthetic Factors in Natural Science* (Lanham: University Press of America).

Snow, C. P. (1959) *The Two Cultures* (Cambridge: Cambridge University Press).

Tauber, A. I. (1992) "The organismal self: Its philosophical context," in L. Rouner (ed.), *Selves, People and Persons* (Notre Dame: Notre Dame University Press), pp. 149–67.

—— (1993) "Goethe's philosophy of science: Modern resonances," *Perspectives in Biology and Medicine*, 36: 244–257.

—— (1994) *The Immune Self. Theory or Metaphor?* (New York and Cambridge: Cambridge University Press).

—— (1996a) "From Descartes' dream to Husserl's nightmare," in A. I. Tauber (ed.), *Aesthetics and Science. The Elusive Synthesis* (Dordrecht: Kluwer Academic Publishers), pp. 289–312.

—— (ed.) (1996b) *Aesthetics and Science. The Elusive Synthesis* (Dordrecht: Kluwer Academic Publishers).

Index